Horst Pagel Christoph Weiss
Wolfgang Jelkmann (Eds.)

Pathophysiology and Pharmacology of Erythropoietin

With 106 Figures and 63 Tables

Springer-Verlag
Berlin Heidelberg New York
London Paris Tokyo
Hong Kong Barcelona
Budapest

Dr. Horst Pagel
Professor Dr. Christoph Weiss
Institut für Physiologie
Medizinische Universität zu Lübeck
Ratzeburger Allee 160, W-2400 Lübeck 1
Germany

Professor Dr. Wolfgang Jelkmann
Physiologisches Institut I
Rheinische Friedrich-Wilhelms-Universität
Nußallee 11, W-5300 Bonn 1
Germany

ISBN-13:978-3-642-77076-0 e-ISBN-13:978-3-642-77074-6
DOI: 10.1007/978-3-642-77074-6

Library of Congress Cataloging-in-Publication Data
Pathophysiology and pharmacology of erythropoietin / Horst Pagel, Christoph Weiss, Wolfgang Jelkmann (eds.).
p. cm. Includes bibliographical references and index.
ISBN-13:978-3-642-77076-0
1. Erythropoietin – Therapeutic use – Congresses, 2. Erythropoietin – Pathophysiology – Congresses, 3. Recombinant erythropoietin – Therapeutic use – Congresses. 4. Recombinant erythropoietin – Pathophysiology – Congresses, 5. Renal anemia – Chemotherapy – Congresses. 6. Anemia – Chemotherapy – Congresses.
I. Pagel, Horst, II. Weiss, Christoph, III. Jelkmann, Wolfgang, 1949-[DNLM: 1. Anemia – drug therapy – congresses. 2. Erythropoietin – physiology – congresses. 3. Erythropoietin-therapeutic use – congress, 4. Kidney Failure, Chronic – drug therapy – congresses. 5. Recombinant Proteins – therapeutic use – congresses. WH 150 P297]
RM666.E834P37 1992 615'.718 – dc20 DNLM/DLC for Library of Congress 92-2312 CIP

Softcover reprint of the hardcover 1st edition 1992

Typesetting: Mitterweger Werksatz GmbH, Plankstadt

21/3130-5 4 3 2 1 0 – Printed on acid-free paper

Foreword

I am honored to be invited to prepare a foreword for the proceedings of the Second International Lübeck Conference on Erythropoietin (Epo). I congratulate Wolfgang Jelkmann, Horst Pagel and Christoph Weiss for their organization of an excellent program for this conference which updated all of us on the advances made in erythropoietin research during the past few years since the first conference in June of 1988. I am sure that Professor Paul Carnot, had he been present at this conference, would be very pleased and proud of the advances made in the field of erythropoietin since his and Madame DeFlandre's seminal finding in 1906 (1) that rabbits produced a humoral substance following bleeding which controls red blood cell production. The reports by Hjort in 1936 (2) and by Erslev in 1953 (3) that large volumes of plasma or serum from rabbits following a bleeding stimulus, when injected into normal donor rabbits, produced a reticulocytosis, were very significant in confirming the existence of a humoral factor which controls erythropoiesis. Reissmann's parabiotic rat experiments in 1950 (4) reawakened interest in erythropoietin when he proved that hypoxia stimulated the production of a factor which regulates red cell production. The studies of several investigators such as Jacobson et al. (5), Fisher and Birdwell (6), Kuratowska et al. (7) and Nathan et al. (8), using some early in vivo and in vitro techniques, were very important in providing the necessary documentation that the kidney produced erythropoietin. Of course, considerable credit should be given to Miyake and Goldwasser in 1977 (9) for purifying erythropoietin to homogeneity; to Fu-Kuen Lin and colleagues at AMGEN in 1985 (10) for their cloning of the gene for Epo; and to Eschbach and Adamson in 1987 (11) for their first clinical trials of recombinant erythropoietin in patients with anemia of end stage renal disease. These are examples of early seminal findings that have had a significant impact on the treatment of anemia. Recombinant human erythropoietin is now approved throughout the world for therapeutic use in patients with anemia of end stage renal disease, and also more recently for anemic patients with the immune deficiency syndrome that are receiving antiviral drug therapy. Certainly in the near future, the largest population of patients who should be treated with erythropoietin is the autologous blood donor group. For example, patients who are awaiting elective heart by-pass surgery or an orthopedic procedure where large amounts of blood may be needed for transfusion at the time of surgery. These patients can bank their own blood prior to surgery and use this blood for transfusion at the time of their surgery. I am delighted to have played a role in the development of erythropoietin and we certainly look forward to resolving some of the controversial topics, such as the cells in the kidney and the liver which produce erythropoietin, at subsequent meetings.

This meeting was certainly provocative and provided us with an excellent opportunity to discuss several important issues in erythropoietin research. There are 40 papers included in this monograph, mostly from European countries, covering several broad topics on the pathophysiology and pharmacology of erythropoietin research such as: structure and action; sites and control of production; blood levels in health and disease; erythropoietin replacement in renal failure; primary hemopoietic disorders and chemotherapy; and the application of erythropoietin in non-renal anemia. The material contained in this monograph should be of value to physiologists, pharmacologists, hematologists, nephrologists, biochemists and other investigators on erythropoietin and the control of erythropoiesis. I hope it will not be too long in the future before another Lübeck International Conference on Erythropoietin will be held, and again I would like to thank personally all of the participants in this symposium for their contributions.

We look forward to reading the articles in the monograph which will include some of these papers and thanks again to the organizers of this conference.

James W. Fisher
Tulane University
School of Medicine
New Orleans, U.S.A.

References

1. Carnot, P, DeFlandre, C (1906) Sur l'activité hématopoiétique des différents organes au cours de la régénération du sang. C. R. Acad. Sci. Paris 143:432–435.
2. Hjort, E (1936) Reticulocyte increase after injection of anemic serum. Norsk Mag F Laegevidensk 97:270–277.
3. Erslev, A J (1953) Humoral regulation of red cell production. Blood 8:349–357.
4. Reissmann, K R (1950) Studies on the mechanism of erythropoietic stimulation in parabiotic rats during hypoxia. Blood 5:372–380.
5. Jacobson, L O, Goldwasser, E, Fried, W, Plzak, L (1957) Role of the kidney in erythropoiesis. Nature 179:633–634.

6. Fisher, J W, Birdwell, B J (1961) The production of an erythropoietic factor by the in situ perfused kidney. Acta Haematol. 26:224–232.
7. Kuratowska, Z, Lewartowski, B, Michalak, E (1961) Studies on the production of erythropoietin by the isolated perfused organs. Blood 18:527–534.
8. Nathan, D G, Schupak E, Stohlman, F, Jr, Merril, J P (1964) Erythropoiesis in anephric man. J. Clin. Invest. 43:2158–2165
9. Miyake, T, Kung, C K-H, Goldwasser, E (1977) Purification of human erythropoietin. J. Biol. Chem. 252:5558–5564.
10. Lin, F-K, Suggs, S, Lin, C H, Browne, J K, Smalling, R, et al (1985) Cloning and expression of the human erythropoietin gene. Proc. Natl. Acad. Sci. USA 82:7580–7584.
11. Eschbach, J W, Egrie, J C, Downing, M R, Browne, J K, Adamson, J W (1987). Correction of the anemia of end-stage renal disease with recombinant human erythropoietin: Results of a combined phase I and II clinical trial. New Engl. J. Med. 316:73–78.

List of Contents

List of Correspondence Authors

Battersby, Rüdiger, IBR Forschungs GmbH & Co, Bioanalytisches Zentrum KG, Feodor-Lynen-Str. 5, W-3000 Hannover 61, Germany

Bechensteen, Anne Grete; Dept. of Pediatrics, Ulleval Hospital, 0407 Oslo 4, Norway

Brune, Thomas; Correspondence should be addressed to Dr. H. Heinrichs, Marburg, Germany

Busch, O. R. C.; Laboratory for Experimental Surgery, Erasmus University, P.O. Box 1738, 3000 DR Rotterdam, The Netherlands

Carozzi, Silvia; Nephrology and Dialysis Unit, St. Paul's Hospital, Monoblocco Valloria, 17100 Savona, Italy

Casati, Stefano; Divisione di Nefrologia e Dialisi, IRCCS Ospedale Maggiore, Via Commenda 15, 20122 Milano, Italy

Castelnovo, Claudia; Divisione di Nefrologia e Dialisi, IRCCS Ospedale Maggiore, Via Commenda 15, 20122 Milano, Italy

Costagliola, Ciro; Eye Clinic, 1rst School of Medicine, University of Naples, Via F. Petrarca 41/a, 80122 Napoli, Italy

Di Benedetto, Attilio; Sez. Nefrologia Emodialisi, Ospedale dei Pellegrini, University of Naples, Via Petrarca 41/A, 80122 Napoli, Italy

Dreyling, K. W.; Abt. Nephrologie, Medizinische Universitätsklinik, Hugstetter Str. 55, W-7800 Freiburg, Germany

Fandrey, Joachim; Hematology Division, Brigham & Women's Hospital, Harvard Medical School, 75 Francis Street, Boston, Massachusetts, 02115, USA

Fisher, James; Dept. of Pharmacology, Tulane University Medical Center, 1430 Tulane Avenue, New Orleans, Louisiana, 70112, USA

Ganser, A.; Zentrum d. Inneren Medizin, Abt. Hämatologie, Klinikum d. Johann Wolfgang Goethe-Universität, Theodor-Stern-Kai 7, W-6000 Frankfurt a. M. 70, Germany

Grossi, Alberto; Division of Hematology, University of Florence, USL 10/D, Ospedale Careggi, 50134 Florence, Italy

GRÜTZMACHER, PETER; II. Medizinische Klinik, St. Markus Krankenhaus, Wilhelm-Epstein-Str. 2, W-6000 Frankfurt a.M. 50, Germany

HEINRICHS, HUBERT; Behringswerke AG, Klinische Forschung, Postfach 1140, W-3550 Marburg/Lahn, Germany

JELKMANN, WOLFGANG; Physiologisches Institut I, Rheinische Friedrich-Wilhelms-Universität, Nußallee 11, W-5300 Bonn, Germany

KALTWASSER, J. P.; Bereich Rheumatologie, Zentrum d. Inneren Medizin, Johann Wolfgang Goethe-Universität, Theodor-Stern-Kai 7, W-6000 Frankfurt a. M. 70, Germany

KOKOT, FRANCISZEK; Dept. of Nephrology, Silesian School of Medicine, Inst. of Intern. Med., Ul. Francuska 20/24, 40-027 Katowice, Poland

LACOMBE, CATHERINE; Inserm U 152 and CNRS UA 628, Hospital Cochin, 27 rue du Faubourg Saint-Jacques, 75014 Paris, France

LAMPERI, SILVANO; Correspondence should be addressed to Dr. S. Carozzi, Savona, Italy

LAPPIN, TERENCE R. J., Dept. of Haematology, Royal Victoria Hospital, Grosvenor Road, Belfast BT12 6BA, Northern Ireland

MÜLLER-WIEFEL, DIRK E.; Nephrologische Abteilung, Universitäts-Kinderklinik Hamburg Eppendorf, Martinistr. 52, W-2000 Hamburg 20, Germany

NEUMANN, HARTMUT; Abt. Innere Medizin IV, Medizinische Universitätsklinik, Hugstetter Straße 55, W-7800 Freiburg, Germany

NIJHOF, WILLEM; Laboratorium voor Fysiol., Chem. der Rijksuniversiteit, Bloemsingel 10, 9712 KZ Groningen, The Netherlands

NOÉ, GUDRUN; Abt. Transfusionsmedizin, Universität Ulm, Helmholtzstraße 10, W-7900 Ulm, Germany

PECHERSTORFER, MARTIN; Krankenhaus der Stadt Wien-Lainz, 5. Medizinische Abteilung mit Onkologie, Wolkersbergenstraße 1, 1130 Wien, Austria

PETTERSSON, TOM; Fourth Dept. of Medicine, Helsinki University Central Hospital, Unioninkatu 38, 00170 Helsinki, Finland

POHL, CHRISTOPH; Medizinische Klinik I, Universität Köln, Joseph-Stelzmann-Str. 9, W-5000 Köln 41, Germany

RICH, IVAN; Experimentelle Hämatologie, Abteilung Transfusionsmedizin der Universität Ulm, Oberer Eselsberg 10, W-7900 Ulm, Germany

ROBAK, TADEUSZ; Dept. of Clinical Pharmacology, Medical Academy; Copernicus Hospital, Ul. Pabianicka 62, 93-513 Lódz, Poland

Schlageter, Marie-H.; Hospital St. Louis, Service de Medicine Nucléaire, 1 Avenue Claude Vellefaux, 75010 Paris, France

Schmidt, Walter; Abteilung f. Sport- u. Arbeitsphysiologie, Medizinische Hochschule Hannover, Konstanty-Gutschow-Str. 8, W-3000 Hannover 61, Germany

Schrezenmeier, Hubert; Abt. Innere Medizin III, Medizinische Universitätsklinik, Robert-Koch-Straße 8, W-7900 Ulm, Germany

Schurek, Hans-Joachim; Innere Medizin, Abt. Nephrologie, St. Bonifatius-Hospital, Wilhelmstraße 13, W-4450 Lingen (Ems), Germany

Schütterle, G.; Med. Universitäts-Klinik II, Zentrum für Innere Medizin, Justus-Liebig-Universität, Klinikstraße 36, W-6300 Gießen, Germany

Stockenhuber, Felix; I. Medizinische Universitätsklinik, Lazarettgasse 14, 1090 Wien 9, Austria

Tan, Chorh-Chuan; Institute of Molecular Medicine, Room 317, John Radcliffe Hospital, Oxford OX3 9DU, England

Vannucchi, Allessandro M.; Dept. of Hematology, University of Florence USL 10/D, Ospedale di Careggi, 50134 Florence, Italy

Wilms, H. W. F.; Dialyse Kuratorium Hamburg e.V., Alte Landstraße 284, W-2000 Hamburg 65, Germany

Wolff, Martin; Physiologisches Institut I, Universität Bonn, Nußallee 11, W-5300 Bonn 1, Germany

Wurnig, C.; Correspondence should be addressed to Dr. F. Stockenhuber, Wien, Austria

I. Structure and Action

Structure-Function Relationships of Erythropoietin

T.R.J. LAPPIN

Introduction

One of the major questions in erythropoietin research centres on the relationship between the structure of the hormone and its mode of action. Erythropoietin is a glycoprotein which is 39% carbohydrate based on the molecular mass of 30 400 daltons, as determined by sedimentation equilibrium [1]. The human erythropoietin gene encodes a 193 amino acid protein. Cleavage of a 27 amino acid leader sequence gives a mature protein which undergoes post-translational processing. N-linked glycosylation occurs at three asparagine sites, residues 24, 38 and 83 [2] and O-linked glycosylation at serine residue 126 [3] (see Fig. 1). In addition, the C-terminal arginine (residue 166) is removed, possibly by an intracellular carboxypeptidase [4].

Carbohydrate Structure and Function

The structure of the N-linked and O-linked oligosaccharides of erythropoietin produced in Chinese hamster ovary (CHO) cells transfected with a cDNA clone for erythropoietin is almost identical to those of human urinary erythropoietin [3, 5, 6]. The major carbohydrate component is sialated tetra-antennary oligosaccharide, either with or without *N*-acetyllactosaminyl repeats.

When the inhibitor 1-deoxymannojirimycin was used to block the formation of the oligosaccharide chains the modified erythropoietin contained 50% less carbohydrate but retained full biological activity [7]. Similarly, expression of erythropoietin in insect cells reduced the carbohydrate to 30% of its normal content, but full biological activity was retained [7]. In contrast tunicamycin, an inhibitor of N-linked glycosylation caused a marked reduction in the secretion of erythropoietin from the human hepatoblastoma cell Hep G-2 [8].

Desialation – the removal of neuraminic acid from erythropoietin – results in increased activity in vitro [9]. After desialation or deglycosylation, activity in vivo was found to be completely lost [10]. Desialation exposes galactose residues and the loss of activity of asialoerythropoietin in vivo has been

Pagel/Weiss/Jelkmann (Eds.)
Pathophysiology and Pharmacology
of Erythropoietin

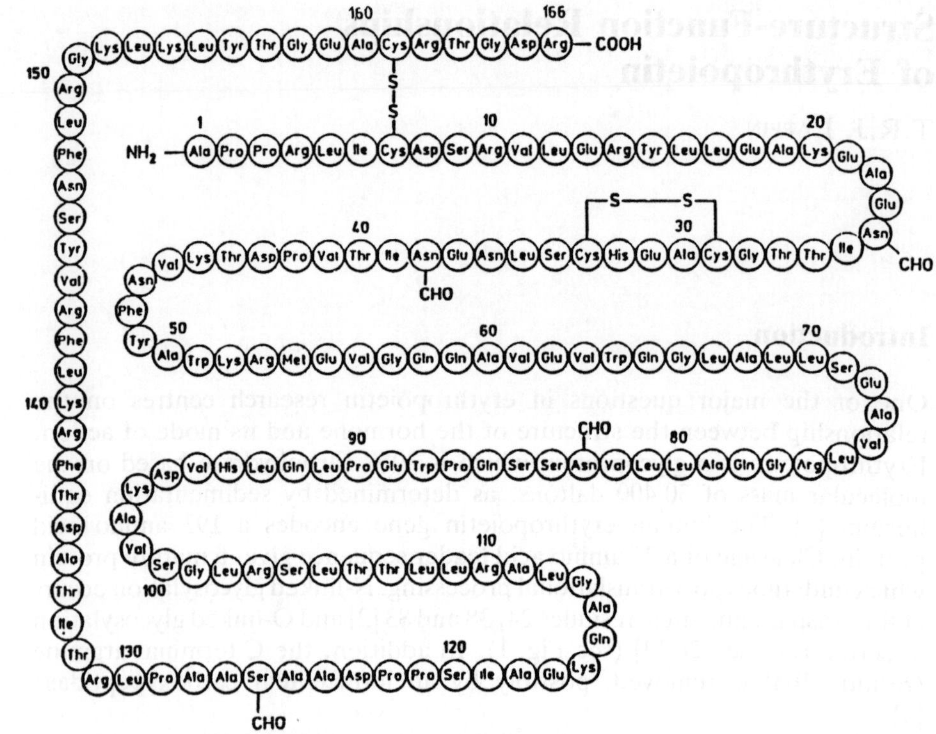

Fig. 1. The primary amino acid sequence of human erythropoietin. The sequence is folded arbitrarily for illustration. Two disulphide bonds link Cys 7 with Cys 161 and Cys 29 with Cys 33. The three N-linked glycosylation sites occur at aspartyl residues 24, 38 and 83. One O-linked glycosylation site is present at Ser 126 (from [32])

explained by its removal from the circulation by galactose receptors in the liver [9]. Imai et al. [11] have examined a series of partially and fully desialated human erythropoietins prepared by controlled neuraminidase digestion. Asialoerythropoietin showed an activity of 220 IU/mg protein in vivo, 1000-fold less than intact erythropoietin. A linear relationship was found between the logarithm of the specific activity and the number of sialic acid groups. Asialoerythropoietin showed a fourfold increase in activity compared with intact erythropoietin in an in vitro assay using mouse marrow cells. Asialoerythropoietin was found to bind to its receptor faster, which may account for its increased activity in vitro.

In vitro mutagenesis is a powerful technique for investigating the relationship between the structure of a protein and its mode of action. Dube et al. [12] have reported that replacement of asparagine by glutamic acid at position 24 led to decreased biological activity of erythropoietin but did not affect its biosynthesis or secretion in stably transfected baby hamster kidney cell lines. Similar mutations at 38 and 83 prevented production of erythropoietin, despite normal levels of mRNA. When the serine 126 residue was replaced by glycine all bioactivity was lost from human erythropoietin [12], which is surprising because mouse erythropoietin has a proline residue at position 126, and is therefore incapable of O-glycosylation at this position.

Protein Structure and Function

Human erythropoietin contains two disulphide bonds, cysteine 29-cysteine 33 and cysteine 7-cysteine 161, which are essential for biological activity. In mouse erythropoietin there is a proline at residue 33 which prevents the formation of a disulphide bond with the cysteine at residue 29. However Lin [13] found that replacement of cysteine by proline at position 33 in human erythropoietin resulted in almost complete loss of bioactivity, in vitro. Taken together the results of these in vitro mutagenesis experiments indicate that it is difficult to predict the effect of small changes in primary structure on biological activity of erythropoietin, with our current knowledge of interspecies differences.

Site-Specific Antibodies

Site-specific antibodies have been used in attempts to identify the functional domains of erythropoietin. This approach is based on the assumption that sequences of amino acids with relatively high hydrophilicity are more likely to be exposed at the surface of the molecule. In general these peptides induce high titres of peptide-specific antibodies in rabbits but their ability to produce antibodies that cross-react with erythropoietin is variable and unpredictable (see Table 1).

Sytkowski and his colleagues [14, 15] used seven peptides corresponding to residues 1–26, 40–59, 80–99, 99–118, 111–129, 131–150 and 147–166 of human erythropoietin. All of these peptides, except one, 147–166, elicited the production of antibodies in rabbits which cross-reacted with native human erythropoietin. Only antibodies to peptide 99–118 and to peptide 111–129 inhibited the biological activity of erythropoietin in vitro. These results suggest that the region 99–129 contains a domain which is functionally important.

In contrast, others (Taylor et al., unpublished data) who have raised antibodies to seven synthetic peptides found that only one, 1–26, elicited production of an antibody which cross-reacted with native human erythropoietin, while six other peptides, 3–26, 8–15, 38–58, 43–62, 111–119, 116–128, failed to produce antibodies which cross-reacted with erythropoietin.

Fibi et al. [16] have raised antibodies to five peptides corresponding to residues 7–23, 52–63, 84–95, 110–123 and 152–166. Three of the peptides – 7–23, 110–123 and 152–166 – elicited production of antibodies which cross-reacted with erythropoietin, whereas peptides 52–63 and 84–95 did not.

Table 1. Cross-reactivity and neutralizing ability of antibodies elicited by synthetic peptides

Study	Peptide	Cross-reactivity with erythropoietin	Inhibition of biological activity
Sytkowski et al. [14, 15]	1– 26	+	–
	40– 59	+	–
	80– 99	+	–
	99–118	+	+
	111–129	+	+
	131–150	+	–
	147–166	+	–
Taylor et al. (unpublished data)	1– 26	+	–
	3– 26	+	–
	8– 15	–	–
	38– 58	–	–
	43– 62	–	–
	111–119	–	–
	116–128	–	–
Fibi et al. [16]	7– 23	+	–
	52– 63	–	–
	84– 95	–	–
	110–123	+	–
	152–166	+	+

Only antisera induced by peptide 152–166 inhibited the biological activity of erythropoietin in an in vitro assay, suggesting that the C-terminal region of the molecule is important in the biological function of the hormone.

Two monoclonal antibodies, raised against intact recombinant human erythropoietin, which neutralized erythropoietin and inhibited binding to its receptor have been described [17]. The antibodies did not precipitate sodium dodecyl sulphate-denatured erythropoietin and did not bind fragments of erythropoietin generated by protease digestion, indicating that the epitopes were conformation dependent, and probably contained some non-contiguous residues.

The interpretation of results from experiments involving site-specific antibodies is complicated because loss of bioactivity does not necessarily imply that the antibody has bound to an epitope which is important for bioactivity. Instead the antibody may bind to an adjacent site and cause loss of activity through steric effects.

Deletion Mutants

Boissel and Bunn [18] have used recombinant DNA techniques to produce a series of seven deletion mutants. Each mutant had a sequence of 7 to 17 consecutive amino acids deleted: 9–21, 48–64, 83–98, 111–119, 142–150,

151–157, and 151–165. Only one mutant, 111–119, retained immunoreactivity and biological activity. It is of interest that this sequence contains the overlapping domain (111–118) of the two sequences previously found to elicit production of antibodies which caused inhibition of erythropoietin activity in vitro [15].

Tertiary Structure of Erythropoietin

The tertiary structure of erythropoietin is unknown, and awaits definition by X-ray crystallography. Although erythropoietin shows little similarity in amino acid sequence with growth hormone (GRH), Bazan [19, 20] has postulated that the tertiary conformation of the two hormones is similar. X-ray crystallography of GRH has revealed a four-membered α-helix bundle with connecting loops [21]. Bazan [19] has proposed a similar globular structure for erythropoietin, in which four anti-parallel α-helices are joined by two long and one short loop connection (see Fig. 2a). This model is consistent with the circular dichroism spectrum of erythropoietin which indicates an α-helical content of 50%, and no evidence for ß-sheet structure in the remainder of the molecule [2].

The notion that α-helical structures are important for bioactivity is supported by data from Boissel and Bunn [18] because mutations within any of the predicted α-helical segments caused loss of bioactivity, but the deletion mutant 111–119 positioned on a long loop outside the predicted helical core did not cause a decrease in bioactivity (see Fig. 2b).

Prolactin (PRL), interleukin-6 (IL-6), and granulocyte colony-stimulating factor (G-CSF) also appear to share the GRH-like helical-folding and loop topology with erythropoietin [19, 20, 22].

The Erythropoietin Receptor

Receptor cross-linking experiments using radio-iodinated erythropoietin indicate that the erythropoietin receptor consists of two polypeptides with molecular masses of 85 kDa and 100 kDa. Sawyer [23] subjected the two cross-linked complexes to digestion by V8 protease and found identical fragments. This result was interpreted as indicating that the primary amino acid sequence of the two polypeptides is similar if not identical. Study of many cell lines and cell types has revealed two classes of receptor based on their affinity for erythropoietin. Some confusion exists in the literature concerning the exact definition for the high and low affinity types, but the value of 0.10 nmol/l represents a reasonable cut-off value for the K_d, below which the receptors should be classified as high affinity. Most cells studied either have only the low-affinity type or both types of receptor. In general the receptor number is low, at less than 4000 receptors per cell [reviewed by 24].

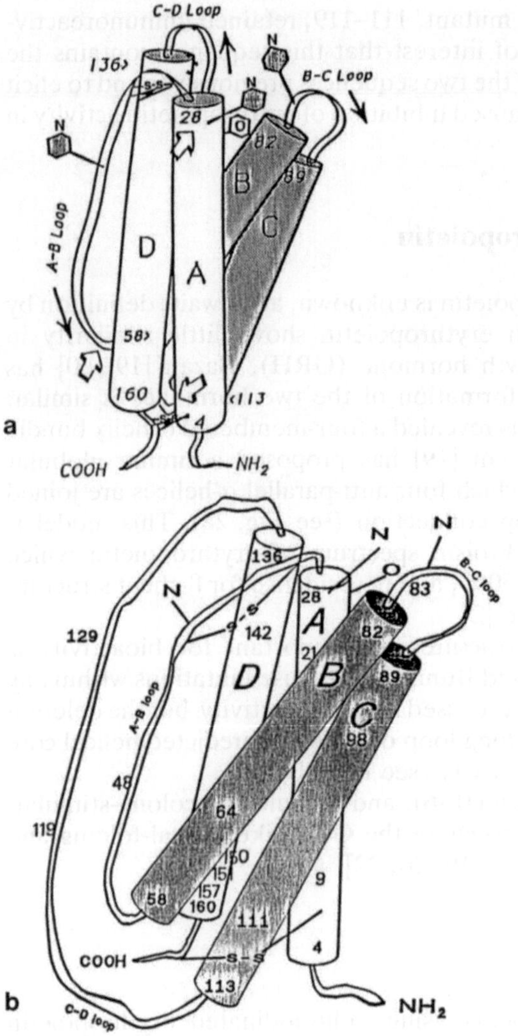

Fig. 2.a The predicted tertiary structure of human erythropoietin. Four anti-parallel α-helices (A-D), each containing 25 amino acids are depicted as cylinders. They are joined by one short (B-C) and two long (A-B and C-D) loops. The helix D may form the primary receptor binding site for erythropoietin to its receptor, by analogy to GRH [20]. Onc disulphide bond (Cys 7–Cys 161) links the A and D helices, while the other (Cys 29 and Cys 33) forms a closed segment of the A-B loop. *N* and *O* mark the sites of N- and O-glycosylation. **b** Reverse view of the tertiary structure of human erythropoietin, exposing the C-D loop. Site-specific antibodies raised against synthetic peptides of residues 99–118 and 111–129 inhibited biological activity of erythropoietin in vitro [15], whereas a mutant erythropoietin from which residues 111–119 had been deleted showed full biological activity in vitro [18]. S-S represents the disulphide bond joining Cys 7 and Cys 161 (modified from [20])

In a major recent advance D'Andrea et al. [17] have cloned the murine erythropoietin receptor. Two independent cDNA clones encoding the erythropoietin receptor were isolated from a pXM expression library made from uninduced murine (MEL) cells. The clones were identified by screening the transfectants for binding and for uptake of radiolabelled erythropoietin, on the assumption that erythropoietin would be internalized by receptor-mediated endocytosis. From the cDNA sequence it may be inferred that the murine erythropoietin receptor is a 507 amino acid polypeptide with a single membrane-spanning domain.

Hormone-Independent Activation of the Erythropoietin Receptor

The successful cloning and expression of the mouse erythropoietin receptor gene has led to valuable insights into the mechanisms involved in the leukaemogenesis induced by Friend virus and in tumourigenicity.

Friend spleen focus-forming virus causes erythroleukaemia in mice and erythroblastosis in bone marrow cultures [25, 26]. A membrane glycoprotein, gp55, encoded by the viral *env* gene, is required both for the induction of leukaemia in vivo and erythroblast proliferation in vitro. Although the mechanism of action of gp55 in these systems is unknown, Li et al. [27] have suggested that gp55 mimics the action of erythropoietin by binding to the erythropoietin receptor. In order to test this idea they co-expressed gp55 and the mouse erythropoietin receptor in a fibroblast cell line. The gp55 was found to bind directly to the erythropoietin receptor in these cells. In addition, co-infection of an IL-3-dependent lymphoid cell line with Friend virus and a virus carrying the murine erythropoietin receptor gene enabled the cell line to grow without IL-3. Thus by binding gp55 to the erythropoietin receptor it was possible to cause prolonged proliferation of infected erythroid cells, which may represent the initial step in Friend virus-induced leukaemogenesis.

Mutant murine erythropoietin receptors were isolated from IL-3-dependent haemopoietic cells infected with retroviruses expressing the receptor cDNA by Yoshimura et al. [28]. Two classes of activating mutations were found. Firstly, a single point mutation in the exoplasmic domain enabled the receptor to induce hormone-independent cell growth and tumourigenesis after expression in non-tumourigenic IL-3-dependent cells. Secondly, a C-terminal truncation of the cytoplasmic domain of the receptor rendered it hypersensitive to erythropoietin, but did not induce hormone-independent growth or tumourigenicity.

The changes caused by the erythropoietin receptor carrying the point mutation are similar to those observed by activation of the erythropoietin receptor by gp55 [27]. These results suggest that tumourigenicity may arise from a variety of mechanisms which override the normal cytokine control systems.

Growth Factor Receptor Homology

Seeking a homology between the erythropoietin receptor and other receptors D'Andrea et al. [29] and Bazan [19, 20, 22] used sequence- and structure-pattern matching techniques to extract common features from the different receptor sequences. Homologous extracellular factor-binding domains were identified in ten receptors – those for erythropoietin, IL-1, IL-4, IL-6, IL-7, GRH, PRL, granulocyte macrophage colony-stimulating factor (GM-CSF), G-CSF and the ß chain of the IL-2 receptor (see Fig. 3). This haemopoietin

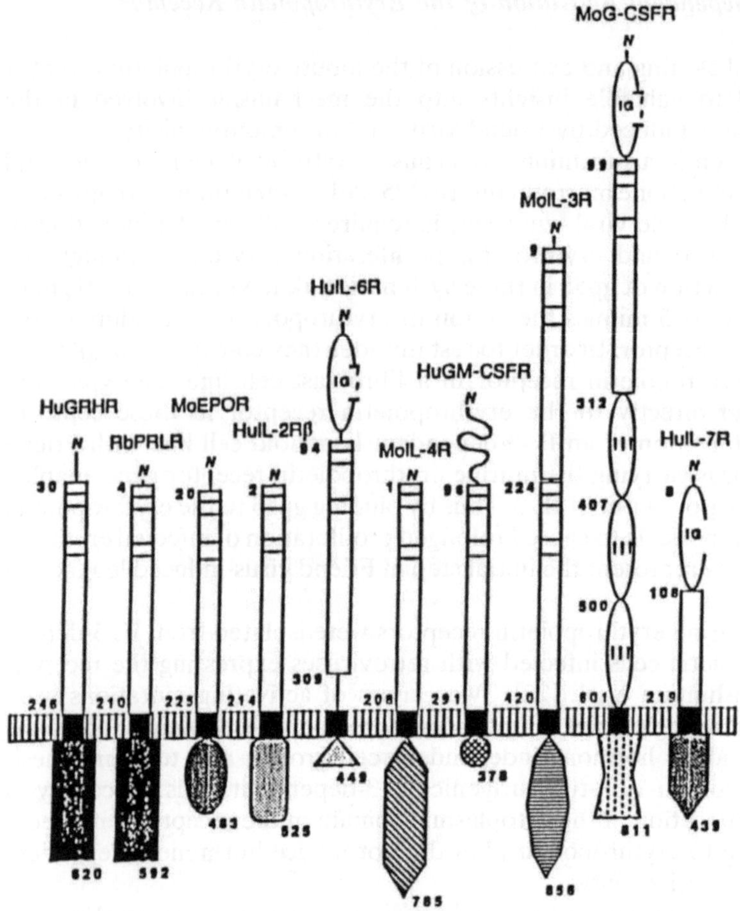

Fig. 3. A schematic comparison of the domain organization of the Class 1 haemopoietin receptor superfamily. The extracelluar cytokine-binding membrane and cytoplasmic domains are shown. Except for the IL-7 receptor, the structures display a characteristic set of Cys residues, depicted here by *dark bands*. *Hu*, *Mo* and *Rb* denote human, mouse and rabbit proteins respectively. (After [20])

receptor superfamily is characterised by an N-terminal set of four conserved cysteine residues except in the case of the IL-7 receptor [20] and a C-terminal collection of spaced aromatic residues, termed the 'WSXWS' box in one letter code notation, with X representing a non-conserved residue [20].

The only ligands for this group of receptor proteins which have a recognised weak similarity in sequence are GRH and PRL. Although the other ligands appear quite different in sequence Bazan [19] suggests that these factors, like their cognate ligands may share hidden symmetries in tertiary fold.

Similarities of structures in disparate ligands suggest the presence of functional analogies. The proposed structural similarities between GRH and erythropoietin suggest that helix D may be involved in the binding of erythropoietin to its receptor [18].

The homologous cytoplasmic domains of GRH receptor (349 residues) and PRL receptor (357 residues) show a weak similarity to those of the erythropoietin receptor (236) and the IL-2 receptor ß chain (286) [19]. The short cytoplasmic extension of the human IL-6 receptor of 81 amino acids was dissimilar to the other molecules. IL-6 receptor has also an N-terminal immunoglobulin(Ig)-like domain [19], and on this basis Yamasaki et al. [30] have suggested that it should be grouped with other multiple-Ig domain receptors for IL-1ß, platelet-derived growth factor and macrophage colony-stimulating factor. In contrast to the similarities noted in the binding domains, little similarity exists between the cytoplasmic extensions of these diverse receptors.

Ulrich et al. [31] have compared the erythropoietic effects of recombinant erythropoietin and recombinant IL-6 on the marrow and peripheral blood in the rat. They found that a single i.v. injection of IL-6 induced a selective erythroid hyperplasia of marrow late normoblasts within 24 h. In contrast, erythropoietin induced a pan-erythroid hyperplasia with successive peaks of pronormoblasts and early normoblasts at 24 h, intermediate normoblasts at 24–48 h and late normoblasts at 72 h.

It is widely assumed that a pyramidal proliferative and differentiative network exists, extending from a progenitor stem cell to a diverse number of mature erythroid, myeloid and lymphoid cell lines. Regulatory growth factors and cell receptors are essential for the maintenance of this system. The structural similarities which are now becoming evident between disparate growth factors, such as erythropoietin and IL-6 on the one hand, and their receptors on the other, suggests that the cytokine network may have arisen from a common primitive system. The wide range of overlapping activities commonly observed for lymphokines and growth factors suggest that common, less specific, receptor-ligand interactions preceded the development of this highly specific and responsive network.

Acknowledgement. I am grateful to Professor J.M. Bridges, Ms. G.E. Elder and Dr M. Afrasiabi for helpful discussions during the preparation of this manuscript.

References

1. Davis JM, Arakawa T, Strickland TW, Yphantis DA (1987) Characterization of recombinant human erythropoietin produced in Chinese hamster ovary cells. Biochemistry 26:2633–2638
2. Lai P-H, Everett R, Wang F-F, Arakawa T, Goldwasser E (1986) Structural characterization of human erythropoietin. J Biol Chem 261:3116–3121
3. Sasaki H, Bothner B, Dell A, Fukuda M (1987) Carbohydrate structure of erythropoietin expressed in Chinese hamster ovary cells by a human erythropoietin cDNA. J Biol Chem 262:12059–12076
4. Recny MA, Scoble HA, Kim Y (1987) Structural characterization of natural human urinary and recombinant DNA-derived erythropoietin. J Biol Chem 262:17156–17163
5. Takeuchi M, Takasaki S, Miyazaki H, Kato T, Hoshi S, Kochibe N, Kobata A (1988) Comparative study of the asparagine-linked sugar chains of human erythropoietins purified from urine and the culture medium of recombinant Chinese hamster ovary cells. J Biol Chem 263:3657–3663
6. Tsuda E, Goto M, Murakami A, Akai K, Ueda M, Kawanishi G, Takahashi N, Sasaki R, Chiba H, Ishihara H, Mori M, Tejima S, Endo S, Arata Y (1988) Comparative structural study of N-linked oligosaccharides of urinary and recombinant erythropoietins. Biochemistry 27:5646–5654
7. Wojchowski DM, Orkin SH, Sytkowski AJ (1987) Active human erythropoietin expressed in insect cells using a baculovirus vector: a role for N-linked oligosaccharide. Biochim Biophys Acta 910:224–232
8. Nielson OJ, Schuster SJ, Kaufmann R, Erslev AJ, Caro J (1987) Regulation of erythropoietin production in a human hepatoblastoma cell line. Blood 6:1904–1909
9. Goldwasser E, Kung CK-H, Eliason JF (1974) On the mechanism of erythropoietin-induced differentiation. XIII. The role of sialic acid in erythropoietin action. J Biol Chem 249:4202–4206
10. Lowy PH, Keighley G, Borsook H (1960) Inactivation of erythropoietin by neuraminidase and by mild substitution reactions. Nature 185:102–103
11. Imai N, Higuchi M, Kawamura A, Tomonoh K, Oh-Eda M, Fujiwara M, Shimonaka Y, Ochi N (1990) Physicochemical and biological characterization of asialoerathropoietin. Eur J Biochem 194:457–462
12. Dube S, Fisher JW, Powell JS (1988) Glycosylation at specific sites of erythropoietin is essential for biosynthesis, secretion, and biological function. J Biol Chem 263:17516–17521
13. Lin F-K (1987) The molecular biology of erythropoietin. In: Rich IN (ed) Molecular and cellular aspects of erythropoietin and erythropoiesis. Springer, Berlin Heidelberg New York, pp 23–36 (NATO ASI series, vol H8)
14. Sytkowski AJ, Fisher JW (1985) Isolation and characterization of an anti-peptide monoclonal antibody to human erythropoietin. J Biol Chem 260:14727–14731
15. Sytkowski AJ, Donahue KA (1987) Immunochemical studies of human erythropoietin using site-specific anti-peptide antibodies. J Biol Chem 262:1161–1165
16. Fibi MR, Stuber W, Hintz-Obertreis P, Luben G, Krumwieh D, Siebold B, Zettlmeissl G, Kupper HA (1991) Evidence for the location of the receptor binding site of human erythropoietin at the carboxyl-terminal domain. Blood 77:1203–1210
17. D'Andrea AD, Lodish HF, Wong GG (1989) Expression cloning of the murine erythropoietin receptor. Cell 57:277–285
18. Boissel J-P, Bunn HF (1990) Erythropoietin structure-function relationships. In: Daniak N, Cronkite EP, McCaffrey R, Shadduck RK (eds) The biology of hematopoiesis. Wiley-Liss, New York, pp 227–232
19. Bazan JF (1989) A novel family of growth factor receptors: a common binding domain in the growth hormone, prolactin, erythropoietin and IL-6 receptors, and the p75 IL-2 receptor β-chain. Biochem Biophys Res Commun 164:788–795

20. Bazan JF (1990) Haemopoietic receptors and helical cytokines. Immunol Today 11:350–354
21. Abdel-Meguid SS, Shieh HS, Smith WW, Dayringer HE, Violand BN, Bentle LA (1988) Three-dimensional structure of a genetically engineered variant of porcine growth hormone. Proc Natl Acad Sci USA 84:6434–6437
22. Bazan JF (1990) Structural design and molecular evolution of a cytokine receptor superfamily. Proc Natl Acad Sci USA 87:6934–6938
23. Sawyer ST (1989) The two proteins of the erythropoietin receptor are structurally similar. J Biol Chem 264:13343–13347
24. Sawyer ST (1990) Receptor for erythropoietin. Distribution, structure, and role in receptor-mediated endocytosis in erythroid cells. In: Harris JR (ed) Blood cell biochemistry, vol. 1. Plenum, New York, p 365
25. Friend C (1957) Cell-free transmission in Swiss adult mice of a disease having the character of a leukemia. J Exp Med 105:307–318
26. Hankins WD, Kost TA, Koury MJ, Krantz SB (1978) Erythroid bursts produced by Friend leukaemia virus in vitro. Nature 276:506–508
27. Li J-P, D'Andrea AD, Lodish HF, Baltimore D (1990) Activation of cell growth by binding of Friend spleen focus-forming virus gp55 glycoprotein to the erythropoietin receptor. Nature 343:762–764
28. Yoshimura A, Longmore G, Lodish HF (1990) Point mutation in the exoplasmic domain of the erythropoietin receptor resulting in hormone-independent activation and tumorigenicity. Nature 348:647–649
29. D'Andrea AD, Fasman GD, Lodish HF (1989) Erythropoietin receptor and interleukin-2 receptor beta chain: a new receptor family. Cell 58:1023–1024
30. Yamasaki K, Taga T, Hirata Y, Yawata H, Kawanishi Y, Seed B, Taniguchi T, Hirano T, Kishimoto T (1988) Cloning and expression of the human interleukin-6 (BSF-2/IFNß 2) receptor. Science 241:825–828
31. Ulrich TR, Castillo J, Yin S, Egrie JC (1991) The erythropoietic effects of interleukin-6 and erythropoietin in vivo. Exp Hematol 19:29–34
32. Lappin TRJ, Maxwell AP (1989) Chemistry and Assays of Erythropoietin. In: Jelkmann W, Gross AJ (eds) Erythropoietin. Springer Berlin Heidelberg New York, p 7–18

Isoforms of Recombinant Human Erythropoietin

R.V. Battersby, B. Ohlrogge, A. Feigler, and C.J. Holloway

Introduction

Recombinant human erythropoietin (rHuEPO) has been made commercially available utilizing a variety of expression systems. Whereas the peptide backbones of these recombinant proteins can be expected to be identical, a polymorphism of these recombinant materials with distinct variations in biological and physicochemical properties is observed. This is related to extensive posttranslational modification processes, which add complex oligosaccharide side chains to the molecule, amounting to approx. 40% of the total molecular mass.

Human EPO has four glycosylation sites; three are N-linked and one is O-linked. Detailed research into the composition of the carbohydrate moiety of rHuEPO from CHO cells [1–9] and BHK cells [10, 11], and comparisons with human urinary EPO [12, 13] have demonstrated the complexity of this glycosylation: The oligosaccharides can be enzymatically cleaved from the rHuEPO molecule and further analyzed. As a result, either bi–, tri–, or tetraantennary oligosaccharide side chains are found, to which in turn sialic acid molecules have been transferred in varying amounts. Especially the differences in the degree of sialization of each molecule allows EPO to be analyzed effectively by isoelectric focusing, thus dividing the recombinant rHuEPO into six or more distinct bands (hereinafter referred to as "isoforms") with pI values ranging from ca. 4.0 to 5.0.

Project Aims

Current research in this field has concentrated on the "entire" carbohydrate portion of EPO. The purpose of this study was to isolate the isoforms of rHuEPO from CHO cells by isoelectric focusing prior to structural analysis, in order to provide more detailed information.

The work described focuses mainly on the characterization of these isoforms with respect to their biological activities. It is demonstrated that the method presented allows the separation and isolation of isoforms of rHuEPO with a high degree of purity, thereby maintaining the biological activity.

Pagel/Weiss/Jelkmann (Eds.)
Pathophysiology and Pharmacology
of Erythropoietin

Methods

Analytical Procedures. The isoelectric focusing analyses were carried out on commercially available (LKB) flat bed electrophoresis systems. The SDS-PAGE analysis was carried out in a vertical system (LKB) using a standard discontinuous buffer system. IEF and SDS-PAGE gels were subsequently stained with either Coomassie blue, or a Western blot/immunostaining was performed employing a murine EPO-specific monoclonal antibody.

Micropreparative Procedures. After IEF separation in 0.5 mm polyacrylamide gels, the gel was fractionated, and the separated isoforms were electroeluted in an Elucon (Biometra) chamber. The quantitation of the isoforms was based on scans of the Coomassie blue-stained gels by two-dimensional laser densitometry, using a LKB Ultroscan.

Test Material. The rHuEPO used in this study was produced in CHO cells and provided by courtesy of Dr. J. Gardner, R.W. Johnson Pharmaceutical Research Institute, Raritan, N.J., USA.

Biological Characterization. The original test material and the subsequently separated isoforms were characterized by radioimmunoassay and by determining their biological activity in the exhypoxic polycythemic mouse bioassay.

Results

Isoelectric Focusing

Prior to the isolation of rHuEPO isoforms, the pI values and relative amounts of the individual isoforms were determined by analytical IEF (Table 1).

Table 1. pI values and relative amounts of individual isoforms determined by analytical IEF

Isoform No.[a]	pI	Relative amount (%)
1	4.70	14.5
2	4.52	21.4
3	4.39	24.8
4	4.25	23.3
5	4.14	14.0
6	4.05	1.8

[a] Isoforms numbered in order of decreasing pI values.

Micropreparative Separation by Isoelectric Focusing

The electrophoretic separation of the rHuEPO isoforms was achieved in a two-step procedure: Initially, using conventional 0.5 mm polyacrylamide IEF gels, rHuEPO raw material (approx. 1.7 mg/ml of protein in a PBS buffer) was separated by isoelectric focusing. Lateral strips were cut from the gel and stained to visualize the EPO bands. These were then replaced onto the gel, support and the bands containing the individual isoforms were fractionated appropriately.

The fractionated gel strips were placed individually into an electroelution chamber and transferred electrophoretically into a TRIS/Glycine buffer medium (pH 8.9). The separated proteins were maintained in this buffer prior to subsequent reanalysis by IEF, SDS-PAGE etc. as described above.

Reanalysis of Isolated Isoforms by Isoelectric Focusing

The degree of purity and the recovery in micrograms of protein for each isoform was determined by one-dimensional laser densitometry (Western blot/immunostaining) and two-dimensional laser densitometry (Coomassie blue staining), respectively. The quantitation was based on the assumption that all rHuEPO isoforms have the same staining intensity both in the Coomassie and the immunostaining procedure. The densitometric values for the individual isoforms were compared with those of the test material. The results are summarized in Table 2.

Reanalysis by SDS-PAGE

The isolated isoforms were also subjected to SDS-PAGE with subsequent Western blotting. The obtained apparent relative molecular masses were evaluated after one-dimensional laser densitometry and are summarized in Table 3.

Table 2. Amount and purity of isolated rHuEPO isoforms

Isoform no.	µg/ml	ml	µg	purity (%)
1	18.2	10.4	189	94.1
2	112.4	6.9	776	93.8
3	66.4	5.7	378	91.2
4	53.9	6.4	345	96.1
5	17.6	5.3	93	100.0
6	–	–	–	–

Table 3. Relative molecular masses of isolated rHuEPO isoforms

Sample	Relative molecular mass (kDa)[a]
Isoform no. 1	35.7
Isoform no. 2	36.2
Isoform no. 3	36.8
Isoform no. 4	37.2
Isoform no. 5	38.7
rHuEPO batch 7915-15	36.4

[a] Mean value.

Biological Characterization

Table 4 summarizes the results of the radioimmunoassays and mouse bioassays performed for the separated rHuEPO isoforms. The calculation of specific activities was based on the protein determination by two-dimensional laser densitometry. The values for the original test material are also given for reasons of comparison.

Summary

A total of five individual isoforms of rHuEPO from CHO cells could be isolated by micropreparative IEF with subsequent electroelution. This procedure does not appear to impair the physicochemical or biological properties of the EPO isoforms, as shown by subsequent reanalysis.

The specific biological activities of the separated isoforms appear to increase with decreasing pI values. This tendency is even more pronounced if related to the ratio of bioassay to radioimmunoassay, which essentially removes the influence of the protein determination on this value. The apparent molecular masses of the isolated isoforms display a similar tendency,

Table 4. Specific activities (U/μg protein) of isolated rHuEPO isoforms

Isoform No.	RIA	Bioassay	Bioassay/RIA
1	155	91	0.59
2	120	88	0.73
3	116	128	1.10
4	128	156	1.22
5	121	185	1.53
rHuEPO batch 7915-15	125	133	1.06

RIA, radioimmunoassay

i.e., to increase (35.7–38.7 kDa) with decreasing pI values, as judged by SDS-PAGE.

The data presented is consistent with a model assuming that a high degree of sialization will protect the circulating EPO molecule from clearance, thus enhancing the in vivo biological activity.

References

1. Imai N, Higuchi M, Kawamura A, Tomonoh K, Oh-Eda M, Fujiwara M, Shimonaka Y, Ochi N (1990) Physicochemical and biological characterization of asialoerythropoietin – Suppressive effects of sialic acid in the expression of biological activity of human erythropoietin in vitro. Eur J Biochem 194:457–462
2. Hokke CH, Bergwerff AA, van Dedem GWK, van Oostrum J, Kamerling JP, Vliegenthart JFG (1990) Sialylated carbohydrate chains of recombinant human glycoproteins expressed in Chinese hamster ovary cells contain traces of N-glycolylneuraminic acid. FEBS Lett 275:9–14
3. Takeuchi M, Takasaki S, Miyazaki H, Kato T, Hoshi S, Kochibe N, Kobata A (1988) Comparative study of the asparagine-linked sugar chains of human erythropoietins purified from urine and the culture medium of recombinant Chinese hamster ovary cells. J Biol Chem 263:3657–3663
4. Takeuchi M, Inoue N, Strickland TW, Kubota M, Wada M, Shimizu R, Hoshi S, Kozutsumi H, Takasaki S, Kobata A (1989) Relationship between sugar chain structure and biological activity of recombinant human erythropoietin produced in Chinese hamster ovary cells. Proc Natl Acad Sci USA 86:7819–7822
5. Takeuchi M, Takasaki S, Shimada M, Kobata A (1990) Role of sugar chains in the in vitro biological activity of human erythropoietin produced in recombinant Chinese hamster ovary cells. J Biol Chem 265:12127–12130
6. Sasaki H, Bothner B, Dell A, Fukuda M (1987) Carbohydrate structure of erythropoietin expressed in Chinese hamster ovary cells by a human erythropoietin cDNA. J Biol Chem 262:12059–12076
7. Sasaki H, Ochi N, Dell A, Fukuda M (1988) Site-specific glycosylation of human recombinant erythropoietin: analysis of glycopeptides or peptides at each glycosylation site by fast atom bombardment mass spectrometry. Biochemistry 27:8618–8626
8. Fukuda MN, Sasaki H, Lopez L, Fukuda M (1989) Survival of recombinant erythropoietin in the circulation: the role of carbohydrates. Blood 73:84–89
9. Spivak JL, Hogans BB (1989) The in vivo metabolism of recombinant human erythropoietin in the rat. Blood 73:90–99
10. Tsuda E, Kaeanishi G, Ueda M, Masuda S, Sasaki R (1990) The role of carbohydrate in recombinant human erythropoietin. Eur J Biochem 188:405–411
11. Dube S, Fisher JW, Powell JS (1988) Glycosylation at specific sites of erythropoietin is essential for biosynthesis, secretion, and biological function. J Biol Chem 263:17516–17521
12. Fukuda M, Sasaki H, Fukuda MN (1989) Structure and role of carbohydrate in human erythropoietin. Adv Exp Med Biol 271:53–67
13. Tsuda E, Goto M, Murakami A, Akai K, Ueda M, Kawanishi G, Takahashi N, Sasaki R, Chiba H, Ishihara H, Mori M, Tejima S, Endo S, Arata Y (1988) Comparative structural study of N-linked oligosaccharides of urinary and recombinant erythropoietins. Biochemistry 27:5646–5654

Activation of the Erythropoietin Receptor in the T3Cl2 Friend Erythroleukemia Cell Line

C. Lacombe, S. Chretien, N. Casadevall, I. Dusanter, S. Gisselbrecht, and P. Mayeux

Introduction

The Friend virus complex consists of a replication competent virus, the Friend murine leukemia virus (F-MuLV) and a replication defective virus, the Friend spleen focus-forming virus (SFFV). The SFFV defective component is responsible for the induction of acute leukemia in adult mice of susceptible strains [1]. Erythroid cells infected with the polycythemia-inducing variant of SFFV (SFFVp) can proliferate and differentiate in the absence of erythropoietin (Epo) whereas the anemia-inducing strain of SFFV renders them hypersensitive to the hormone. The envelope glycoprotein (*Env*) encoded by SFFV has been shown to be responsible for its biological effect [2]. The majority of the SFFVp *Env* protein remains in the endoplasmic reticulum; however, a small part (3%–5%) is fully glycosylated and reaches the plasma membrane. This fraction transported to the cell surface seems to be essential for erythroblastosis [3, 4].

The murine Epo receptor (Epo-R) cDNA has been cloned recently [5]. It encodes a 507 amino acid polypeptide chain with a single membrane-spanning domain and shares structural homologies with several hemopoietin receptors, defining a new receptor gene family [6]. It has been shown that the SFFVp *Env* glycoprotein gp55 directly interacts with the Epo-R, and that this association is responsible for the Epo-independent growth of infected erythroid progenitors [7]. This interaction was shown to occur mostly in the endoplasmic reticulum and suggested to be the biologically active form [8].

We have analyzed the Epo-R expression and the Epo-R gene structure of a murine erythroleukemia cell line, T3Cl2, which originated from a polycythemic Friend virus-induced leukemia in a DDD mouse [9]. We show that a SFFV long terminal repeat (LTR) was inserted in the first exon of the Epo-R gene. Moreover, cross-linking experiments revealed that Epo was mainly cross-linked with a protein of 63 kDa [10]. This protein which is the membrane form of the *Env* glycoprotein gp55, is non-covalently associated with the cloned chain of the Epo-R.

Pagel/Weiss/Jelkmann (Eds.)
Pathophysiology and Pharmacology
of Erythropoietin

LTR Insertional Activation of the Epo-R Gene in the T3Cl2 Cell Line

By PCR analysis, we showed that a viral LTR sequence was inserted in the first exon of one allele of the Epo-R gene in the T3Cl2 cell line. This genomic alteration was confirmed by Southern analysis performed with genomic DNA isolated either from T3Cl2 or from normal DDD mouse liver. A rearranged fragment 0.5 kb longer than the germ line allele, corresponding to the LTR insertion, could be seen in T3Cl2 with three restriction enzymes tested.

A the mRNA level, an overexpression of Epo-R mRNA was detected by Northern blot analysis, and primer extension studies showed that the major Epo-R transcriptional cap site was located 37 bases upstream of the normal cap site. This abnormal initiation originated from the R region of the LTR, and the T3Cl2 Epo-R mRNA was a fusion product whose 5′ end was derived from the LTR sequence. By sequencing the amplified T3Cl2 genomic DNA, we reached the conclusion that the site of insertion of the LTR was in the untranslated part of the first exon, 45 bp upstream of the translational ATG initiation codon (Fig. 1). Thus, this LTR insertion did not interrupt the reading frame and the Epo-R protein was expected to be normal. A very weak initiation was found which corresponded to the promoter of the normal Epo-R allele; consequently the Epo-R mRNA overexpression was due to the abnormal allele modified by LTR insertion. However, the number of binding sites on the T3Cl2 cell line did not differ significantly from other erythroleukemia cell lines. The LTR sequence inserted in T3Cl2 had a deletion in the second direct repeat of the U3 region strongly suggesting that this LTR derived from the SFFV component of the Friend virus [12].

Evidence That in SFFVp-Induced Erythroleukemia Cell Lines, Epo is Cross-Linked with the SFFV Env Protein

We have previously shown that the Epo-R can be solubilized by Triton X 100 as a high molecular weight complex [13].

When cross-linking experiments were performed with disuccinimidyl suberate (DSS) in the T3Cl2 cell line as well as in other SFFVp-induced erythroleukemia cell lines, Epo was mainly cross-linked with a protein of 63 kDa molecular mass. Epo can also be cross-linked with 85 and 100 kDa proteins, as previously described for normal erythroid cells [14], but to a much lesser extent (Fig. 2).

In order to identify these different components, we derived anti-Epo antibodies by immunizing rabbits with purified human recombinant Epo, and also antibodies against the cloned chain of the Epo-R (p 66). Anti-Epo-R antibodies were produced by immunizing rabbits with a fusion protein consisting of the maltose transporter (Mal-E) of *Escherichia coli* [15] and most of the intracellular part of p 66. This construction was cloned in the

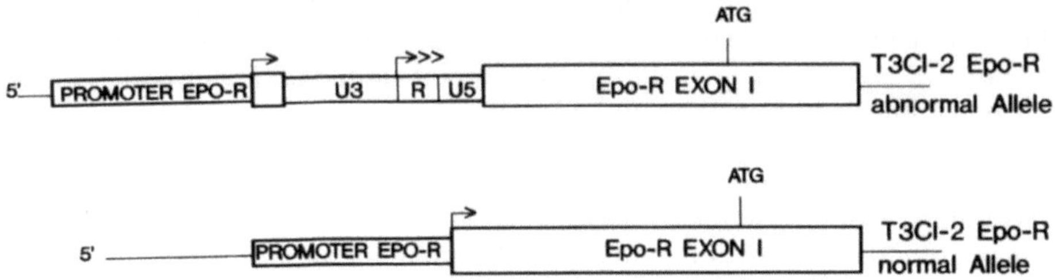

Fig. 1. Schematic representation of the LTR insertion in the T3Cl2 cell line showing structure of the Epo-R allele in T3Cl2 cell line and localization of the SFFV LTR within the first exon of the Epo-R gene

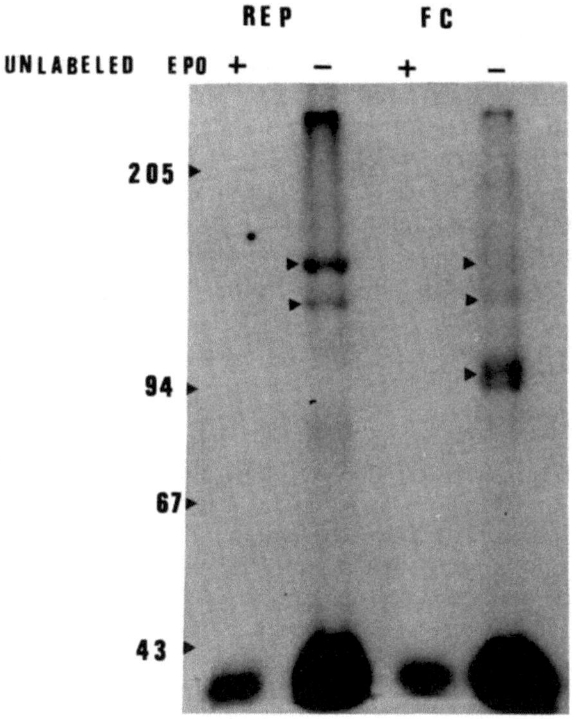

Fig. 2. Cross-linking of Epo to normal rat erythroid progenitors (*REP*) and to T3Cl2 Friend cells (*FC*). ^{125}I-Epo (500 p*M*) was incubated with erythroid cells in the presence or absence of 50 n*M* unlabelled Epo, and cross-linked with 0.5 m*M* DSS. The cross-linked proteins were separated on a 7.5% polyacrylamide slab gel under reducing conditions. Molecular masses of unlabelled protein markers run in adjacent lanes which are revealed by Comassie blue staining. These are indicated with *arrows*

CIMEC plasmid [16] and the fusion protein was purified by affinity chromatography through cross-linked amylose [17]. We also used an antiserum to purified Rauscher *Env* protein which was a gift from Dr. Gruber, National Cancer Institute, Bethesda.

The native Epo-R, after solubilization, could be precipitated by antibodies directed against Epo, p 66, and *Env*, whereas control antibodies directed against the bacterial part of the fusion protein, (anti-Mal E antibodies) did not precipitate any Epo cross-linked protein.

In another set of experiments, solubilized native Epo-R complexes were first immunoprecipitated with anti-Epo antibodies. Then denaturation was performed by boiling in the presence of sodium dodecyl sulphate (SDS) and ß-mercaptoethanol. After denaturation, a second immunoprecipitation with anti-Epo antibodies resulted in the precipitation of all the cross-linked proteins. In striking contrast, anti-p 66 antibodies did not precipitate any Epo cross-linked protein, suggesting that p 66 was not itself cross-linked to Epo. However, the 100 kDa complex, corresponding to the 63 kDa protein when the 34 kDa Epo molecular mass was subtracted, was precipitated by anti-*Env* antibodies. This result strongly suggests that the 63 kDa protein cross-linked to Epo was the SFFVp *Env* protein. Moreover, enzymatic deglycosylation analyses confirmed that the main Epo cross-linked protein was the SFFVp *Env* protein [18].

It has been shown that SFFVp *Env* proteins dimerize by disulfide bonds before reaching the plasma membrane [19]. Immunoprecipitation of the native Epo-R also showed a less abundant 160 kDa complex which corresponds to the dimerized form of SFFVp protein.

Discussion

Our results strongly suggest that the SFFV can play a role in the oncogenic transformation of erythroleukemia cells by LTR insertional activation of the Epo-R gene. This viral LTR insertion resulted in Epo-R mRNA overexpression in T3Cl2 cells. Nevertheless, there was no enhancement of the number of Epo-binding sites at the T3Cl2 cell surface. This absence of correlation could be explained by the expression at a low level of a putative second receptor subunit that would hamper the formation of binding sites at the cell surface. In addition, since gp55 also interacts with the Epo-R in the rough endoplasmic reticulum, resulting in intracellular retention and a decrease in cell surface expression of the Epo-R [7, 20], this may also account for the reduced number of Epo-binding sites.

Many cases of retroviral insertional mutagenesis have been reported as a mechanism to increase expression of oncogenes in several murine models. Recently, constitutive interleukin-2-receptor (IL2-R) and IL6-R gene expression due to the insertion of an intracisternal A particle LTR have been described in murine lymphoma and plasmacytoma cell lines, respectively [21, 22]. This up-regulation of receptor expression can provide a positive pressure in the development of transplantable and tumorigenic leukemia.

Yoshimura et al. [7] previously demonstrated that an IL3-dependent cell line, Ba/F3, transfected with both Epo-R and gp55 cDNAs could grow in the absence of added growth factor. They also showed that the interaction of

Epo-R and gp55 occurred in the rough endoplasmic reticulum, and that co-transfection of gp55 and Epo-R in these cells caused marked reduction in the number of cell surface Epo-binding sites, probably due to retention of Epo-R within the endoplasmic reticulum.

Using cross-linking experiments, and immunoprecipitation of the native or denatured Epo-R complexes in T3Cl2 cells, we confirmed the specific association between gp55 *Env* protein and the Epo receptor.

These experiments were performed with intact cells, and Epo could only be cross-linked with proteins present on the cell surface. Thus, we can assume that the p66-gp55 SFFVp *Env* complexes are able to reach the cell surface, and these data are in good agreement with previous studies showing a strong correlation between the ability of SFFVp *Env* protein to reach the cell surface and its capacity to induce erythroleukemia [4].

Furthermore, we used cross-linking experiments in which DSS was cross-linked to Epo to reveal the presence of the 63 kDa protein. These results show that the p66-gp55 SFFVp *Env* complexes are still able to bind Epo, and that therefore the Epo-R has distinct binding sites for the SFFVp *Env* protein and Epo.

Conclusion

In the T3Cl2 cell line, the Epo receptor expression is up-regulated by LTR insertional activation. We hypothesize that, as a consequence of Epo-R overexpression, a large amount of Epo-R protein is present in the endoplasmic reticulum and able to interact in turn with gp55 *Env* protein. Thus SFFV can play a role in the oncogenic transformation of these cells both by LTR insertional activation of the Epo-R and by gp55 association with this receptor.

Acknowledgements. We acknowledge Jennifer Richardson for critical reading of the manuscript.

References

1. Ruscetti S, Wolff L (1984) Spleen focus forming virus: relationship of an altered envelope gene to the development of a rapid erythroleukemia. Curr Top Microbiol Immunol 112:21–44
2. Wolff L, Ruscetti S (1988) The spleen focus-forming virus (SFFV) envelope gene induces acute erythroleukemia. J Virol 62:2158–2163
3. Machida CA, Bestwick RK, Boswell BA, Kabat D (1985) Role of a membrane glycoprotein in Friend virus-induced erythroleukemia: studies of mutant and revertant viruses. Virology 144:158–172
4. Li JP, Bestwick RK, Spiro C, Kabat D (1987) The membrane glycoprotein of Friend spleen focus-forming virus: evidence that the cell surface component is required for pathogenesis and that it binds to a receptor. J Virol 61:2782–2792

5. D'Andrea AD, Lodish HF, Wong GG (1989) Expression cloning of the murine erythropoietin receptor. Cell 57:277–285
6. Bazan JF (1990) Haematopoietic receptors and helical cytokines. Immunol Today 11:350–354
7. Yoshimura A, D'Andrea AD, Lodish HF (1990) Friend spleen focus-forming virus glycoprotein gp55 interacts with the erythropoietin receptor in the endoplasmic reticulum and affects receptor metabolism. Proc Natl Acad Sci USA 87:4139–4143
8. Li JP, D'Andrea AD, Lodish HF, Baltimore D (1990) Activation of cell growth by binding of Friend spleen focus-forming virus gp55 glycoprotein to the erythropoietin receptor. Nature 343:762–764
9. Ikawa Y, Aida M, Inoue Y (1976) Isolation and characterization of high and low differentiation-inducible Friend leukemia lines. Gan 67:767–770
10. Sasaki R, Yanagawa SI, Hitomi K, Chiba H (1987) Characterization of erythropoietin receptor of murine erythroid cells. Eur J Biochem 168:43–48
11. Todoroko K, Kanazawa S, Amanuma H. Ikawa Y (1987) Specific binding of erythropoietin to its receptor on responsive mouse erythroleukemia cells. Proc Natl Acad Sci USA 84:4126–4130
12. Lacombe C, Chrétien S, Lemarchandel V, Mayeux P, Romeo PH, Gisselbrecht S, Cartron JP (1991) Spleen focus-forming virus long terminal repeat insertional activation of the murine erythropoietin receptor gene in the T3Cl-2 Friend leukemia cell line. J Biol Chem 266:6952–6956
13. Mayeux P, Casadevall N, Lacombe C, Muller O, Tambourin P (1990) Solubilization and hydrodynamic characteristics of the erythropoietin receptor. Evidence for a multimeric complex. Eur J Biochem 194:271–278
14. Mayeux P, Casadevall N, Muller O, Lacombe C (1990) Glycosylation of the murine erythropoietin receptor. FEBS Lett 269:167–170
15. Duplay P, Bedouelle H, Fowler A, Zabin I, Saurin W, Hofnung M (1984) Sequences of the malE gene and its product, the maltose-binding protein of Escherichia coli K 12. J Biol Chem 259: 10606–10613
16. Clement JM, Szmelcman S, Jehanno M, Martineau P, Schwartz O, Hofnung M (1989) Propriétés neutralisantes pour le virus HIV d'une protéine hybride MalE-CD4 exprimée chez E. coli et purifiable en une étape. C R Acad Sci (III) 308:401–406
17. Bedouelle H, Duplay P (1988) Production in Escherichia coli and one-step purification of bifunctional hybrid proteins which bind maltose. Export of the klenow polymerase into the periplasmic space. Eur J Biochem 171:541–549
18. Casadevall N, Lacombe C, Muller O, Gisselbrecht S, Mayeux P (1991) Multimeric structure of the membrane erythropoietin receptor of murine erythroleukemia cells (Friend cells): cross-linking of erythropoietin with the spleen focus-forming virus envelope protein. J Biol Chem 266: 16015–16020
19. Gliniak BC, Kabat D (1989) The leukemogenic membrane glycoprotein encoded by Friend spleen focus-forming virus: transport to cell surface and shedding are controlled by disulfide bonded dimerization and by cleavage of a hydrophobic membrane anchor. J Virol 63:3561–3568
20. Ruscetti SK, Janesh NJ, Chakraborti A, Sawyer ST, Hankins WD (1990) Friend spleen focus-forming virus induces factor independence in an Erythropoietin-dependent erythroleukemia cell line. J Virol 64:1057–1062
21. Kono T, Doi T, Yamada G, Hatakeyama M, Minamoto S, Tsudo M, Miyasaka M, Miyata T. Taniguchi T (1990) Murine interleukin 2 receptor ß chain: Dysregulated gene expression in lymphoma line EL-4 caused by a promoter insertion. Proc Natl Acad Sci USA 87:1806–1810
22. Sugita T. Totsuka T, Saito M, Yamasaki K, Taga T, Hirano T, Kishimoto T (1990) Functional murine interleukin 6 receptor with the intracisternal A particle gene product at its cytoplasmic domain. Its possible role in plasmacytomagenesis. J Exp Med 171: 2001–2009

II. Sites and Control of Production

The Role of Adenosine in Hypoxic Regulation of Kidney Production of Erythropoietin

J. FISHER and J. NAKASHIMA

Introduction

Hypoxia is the primary stimulus for erythropoietin (Ep) production [1, 2]. We postulate that there is a sensitive oxygen sensor mechanism which is involved in the regulation of both renal and extrarenal Ep production. Hypoxia and cobalt treatment have been demonstrated to increase [3] and decrease [4] Ep biosynthesis in two hepatoblastoma (HepG2 and Hep3B) cell lines. Goldberg et al. [5] have presented an interesting model for the hypoxic regulation of Ep production in the kidney in which they postulate that the renal oxygen sensor is a heme protein. These investigators [5] have proposed from their studies in Hep3B cells that Ep production is regulated by the conformation of this heme protein which depends upon the oxygen tension. Our present studies consider the possibility that hypoxia releases adenosine from an oxygen-sensing cell which acts as a messenger substance to activate adenylate cyclase to trigger a cascade of events leading to an increase in Ep production and/or secretion.

Model for the Control of Ep Production

Our model for the control of Ep production in the kidney is shown in Fig. 1. We postulate that there are several transducer substances that we have identified primarily as adenosine, eicosanoids, or beta-2-adrenergic agonists which are released during hypoxia that trigger Ep production. During reoxygenation oxygen-derived metabolites, such as hydrogen peroxide, may trigger Ep production. As noted in this model, we postulate that hypoxia either triggers an oxygen sensor in the cell which produces Ep, or from a closely adjacent cell, which triggers the release of external messengers such as adenosine, eicosanoids, or beta-2-adrenergic agonists. These messenger substances activate a receptor on the Ep-producing cell which is coupled to adenylate cyclase causing the generation of cyclic AMP. The latter leads to the activation of protein kinases which increase the production of phosphoproteins that are involved in transcription and/or translation and release of the final 165 amino acid Ep molecule [6]. The stimuli that we have studied which primarily affect Ep secretion may act through the activation of adenylate

Pagel/Weiss/Jelkmann (Eds.)
Pathophysiology and Pharmacology
of Erythropoietin

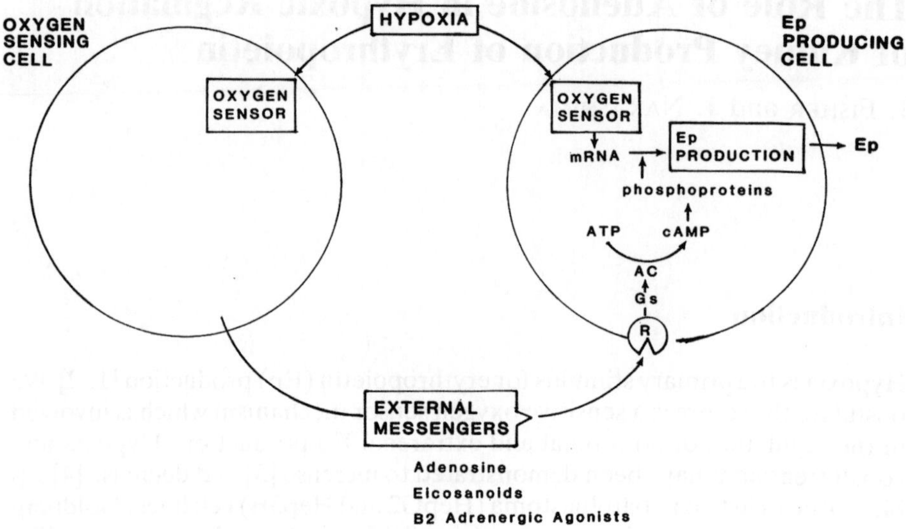

Fig. 1. Schematic model for renal oxygen sensing and erythropoietic production. Ep, erythropoietin; R, receptor; Gs, G stimulatory; Ac, adenylate cyclase; ATP, adenosine triphosphate; cAMP, 3',5'-adenosine cyclic monophosphate

cyclase or adenosine [7]; the eicosanoids PGE_2, PGI_2, 6-keto-PGE_1 [1]; hydrogen peroxide which is generated from superoxide ion [8–10] during hypoxia and reoxygenation; and beta-2-adrenergic agonists [11–13]. Therefore, we postulate that the increase in Ep production in response to hypoxia in vivo is most likely due to the release of several transducer molecules. A mild to moderate hypoxic stimulus probably increases the generation of adenosine from intracellular ATP and extracellular nucleotidases which break down ATP to adenosine. A more severe hypoxic stimulus could produce the eicosanoids and beta-2-adrenergic agonists, and with reoxygenation perhaps hydrogen peroxide, which could also modulate Ep production. We do know that our cloned HepG2 and Hep3B cell will respond to hypoxia in vitro to increase Ep mRNA and Ep production. Therefore, the possibility also exists that the Ep-producing cell may have an oxygen-sensing mechanism of its own to trigger Ep production which does not depend upon the adenylate cyclase/cyclic AMP system. The phosphoproteins generated by adenosine probably act at a post-transcriptional stage because hypoxia alone in these cloned cells will raise the level of Ep messenger RNA, but does not increase cyclic AMP. Thus, the cyclic AMP mechanism is probably more likely related to the external messenger substances that are produced in closely adjacent cells in vivo following a hypoxic stimulus and may act to supplement Ep production following the increase in Ep mRNA in the Ep-producing cells.

In order to characterize the in vivo effects of hypoxia in Ep production, we determined the plasma levels of Ep in normal mice exposed to 0.42 atm hypoxia for 22 h per day for 14 days and then the mice were placed at ambient pressure for the next 10 days [14]. Note in Fig. 2 that when normal mice were exposed to hypoxia for 14 days followed by 10 days at ambient pressure, the peak level of serum Ep occurs on the 3rd and 4th days of hypoxia which is exceedingly high and in the range of 450 mU/ml. Note however that with

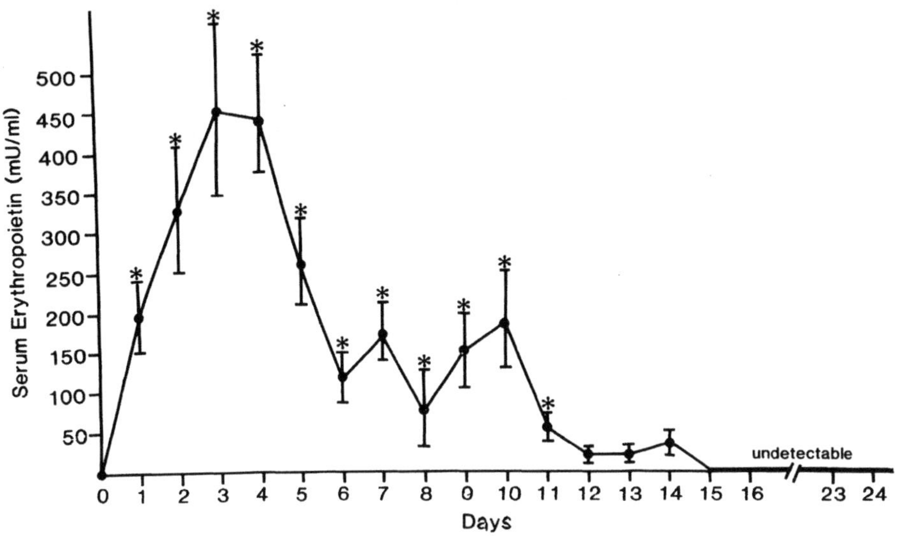

Fig. 2. Serum erythropoietin concentrations in mice exposed to hypoxia (0,42 atm, 22 h/day) for 14 days and during a 10-day posthypoxic period. *Lines* over each point represent ± SE; n=5–11 mice at each time interval. *, Significantly different from day 0 control value (P<0.05) (from [14])

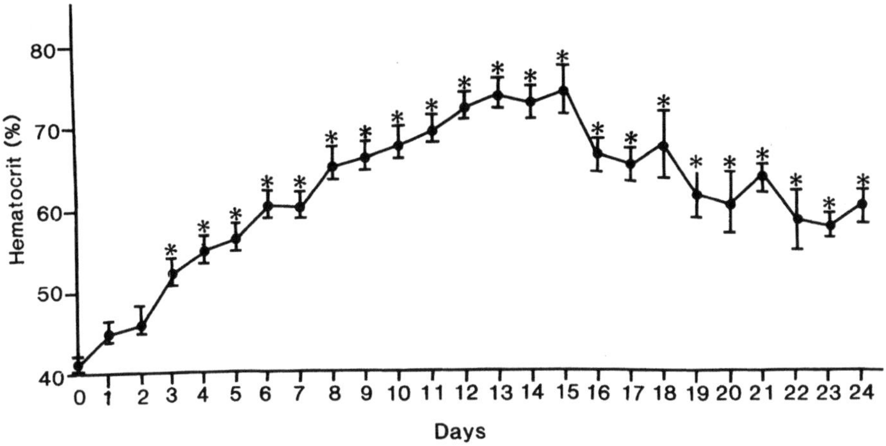

Fig. 3. Hematocrit values of mice exposed to hypoxia (0.42 atm, 22 h/day for 14 days followed by 10 posthypoxic days at ambient pressure. *Lines* over each point represent ± SE; n=5–11 mice at each time interval. *, Significantly different from day 0 control (P<0.05) (from [14])

continued exposure to 22 h per day, the Ep titers in plasma decreased toward baseline levels between 12 and 14 days and reached levels which are still above controls, but significantly less than that seen at 3–4 days. When the mice were removed from the hypobaric chamber and placed at normal atmospheric pressure, the Ep titers in serum were undetectable indicating some type of hyperoxic negative feedback on kidney production of Ep. Note in Fig. 3 that the hematocrit also rises over the 14 days that the mice were exposed to discontinuous hypoxia. Note the significant decline in hematocrit beginning on the 14th day when the mice were removed from the hypobaric chamber and placed at ambient pressure. The negative feedback mechanism which controls kidney production of Ep has not been elucidated.

The Role of Adenosine in Hypoxic Regulation of Ep Production

In considering possible mechanisms of hypoxic regulation of Ep production, we were impressed with our early findings shown in Fig. 4 in which we were

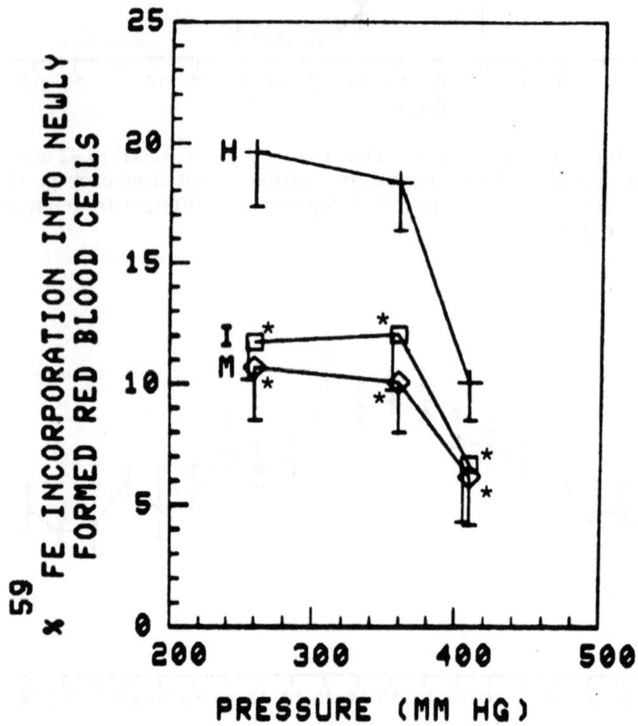

Fig. 4. ^{59}Fe incorporation in exhypoxic polycythemic mice exposed to 260, 360, and 410 mmHg (hypobaric hypoxia) after pretreatment with Na_2CO_3 vehicle (*H*), indomethacin (*I*; 5 mg/kg i.p.), or meclofenamate (*M*; 5 mg/kg i.p.). Each value expresses mean ± SEM for 8–15 mice per group. *, Significantly ($P<.05$) different from the respective pressure for the hypoxia (*H*) vehicle-treated controls (from [22])

able to demonstrate that meclofenamate and indomethacin, both potent cyclooxygenase inhibitors, partially suppressed the rise in Ep in serum of exhypoxic polycythemic mice as indicated by radioiron incorporation. However, an approximately 50% increase in Ep production still occurred. This led us to believe that there were other mechanisms for Ep regulation other than eicosanoids.

Adenosine is a nucleoside which is involved in the regulation of a number of biological processes. The chemical structure is shown in Fig. 5. We became interested in adenosine because adenosine production in the kidney significantly increases very early following ischemic hypoxia [15]. There is a primary oxygen-sensing reaction which may be triggered by secondary biochemical changes such as a decrease in cellular ATP, increase in ADP and AMP, and stimulation of adenosine and hypoxanthine production [16]. This metabolic pathway for adenosine production during hypoxia is shown in Fig. 6. There are two independent routes for the production of extracellular adenosine. When energy demand exceeds supply, cytoplasmic ATP is converted to ADP and AMP, and intracellular 5′-nucleotidase dephosphorylates AMP to yield adenosine which is transported out of the cell by the symmetrical nucleoside transporter. Adenosine has a "retaliatory" action against stimuli that deplete intracellular ATP. For example, in many tissues adenosine acts to improve energy supply by vasodilation and stimulation of an insulin-dependent glucose substrate. Extracellular adenosine can also be formed by a second route from ATP release into the extracellular space, for example, from the exocytosis of vesicles containing the nucleotide. Ectonucleotidases then degrade ATP to adenosine. This extracellular pathway for adenosine synthesis acts to antagonize many of the extracellular effects of ATP. Please note that ATP metabolism is initiated in hypoxic tissue causing the depletion of ATP through the limited availability of oxygen for ATP production.

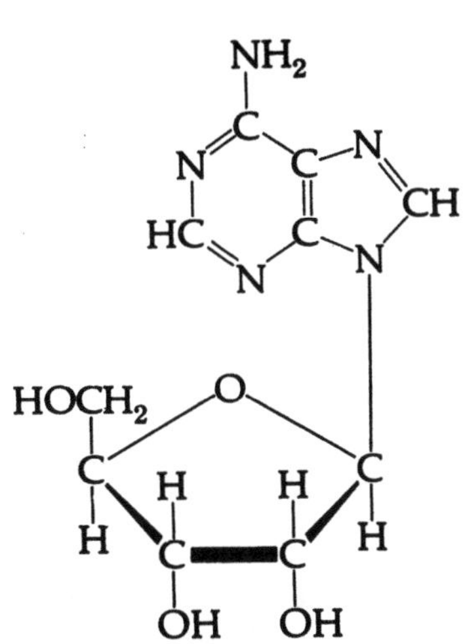

Fig. 5. Chemical structure of adenosine

Adenosine

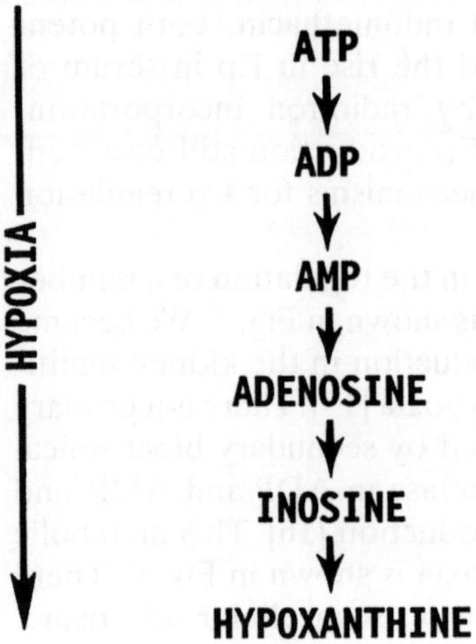

Fig. 6. Model for the metabolic degradation of AP in hypoxia tissues

In order to prove that adenosine is involved in hypoxic regulation in Ep production, several criteria were set forth for these studies. First, the synthesis and release of adenosine should be correlated with a rise in kidney levels of Ep. Secondly, mimicry must be proven which means that adenosine itself is capable of directly stimulating Ep production. Thirdly, adenosine receptor agonists, which are classified as A_1 and A_2, should increase or decrease Ep production. Fourthly, adenosine receptor antagonists, such as theophylline and 8-phenyltheophylline (8-PT), should inhibit the A_1 and A_2 effects of Ep production. Finally, there should be a correlation between the binding of adenosine analogues to adenosine receptors and increased Ep production. In studying these various criteria, the first goal was to prove that adenosine may be a mediator of Ep secretion by determining whether hypoxia itself increases the synthesis and/or release of adenosine by the kidney. These studies have already been carried out previously by Miller et al. [15] in which they have shown in the dog that as early as 5–10 min following exposure to renal artery occlusion to reduce the oxygen supply to the kidney, a significant increase in adenosine, inosine and hypoxanthine levels in the kidney occur (Table 1). In addition, hypoxemic perfusion of the dog kidney in our laboratory (unpublished observations) produces a significant elevation in adenosine levels in the renal tissue after 1 h perfusion with hypoxemic blood.

The next criteria to provide support for the hypothesis that adenosine may be a mediator of Ep production is mimicry. For these studies, we have demonstrated a direct effect of exogenous adenosine on Ep secretion in mice after intravenous infusion of adenosine. Note in Table 2 the rise in plasma levels of Ep, determined by radioimmunoassay, in mice exposed to hypoxia and injected i.v. with 1 and 10 μmol of adenosine per kilogram per day for 4 days. A significant increase in Ep levels in plasma was seen, which was 184.3 ± 24.8 mU/ml with 1 μmol and 274.0 ± 60.6 mU/ml with 10 μmol adenosine

Table 1. Adenosine, inosine, and hypoxanthine in wet weight (nmol/±SE) in normal dog kidneys after renal artery occlusion. (From [15]

	Preocclusion wet weight[a] (nmol/g)	Wet weight after 5 min[a] (nmol/g)	Wet weight after 10 min[b] (nmol/g)
Adenosine	7.6±3.4	18.8± 5.0	25.29± 8.1 $p<0.05$
Inosine	4.3±1.9	20.5± 6.6	84.59±28.7 $p<0.005$
Hypoxanthine	11.0±6.0	75.3±21.6	281.8±88.6 $p<0.001$

[a] $n = 6$.
[b] $n = 2$.

Table 2. Effects of adenosine and hypoxia on plasma levels of Ep in mice

Group	No.	Plasma Ep levels (mU/ml)		
		Control[a]	Adenosine[b] (μmol/kg)	
			1.0	10.0
Hypoxia	6	35.6±4.4	184.3±24.8*	274.0±60.6*

* Significantly ($p<0.05$) higher than control mice (Tukey's test).
[a] For 4 h, 0.42 atm on 4th day of injection.
[b] Injected i.v. per day for 4 days.

per kilogram per day. Thus, it would appear that adenosine itself when used in conjunction with hypoxia significantly increases plasma levels of Ep.

The next criteria for proving adenosine is a mediator of Ep production is the effects of specific adenosine A_1 and A_2 receptor agonists on Ep production. Two subclasses of cell membrane adenosine receptors have been proposed (Fig. 7) [17–19]. These receptors have been characterized physiologically and pharmacologically. A_1 adenosine receptors exhibit high affinity in binding studies (nanomolar) and are coupled to, and inhibit, adenylate cyclase. On the other hand, A_2 adenosine receptors exhibit lower affinity (micromolar) and are coupled to, but stimulate, adenylate cyclase. Adenosine could stimulate Ep production at high concentrations through adenosine A_2 receptor activation and could inhibit Ep production at lower concentrations through adenosine A_1 receptor activation. A_1 receptor stimulation may lead to the activation of a G inhibitory protein which inhibits adenylate cyclase, whereas adenosine A_2 activation may increase a G stimulatory protein which increases adenylate cyclase. Intracellular P binding sites for adenosine have also been identified [20] which may be important in modulating the effects of adenosine on biological processes. An increase in

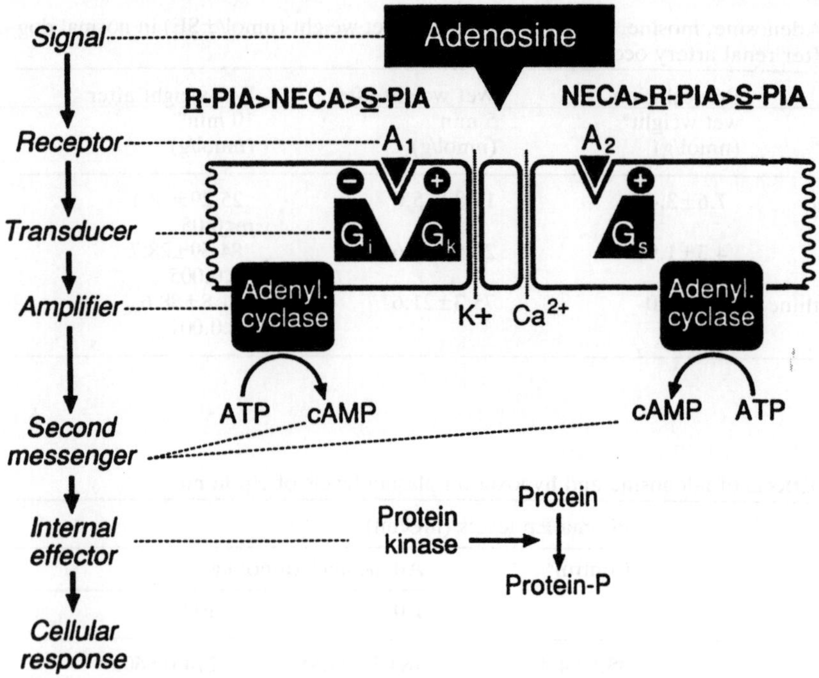

Fig. 7. Schematic description of possible post-receptor reactin cascades which mediate the cellular response following adenosine's receptor activation. A_1 and A_2 adenosine receptor subtypes are depicted as well as an unnamed subtype. A_1 and A_2 receptors are coupled via G-proteins to the adenylate cyclase system. Agonist binding to A_1 receptor subtype can result in inhibition of adenylate cyclase and reduced intracellular cAMP level. In contrast, agonist binding to A_2 receptor subtype can result in stimulation of adenylate cyclase and increased intracellular level of cAMP (Modified from [19])

adenylate cyclase activity leads to the generation of cyclic AMP, the activation of protein kinase A, and the phosphorylation of important nuclear proteins that may be important in the transcriptional and/or the translational stage of Ep biosynthesis in the kidney.

Adenosine itself is degraded both in vivo and in vitro very rapidly by adenosine deaminase and other enzymatic processes making it difficult to use in studies on adenosine as an endogenous mediator of Ep production. Therefore, the two classes of A_1 and A_2 agonists are synthetic analogues of adenosine which have a longer plasma half-life than adenosine itself (chemical structures shown in Fig. 8). The classical A_2 receptor agonist is 5'-*N*-ethylcarboxamidoadenosine (NECA) which has an ethylcarboxamide attachment in the 5' carbon of adenosine. On the other hand, N^6-substituted analogues of adenosine bestow both A_1 and A_2 activity depending upon the substituent group such as with N^6-cyclohexyladenosine (CHA), N^6-cyclopentyladenosine (CPA), and N^6-phenylisoprophyladenosine (PIA).

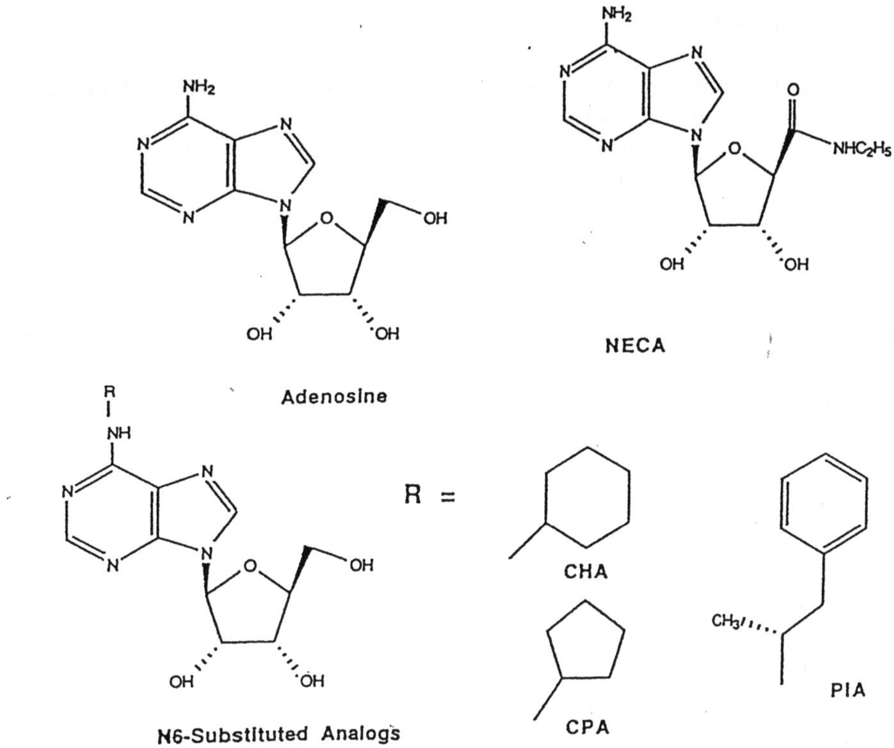

Fig. 8. Chemical structures of adenosine, 5'-*N*-ethylcarboxamidoadenosine (*NECA*) and N^6-cyclohexyladenosine (*CHA*), N^6-cyclopentyladenosine (*CPA*) and N^6-phenylisoprophyladenosine (*PIA*)

Methods Used for Assessing Receptor Agonists and Antagonists

In vivo Ep Studies

Exhypoxic polycythemic mice were employed to determine the in vivo effects of adenosine analogues on Ep production. Adenosine analogues were administered i.v. as single doses in a volume of 0.2 ml on posthypoxic days 3, 4, 5, and 6 and the mice were exposed to reduced atmospheric pressure (0.42 atm) in a hypobaric chamber for 4 h on the 6th posthypoxic day. Blood was removed via cardiac puncture for the determination of microhematocrit and serum Ep (radioimmunoassay).

In vitro Ep Studies

Aliquots of Hep3B cells were dispensed in 24 multi-well plates and incubated with adenosine analogues for 20 h in the presence of hypoxia following 24 h preincubation in a normoxic atmosphere. Supernatants were harvested for Ep determinations in our radioimmunoassay [21].

Radioimmunoassay of Ep

The radioimmunoassay technique has been previously reported in our laboratory by Mason-Garcia et al. [21]. The medium levels of Ep were determined with a sensitive radioimmunoassay using purified human recombinant Ep labelled with I^{125} and an antibody raised to purified Ep.

Cyclic AMP Determination

Hep3B cells were incubated with the adenosine analogues for 1 h in the presence of hypoxia (1% 0_2). The Hep3B cell extracts were assayed for cyclic AMP by radioimmunoassay using a cyclic AMP assay kit.

Tritium-Labelled NECA Binding Studies

Duplicate samples of Hep3B cell membrane were incubated with varying concentrations of tritium (^{3}H)-labelled NECA, 50 µmol CPA (to eliminate A_1 receptor binding), and 1 µmol of adenosine deaminase in the presence or absence of 100 µmol unlabelled NECA for 120 min at 25°C in a final volume of 200 µl. Incubations were terminated by filtration and the filters counted with a liquid scintillation counter.

Results and Discussion

Please note in Fig. 9 that using our radioimmunoassay for Ep, the serum levels of Ep were increased up to a level of almost 400 mU/ml utilizing 0.1 µmol/kg/per day NECA i.v. for 4 days and subsequently exposing the mice to hypoxia. Fig. 10 shows that NECA in very similar concentrations (0.05 – 0.1 µmol/kg/per day) produced a marked increase in kidney levels of Ep reaching levels of 100 mU/gm kidney (hypoxia control: 20 mU/gm kidney). We were very interested to know if NECA was capable of activating Ep secretion in our Ep-producing Hep3B cell line in response to hypoxia (1% 0_2, 5% CO_2, and 94% N_2). Please note in Fig. 11 that NECA produced an exponential rise in Ep levels in our Ep-producing Hep3B cells with increasing

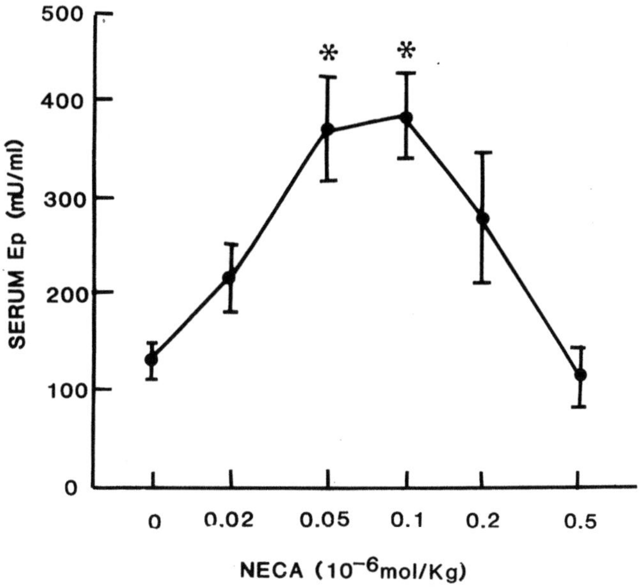

Fig. 9. Effects of NECA on serum Ep levels in exhypoxic polycythemic mice in response to 4 h hypoxia on the 6th posthypoxic day. NECA at a dose range of 0.02–0.5 μmol/kg/per day was administered via the tail vein as a single dose in a 0.2 ml volume on posthypoxic days 3, 4, 5, and 6. Each value represents the mean ± SEM and the *asterisk* indicates significantly different from control ($P<0.05$)

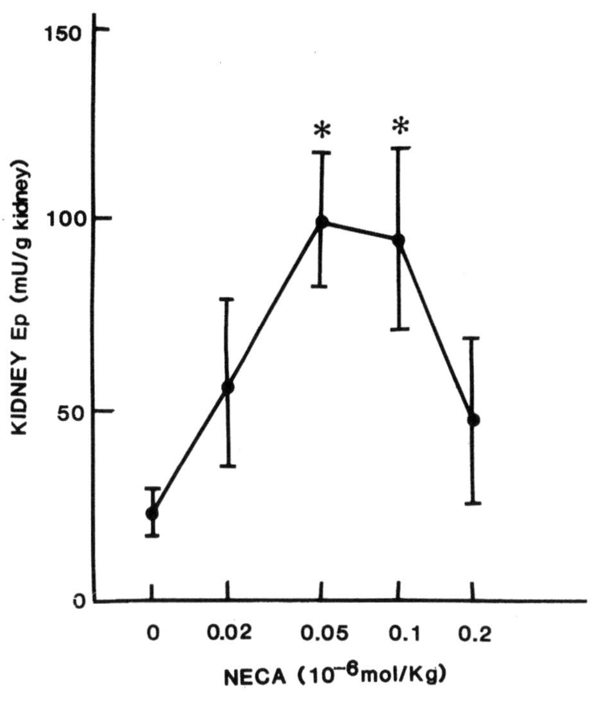

Fig. 10. Effects of NECA on kidney Ep levels in exhypoxic polycythemic mice in response to 4 h hypoxia on the 6th posthypoxic day. NECA in concentrations of 0.02–0.5 μmol/kg/per day was administered via the tail vein on posthypoxic days 3, 4, 5, and 5. Each value represents mean ± SEM and the *asterisk* indicates significantly different from control ($P<0.05$)

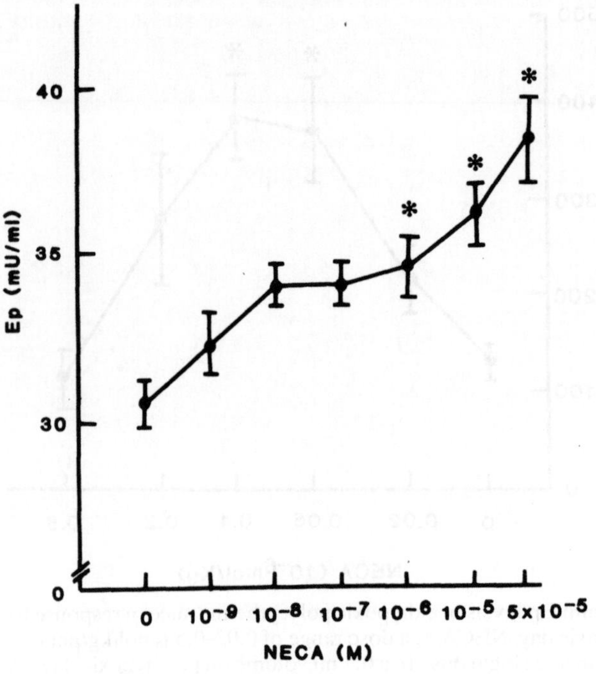

Fig. 11. Effects of NECA on Ep secretion in Hep3B cells exposed to hypoxia Hep3B cells were incubated with NECA at a concentration range of $10^{-8} - 5\times10^{-5}M$ for 20 h under hypoxic (1% O_2) conditions. Each value represents mean ± SEM of 12 samples and the *asterisk* indicates significantly different from hypoxia control ($P<0.05$)

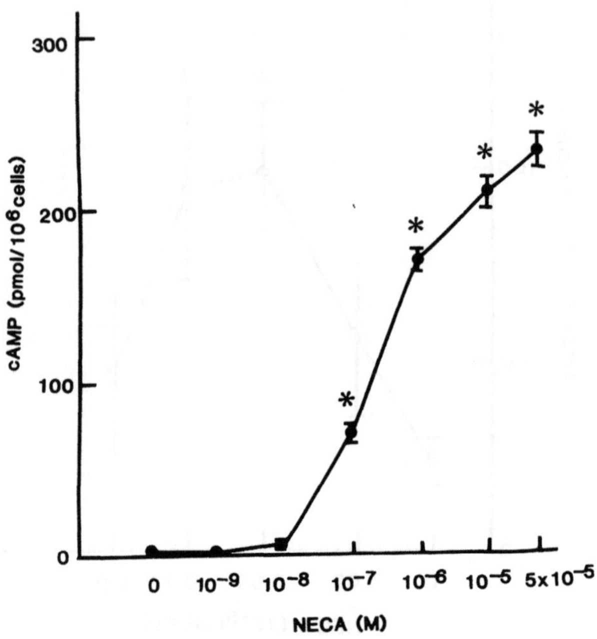

Fig. 12. Effects of NECA on cyclic AMP accumulation in Hep3B cells exposed to hypoxia. Hep3B cells were incubated with NECA in concentrations of $10^{-9} - 5\times10^{-5}M$ for 60 min under hypoxia conditions. Each value represents mean ± SEM of eight samples and the *asterisk* indicates significantly different from hypoxia control ($P<0.05$)

concentrations: For example, at $5 \times 10^{-5} M$ NECA Ep titers were in the range of 38 mU/ml when compared to hypoxia controls of around 30 mU/ml. In order to determine whether the rise in Ep secretion in Hep3B cells in response to NECA was correlated with an increase in cellular levels of the second messenger cyclic AMP, increasing concentrations of NECA on cyclic AMP accumulation in Hep3B cells in response to hypoxia was assessed. Please note in Fig. 12 the exponential rise in cyclic AMP in response to hypoxia and NECA in Hep3B cells.

We next turned to characterizing the adenosine A_1 receptor response utilizing the classical A_1 agonist CHA. In order to study further the effect of CHA on Ep secretion/production, we used the Ep-producing Hep3B cell line exposed to hypoxia (1% 0_2). Please note in Fig. 13 that CHA at a concentration of $10^{-7} M$ slightly suppressed Ep which may be related to its A_1 receptor effects. However, with exceedingly high dosages of CHA, in a range of 10^{-5} and $5 \times 10^{-5} M$, an A_2 agonist effect was seen, thus increasing Ep secretion. This would indicate a biphasic effect of CHA, e.g., a mild suppression of Ep secretion at low concentrations and an increase in Ep secretion at higher concentrations. In order to determine if the high

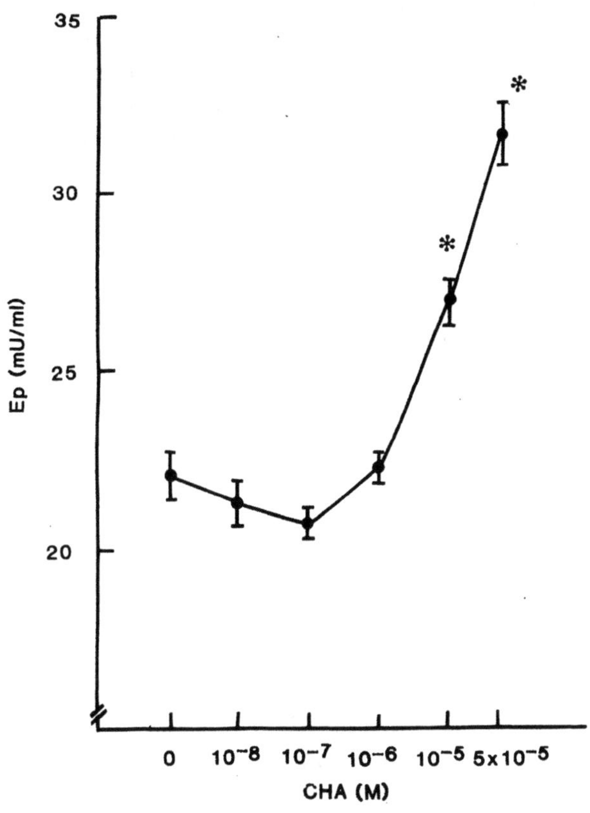

Fig. 13. Effects of N^6-cyclohexyladenosine (CHA) on Ep secretion in Hep3B cells exposed to hypoxia. Hep3B cells were incubated with CHA at a concentration range of 10^{-8} – $5 \times 10^{-5} M$ for 20 h under hypoxia (1% O_2) conditions. Each value represents mean ± SEM of 12–18 samples. *, Significantly different from hypoxia alone ($P<0.05$)

concentrations of CHA which induced Ep secretion were related to changs in intracellular cyclic AMP, the levels of cyclic AMP in the Hep3B cells in response to increasing concentrations of CHA were studied. Please note in Fig. 14 that concentrations of 10^{-5} and $5 \times 10^{-5} M$ CHA, which were the concentrations that increased Ep secretion, produced significant increases in cyclic AMP levels in the Hep3B cells which reached levels exceeding 200 pmol/10^6 cells (control: 3.4 pmol/10^6 cells). Therefore, the A_2 agonist effects of CHA were related to increased Ep secretion and were correlated with an increase in intracellular Hep3B cell levels of cyclic AMP.

The next criteria used in the present studies to prove that adenosine may be a mediator of Ep secretion was the effects of adenosine receptor antagonists on Ep secretion. For these studies, we used theophylline, a xanthine derivative that is an antagonist at both A_1 and A_2 adenosine receptors. For the in vivo studies, we used the exhypoxic polycythemic mouse model to assess a possible inhibitory role of theophylline on NECA-induced Ep secretion. Please note in Fig. 15, that when compared with serum levels of Ep of approx. 140 mU/ml in control mice in response to 4 h hypoxia alone, NECA in a concentration of 0.1 μmol/kg/per day for 4 days produced an elevation of serum Ep to near 500 mU/ml. Note that Ep serum levels with theophylline (20 mg/kg/per day, i.p.) alone were very similar to control levels. On the other hand, theophylline in a concentration of 20 mg/kg significantly inhibited the NECA-induced rise in serum levels of Ep. This data provides evidence that in vivo secretion of Ep in response to NECA is significantly antagonized by the nonselective A_1, A_2 antagonist theophylline.

In further studies, we assessed another xanthine analogue antagonist 8-PT, which has minimal phosphodiesterase inhibitory activity on Ep secretion

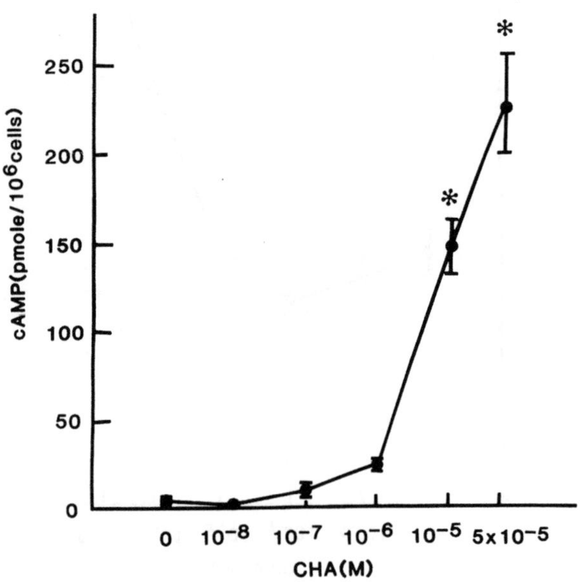

Fig. 14. Effects of CHA on cyclic AMP accumulation in Hep3B cells exposed to hypoxia. Hep3B cells were incubated with CHA at a concentration range of $10^{-8} - 5 \times 10^{-5} M$ for 60 min under hypoxic conditions (1% oxygen). Each value represents mean ± SEM of 11 samples. *, Significantly different from hypoxia controls ($P<0.05$)

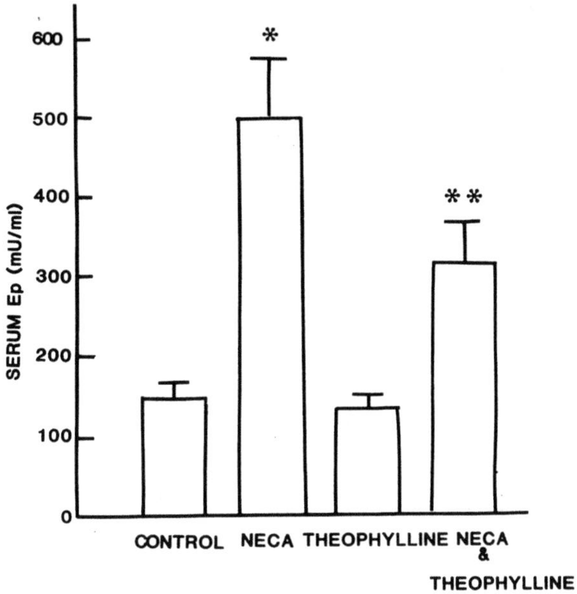

Fig. 15. Effects of theophylline on NECA enhancement in serum Ep levels in exhypoxic polycythemic mice in response to 4 h hypoxia on the 6th posthypoxic day. NECA (0.1 µmol/kg/per day, i.v.) and theophylline (20 mg/kg/per day, i.p.) were administered on posthypoxic days 3, 4, 5 and 6. Each value represents mean ± SEM. *, Significantly different from hypoxia control alone ($P<0.05$); **, significantly different from NECA alone ($P<0.05$)

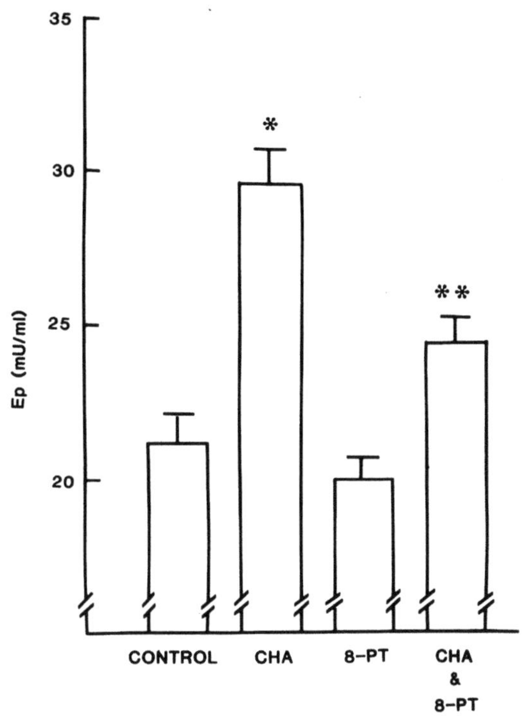

Fig. 16. Effects of 8-phenyltheophylline (8-PT) on CHA-stimulated Ep secretion in Hep3B cells exposed to hypoxia. Hep3B cells were incubated with CHA ($10^{-5}M$) and/or 8-PT ($5\times10^{-7}M$) for 20 h in a hypoxic atmosphere (1% O_2). Each value represents mean ± SEM for 12 samples

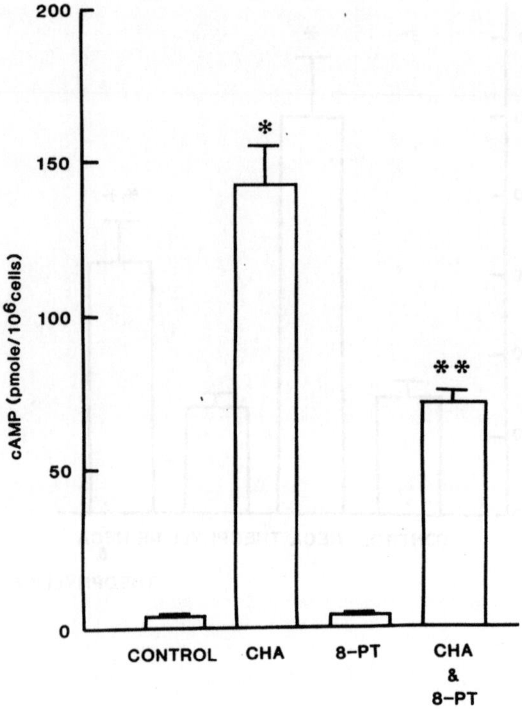

Fig. 17. Effects of 8-PT on CHA-stimulated cyclic AMP accumulation in Hep3B cells exposed to hypoxia. Hep3B cells were incubated with CHA ($10^{-5}M$) and/or 8-PT ($5\times10^{-7}M$) for 60 min in a hypoxic atmosphere (1% oxygen). Each value represents mean ± SEM of eight samples. *, Significantly different from hypoxia alone ($P<0.05$); **, significantly different from CHA alone ($P<0.05$)

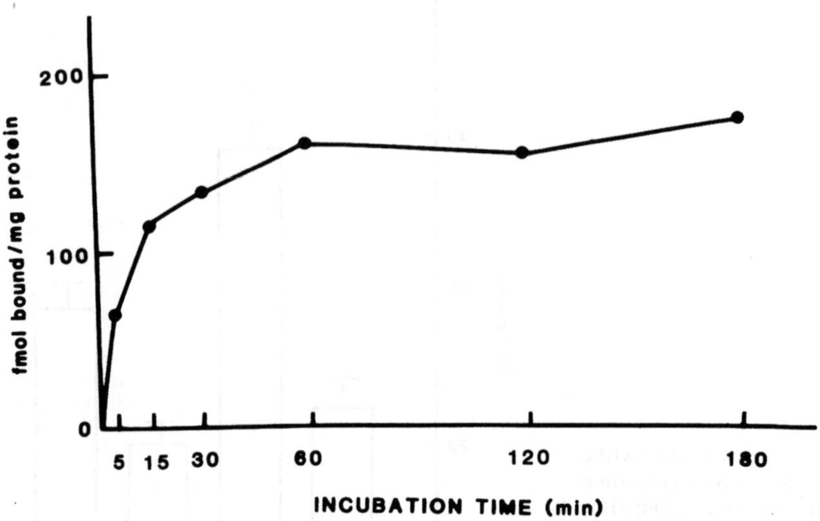

Fig. 18. Time course of specific binding of [^{3}H]NECA to Hep3B cell membranes. Hep3B cell membrane preparations were incubated with 100 nM [^{3}H] NECA, 50 n*M* CPA and 0.1 U/ml of adenosine deaminase in the presence or absence of 100 μ*M* unlabelled NECA for the indicated periods of time at 25°C. Specific binding is defined as the difference between total and nonspecific binding

(Fig. 16) and on cyclic AMP accumulation (Fig. 17) in Hep3B cells. Please note that a concentration of $5 \times 10^{-5} M$ CHA significantly increased Ep secretion (Fig. 16) under hypoxic conditions and also produced a significant increase in cyclic AMP (Fig. 17) in the Hep3B cells. This effect of CHA was significantly inhibited by the xanthine derivative 8-PT. These data further indicate that in our in vitro Hep3B cell line, the A_2 agonist effects of CHA at high concentrations on cyclic AMP accumulation and Ep secretion were significantly reduced by 8-PT.

As a final criterion, we were very interested to determine whether there were A_2 type receptors on Hep3B cell membranes. For these studies, we used tritium-labelled NECA binding to Hep3B cells. Note in Fig. 18 the specific binding curve for ^{3}H-NECA. When we increased the incubation time a plateau

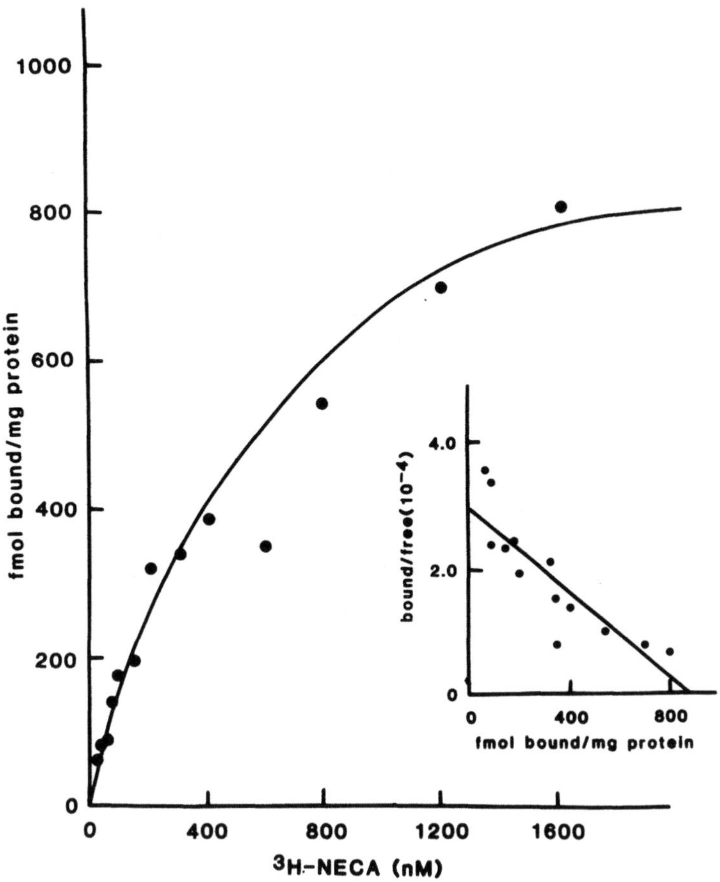

Fig. 19. [^{3}H]NECA binding to Hep3B cell membranes. Binding of [^{3}H]NECA was measured as described in the Materials and Methods section. The Scatchard plots generated from the specific binding data by means of the LIGAND program are shown in *inset* in this figure. Each value represents the mean of 3–5 different experiments

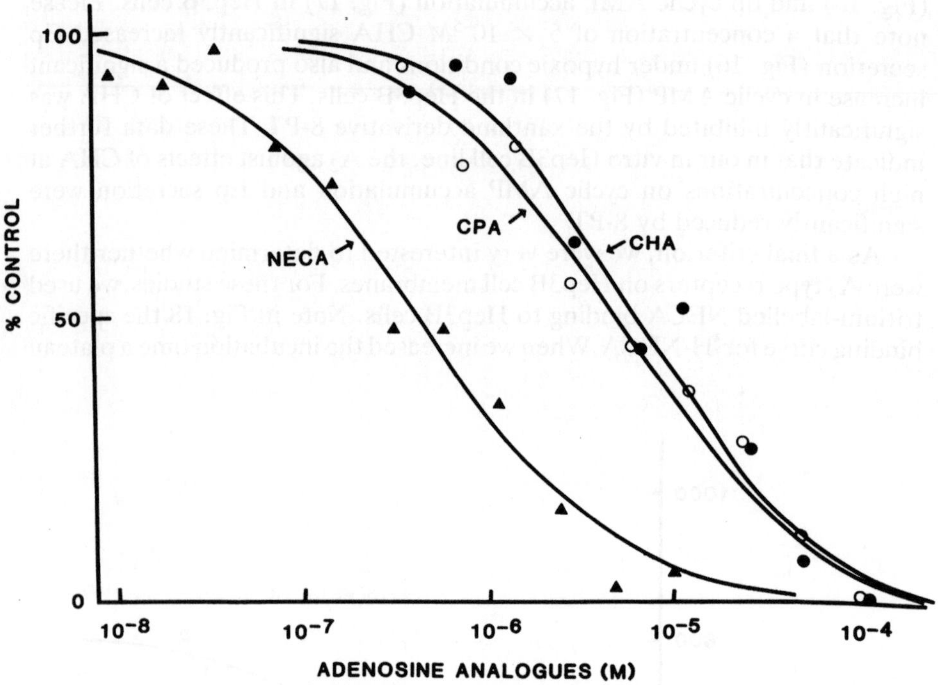

Fig. 20. Displacement curves for adenosine analogues versus [^{3}H]NECA in Hep3B cells. Inhibition of specific [^{3}H]NECA binding by NECA (▲), CHA (●), and CPA (○) was measured as described in the Materials and Methods section. Fach value represents the mean of 3–4 different experiments

for NECA binding to Hep3B cells was reached after 60 min. Please note in Fig. 18 using a 60-min incubation time, the exponential rise in binding of tritium-labelled NECA to Hep3B cell membranes reached near saturation at 1600 nmol. Note the Scatchard analysis (Fig. 19, inset); there is probably a single class of binding sites with a dissociation constant (Kd) of 0.43 nmol and maximal binding capacity (Bmax) of 863 fmol/mg protein ($n=5$). We then carried out binding displacement studies with some adenosine analogues and tritium-labelled NECA. Note in Fig. 20 the binding displacement curves

Table 3. Correlation between the Ki value and the Ec50 for adenosine analogues on cAMP accumulation in Hep3B cells

	Ki (*M*)	Ec50 for cAMP (*M*)
NECA	0.38	0.33
CHA	6.3	4.2
CPA	3.9	2.2

Ki, inhibitory constant; Ec50, 50% effective concentration

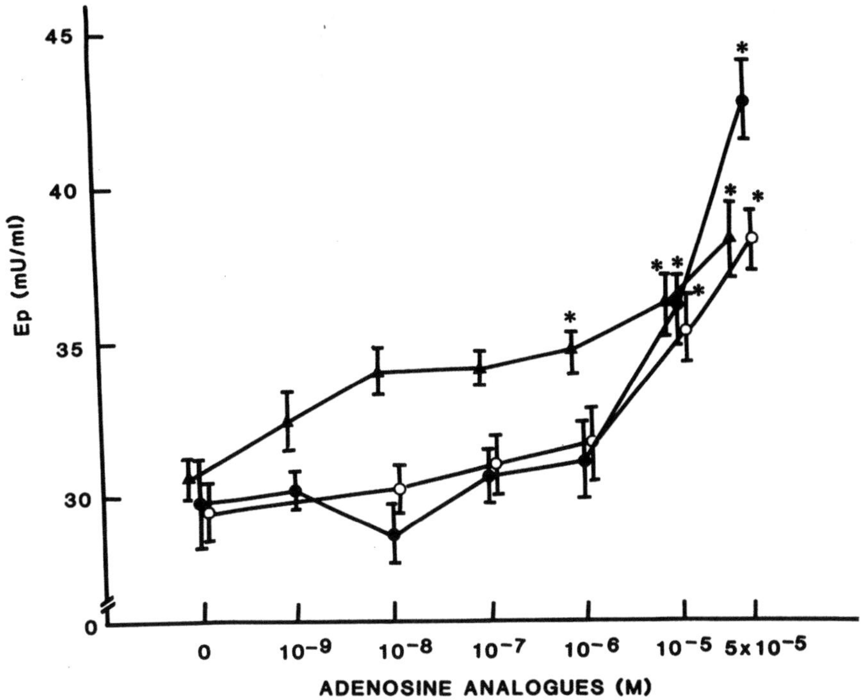

Fig. 21. Effects of adenosine analogues on Ep secretion in Hep3B cells exposed to hypoxia. Hep3B cells were incubated with NECA (▲), CHA (●), and CPA (○) in the presence of 1 U/ml adenosine deaminase for 20 h under hypoxic (1% O_2) conditions. Each value represents the mean ± SEM of 12 samples and the *asterisk* indicates significantly different from respective hypoxia control ($P<0.05$)

with ^{3}H-NECA vs cold NECA, CPA, and CHA. CHA and CPA at these concentrations displaced tritium-labelled NECA from binding to Hep3B cell membranes to a significantly lesser extent than cold NECA (Fig. 20). The effects of these adenosine analogues on Ep-production in Hep3B cells in response to hypoxia were then evaluated. It is of interest to note that NECA was significantly more potent in the range of 10^{-9}–10^{-6} mol than either CPA or CHA at these concentrations (Fig. 21). However, at higher concentrations all three compounds were demonstrated to have A_2 receptor activity and increased Ep secretion. Please note in Table 3 the significantly lower inhibitory constant (Ki) and 50% effective concentration (Ec50) of NECA on cyclic AMP in Hep3B cells than that of CHA and CPA.

In order to follow up the role of the second messenger cyclic AMP in the effects of adenosine agonists on Ep production, we assessed directly the effects of 8 bromo cyclic AMP (8 bcAMP) on Ep secretion in Hep3B cells. Note in Fig. 22 that concentrations of 8bcAMP at 10^{-5}–10^{-4} mol produced significant increases in Ep secretion after 20 h exposure. Note in Fig. 23 that

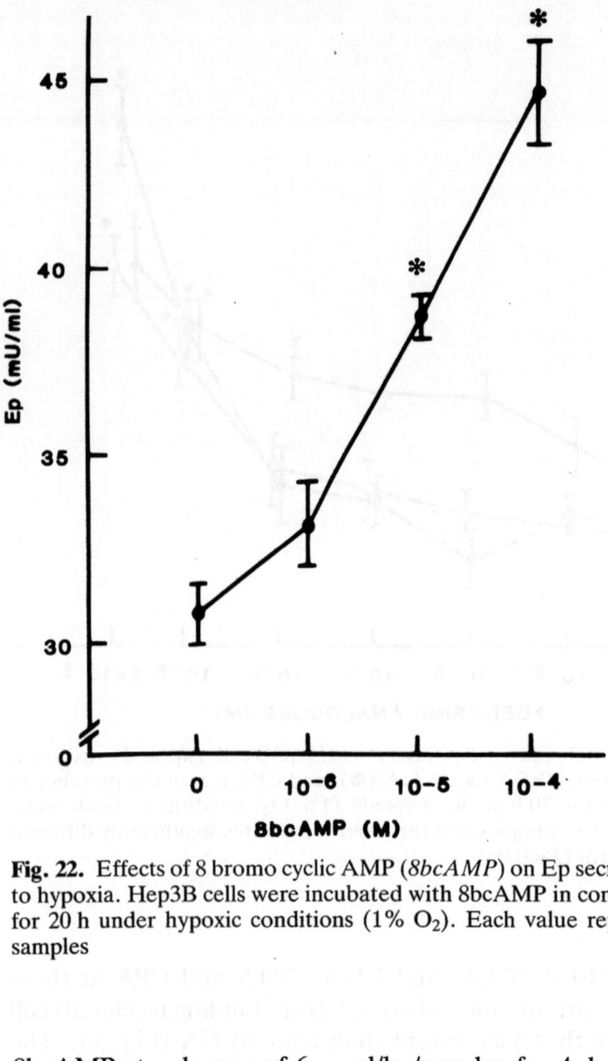

Fig. 22. Effects of 8 bromo cyclic AMP (*8bcAMP*) on Ep secretion in Hep3B cells exposed to hypoxia. Hep3B cells were incubated with 8bcAMP in concentrations of 10^{-7} – 10^{-4} mol for 20 h under hypoxic conditions (1% O_2). Each value represents mean ± SEM of 12 samples

8bcAMP at a dosage of 6 μmol/kg/per day for 4 days produced a significant increase in plasma levels of Ep in exhypoxic polycythemic mice. Therefore, it seems clear that adenosine induced regulation and Ep secretion is involved with the second messenger cyclic AMP.

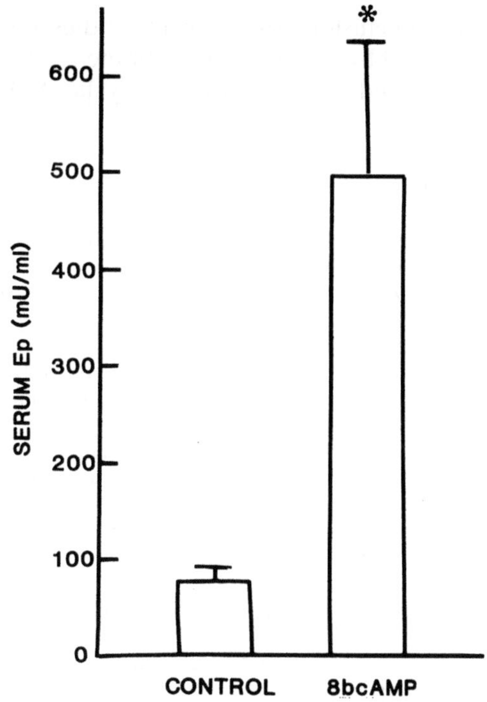

Fig. 23. Effects of 8bcAMP on serum Ep levels in exhypoxic mice in response to 4 h hypoxia on the 6th posthypoxic day. 8bcAMP in a dosage of 6 μmol/kg/per day was administered via the tail vein on posthypoxic days 3, 4, 5, and 6. Each value represents the mean ± SEM and the asterisk indicates significantly different from control ($P<0.05$)

Summary

We have presented a model for the role of external messenger substances in hypoxic stimulation of kidney production of Ep (Fig. 1). NECA produced significant increases in serum Ep levels in exhypoxic polycythemic mice following a 4 h exposure to hypoxia. Theophylline, an adenosine receptor antagonist, significantly inhibited the stimulatory effects of NECA on serum Ep levels. A significant increase in serum levels of Ep was produced by 8bcAMP in exhypoxic polycythemic mice after 4 h exposure to hypoxia. In vitro culture of Hep3B cells with NECA under hypoxic conditions produced a significant increase in medium levels of Ep. NECA also significantly increased cyclic AMP levels in Hep3B cells exposed to hypoxia. 8bcAMP also increased medium levels of Ep in Hep3B cell cultures. Binding studies with tritium-labelled NECA demonstrated that Hep3B cells contain sensitive binding sites for NECA. The rank order of potency with stimulation of Ep production and cyclic AMP accumulation in Hep3B cells was NECA > CPA = CHA. In studies of several adenosine analogues, there was a correlation found between binding of these analogues to A_2 receptors, Ep production and cyclic AMP accumulation in Hep3B cells.

In conclusion, the present studies indicate that Ep production may be regulated by adenosine membrane A_2 receptor-coupled activation of adenyl cyclase and the generation of cyclic AMP.

References

1. Fisher JW (1988) Pharmacologic modulation of erythropoietin production. Annu Rev Pharmacol Toxicol 28:101–122
2. Kurtz AW, Jelkmann A, Pfuhl K, Malmstrom K, Bauer C (1986) Erythropoietin production by fetal mouse liver cells in response to hypoxia and adenylate cyclase. Endocrinology 118:567–572
3. Goldberg MA, Glass A, Cunningham JM, Bunn HF (1987) The regulated expression of erythropoietin by two human hepatoma cell lines. Proc Natl Acad Sci USA 84:7972–7976
4. Nielsen OJ, Schuster SJ, Kaufman R, Erslev AJ, Caro J (1987) Regulation of erythropoietin production in a human hepatoblastoma cell line. Blood 70:1904–1909
5. Goldberg MA, Dunning SP, Bunn HF (1988) Regulation of the erythropoietin gene: evidence that the oxygen sensor is a heme protein. Science 242:1412–1415
6. Recny MA, Scoble HA, Kim Y (1987) Structural characterization of natural human urinary and recombinant DNA-derived erythropoietin. J Biol Chem 262(35):17156–17163
7. Ueno M, Brookins J. Beckmann B, Fisher JW (1988) A_1 and A_2 adenosine receptor regulation of erythropoietin production. Life Sci 43:229–237
8. Ueno M, Brookins J, Beckman BS, Fisher JW (1988a) Effects of reactive oxygen metabolites on erythropoietin production in renal carcinoma cells. Biochem Biophys Res Commun 154:773–780
9. Shah SV (1984) Effect of enzymatically generated reactive oxygen metabolites on the cyclic nucleotide content in isolated rat glomeruli. J Clin Invest 74:393–401
10. McCord JM (1985) Oxygen derived free radicals in postischemic tissue injury. N Engl J Med 312:159–163
11. Jelkmann W, Beckman BS, Fisher JW (1979) Enhanced effects of hypoxia on erythropoiesis in rabbits following beta-2 adrenergic activation with albuterol. J Pharmacol Exp Ther 211:99–103
12. Jelkmann W, Brookins J, Fisher JW (1979) Indomethacin blockade of albuterol-induced erythropoietin production in isolated perfused dog kidneys. Proc Soc Exp Biol Med 162:65–70
13. Gross DM, Fisher JW (1978) Effects of terbutaline, a synthetic beta adrenergic agonist, on in vivo erythropoietin production. Arch Int Pharmacodyn Ther 236(2):192–201
14. Seferynska I. Brookins J. Rice JC, Fisher JW (1989) Erythropoietin production in exhypoxic polycythemic mice. Am J Physiol 256:C925–C929
15. Miller WL, Thomas RA, Berne RM, Rubia R (1978) Adenosine production in the ischemic kidney. Circ Res 43:390–397
16. Jones DP (1986) Renal metabolism during normoxia, hypoxia and ischemic injury. Annu Rev Physiol 48:33–50
17. Londos C, Cooper DMF, Wolff J (1980) Subclasses of external adenosine receptors. Proc Natl Acad Sci USA 77:2551–2554
18. Daly JW (1982) Receptors: Targets for future drugs. J Med Chem 25:197–207
19. Pelleg A, Porter S (1990) The pharmacology of adenosine. Pharmacotherapy 10(3):157–174
20. Londos C. Wolff J (1988) Two distinct adenosine-sensitive sites on adenylate cyclase. Proc Natl Acad Sci USA 74:5482–5486

21. Mason-Garcia M, Beckman BS, Brookins JW, Powell JS, Lanham W. Blaisdell S, Keay L, Chen, I-Li, Fisher JW (1990) Development of a new radioimmunoassay for erythropoietin using recombinant erythropoietin. Kidney Int 38:969–975
22. Nelson PK, Brookins J, Fisher JW (1983) Erythropoietin effects of prostacyclin (PGI_2) and its metabolite 6-keto-prostaglandin E_1. J Pharmacol Exp Ther 226:493–499

Oxygen Shunt Diffusion in Renal Cortex and Its Physiological Link to Erythropoietin Production

H.-J. Schurek

Oxygen Deficiency in Renal Cortex and Medulla

Oxygen supply in the kidney is at the brink of oxygen deficiency, though blood flow to the kidney is tremendously high. Some of the reasons for this paradox are the recently demonstrated preglomerular oxygen diffusion shunt [13], which may contribute to the high amount of low cortical pO_2 values measured by invading electrodes [3, 13] and the construction of the capillary network within the renal cortex, the width of which is defined by the tubular diameters. The analysis of cytochrome aa_3, 30% of which is in the reduced form under physiological conditions, as measured by transmission spectrophotometry [1], indicated that 70% of the oxygen-deficient area was in the renal cortex, as only 30% was convertible by furosemide and thereby could mide be attributed to (TAL) segments of Herle's loop within the outer medulla [5]. Within the renal medulla, oxygen shunt diffusion in the vascular bundles is the main reason for oxygen deficiency [11]. This critical oxygen supply may induce an oscillation (OSC) between aerobic and anaerobic energy production, a hypothesis [9] derived from a complementarily observed OSC of proximal tubular pressures [6–8], a phase lag OSC of distal tubular fluid conductivity (Gutsche 1988, personal communication), and from recent experimental evidence for an OSC of the pO_2 at the renal surface with the same frequency [10], representing an OSC of tubuloglomerular feedback (TGF) activity. There is evidence [13] that the locus where oxygen supply is limited in the renal cortex is also the locus of erythropoietin production, where an oxygen sensor may be active. The aim of our study was to find a common denominator for the high renal susceptibility to hypoxia and for an oxygen sensor used for the regulation of erythropoietin production in the renal cortex by analyzing oxygen pressures along the cortical vasculature. The availability of rats (MWF strain) exhibiting a large number of glomeruli at the kidney surface allowed us to use these glomeruli as an anatomical landmark within the cortical circulation. Newly developed microcoaxial needle sensors for oxygen were used to analyze glomerular pO_2 at the renal surface during inspiration of air or alternatively pure oxygen [13].

Systemic arterial pO_2 was measured via a femoral arteriovenous shunt inline with temperature-adjusted pO_2 Clark type electrodes. At a mean

Pagel/Weiss/Jelkmann (Eds.)
Pathophysiology and Pharmacology
of Erythropoietin

systemic pO_2 of 90±8 mmHg ($\bar{x}\pm SD$, $n=54$) pO_2 at superficical glomeruli averaged 46±13 mmHG ($\bar{x}\pm SD$, $n=71$) with only one value overlap ($p\leq 0.001$). After switching from inspiration of air to pure oxygen by a mechanical respirator, systemic arterial pO_2 was increased from 90±8 to 593±27 mmHg; however, glomerular pO_2 increased from 46±13 to only 80±28 mmHG ($\bar{x}\pm SD$, $n=71$, $p<0.001$). From these data we conclude that a highly effective preglomerular shunt diffusion exists for blood gases especially O_2 and CO_2 [10, 12, 13]. Trapping of CO_2 with the highest values (65 mmHg) in the outermost subcapsular zone [4, 10] facilitates desaturation of oxyhemoglobin by shifting the oxygen-binding curve to the right (Bohr effect) most effectively. An additional quality of information could be obtained by using multiwire gold electrodes and halothane anesthesia to measure tubular and glomerular pO_2 at the renal surface. The same frequency of OSC of proximal tubular pressures, distal tubular conductivities, and surface pO_2 values support our suggestion that a limited oxygen supply of the nephron makes TAL segments oscillate [10] between aerobic and anaerobic energy supply [9]. Glycolysis in TAL segments reduces the filtered load of the corresponding nephrons via macula densa and thus also prevents permanent oxygen deficiency in proximal tubules. The latter seems to be important as proximal convoluted tubules cannot withstand oxygen deficiency for an extended period. Their energy reserves (ATP) break down after some few seconds of ischemia, whereas thick ascending limb segments of Henle's loop hold their ATP levels for some 60 sec by switching metabolism to anaerobic glycolysis. This difference in tubular transport capacity is due to the different enzymatic pattern for glycolysis, which is sufficient in TAL segments but insufficient in proximal convoluted tubules [2].

This represents a new dimension for the significance of the TGF and a prerequisite for regulation of erythropoietin production within the renal cortex, which is far from being luxuriously supplied with oxygen.

Summary

The low pO_2 at superficial glomeruli is thought to be evidence of a highly effective shunt diffusion for blood gases mainly between interlobular arteries and veins within the renal cortex. With a limited oxygen supply to the nephron (including renal cortex and medulla), the thick ascending TAL segment oscillates between aerobic and anaerobic energy production. Thus renal cortex seems to be an attractive locus for an oxygen sensor to regulate effectively the production of erythropoietin.

Acknowledgement. This work includes material from thesis of Uwe Jost, Harald Bertram and Oliver Johns. For excellent technical assistance I am grateful to Heike Haas and Jörg-Dieter Biela.

This work has been supported by the Deutsche Forschungsgemeinschaft Grant Schu 343/6-1. The group in Hannover was also supported by the State of Lower Saxony, Department of Science and Arts.

References

1. Balaban RS, Silvia AL (1981) Spectrophotometric monitoring of O_2-delivery to the exposed rat kidney. Am J Physiol 241:F257-F262
2. Bastin J, Cambon N, Thompson M, Lowry OH, Burch HB (1987) Change in energy reserves in different segments of the nephron during brief ischemia. Kidney Int 31:1239–1247
3. Baumgärtl H, Leichtweiss HP, Lübbers DW, Weiss C, Huland H (1972) The oxygen supply of the dog kidney: measurements of intrarenal pO_2. Microvasc Res 4:247–257
4. DuBose TD, Caflisch CR, Bidani A (1984) Role of metabolic CO_2 production in the generation of elevated renal cortical pCO_2. Am J Physiol 246:F592–599
5. Epstein FH, Balaban RS, Ross BD (1982) Redox state of cytochrom aa_3 in isolated perfused rat kidney. Am J Physiol 243:F356–363
6. Holstein-Rathlou NH (1987) Synchronization of proximal intratubular pressure oscillations: evidence for interaction between nephrons. Pflügers Arch 408:438–443
7. Leyssac PP (1986) Further studies on oscillating tubuloglomerular feedback responses in the rat kidney. Acta Physiol Scand 126:271–277
8. Leyssac PP, Baumbach L (1983) An oscillating intratubular pressure response to alterations in Henle loop flow in the rat kidney. Acta Physiol Scand 117:415–419
9. Schurek HJ (1988) Die Nierenmarkhypoxie: Ein Schlüssel zum Verständnis des akuten Nierenversagens? Klin Wochenschr. 66:828–835
10. Schurek HJ, Johns O (1990) Tubulo-glomerular feedback prevents nephron oxygen deficiency via TAL segments. JASN 1:603
11. Schurek HJ, Kriz W (1985) Morphologic and functional evidence for oxygen deficiency in the isolated perfused rat kidney. Lab Invest 53:145–155
12. Schurek HJ, Jost U, Bertram H, Baumgärtl H (1989) Preglomerular cortical oxygen diffusion shunt: a prerequisite for effective erythropoietin regulation? Contrib Nephrol 76:57–66
13. Schurek HJ, Jost U, Baumgärtl H, Bertram H, Heckmann U (1990) Evidence for a preglomerular oxygen diffusion shunt in rat renal cortex. Am J Physiol 259:F910–915
14. Sohtell M, Karlmark B (1976) In vivo micropuncture pCO_2 measurements. Pflügers Arch 363:179–180

Oxygen Sensing and Erythropoietin mRNA Production in Isolated Perfused Rat Kidneys*

C.C. Tan and P.J. Ratcliffe

Introduction

In the adult, at the kidney is the major site of erythropoietin (Epo) synthesis (Jacobson et al. 1957), and recent studies showing that oxygen-modulated production of Epo mRNA and hormone occurs in isolated perfused kidneys (Ratcliffe et al. 1990; Pagel et al. 1990; Scholz et al. 1991) demonstrate that an oxygen-sensing system operates intrarenally. As increases in Epo levels can be achieved by stimuli in which systemic arterial oxygen tensions are normal, such as anaemia (Bondurant and Koury 1986; Erslev et al. 1987) or increased erythrocyte haemoglobin-oxygen affinity (Adamson and Finch 1975), it is likely that it is the changes in local tissue oxygen tension which activate the Epo oxygen-sensing machanism.

Two important questions follow: firstly, how do changes in overall renal oxygen delivery affect tissue oxygen levels, and, secondly, how is the tissue oxygen level sensed. Several authors have suggested that it is the balance between renal oxygen delivery and consumption that determines the level of renal Epo synthesis (Fried et al. 1957; Krantz and Jacobson 1970; Eckardt et al. 1989a), through their effects on local oxygen tensions within the kidney. It has been further proposed that the kidney is uniquely suited to detect decreases in blood oxygen availability as renal oxygen consumption is determined largely by tubular sodium reabsorption (Kiil et al. 1961), which in turn depends on the glomerular filtration rate, and is regulated in parallel with renal blood flow. Hence, changes in renal perfusion which occur independently of alterations in blood oxygen content are envisaged to lead to changes in renal oxygen supply which are balanced by parallel changes in renal oxygen consumption (Erslev et al. 1985). While a large body of data has established that blood oxygen availability is the major determinant of Epo production (Cotes 1982; Erslev et al. 1987; Eckardt et al. 1989b), the role of renal oxygen consumption is still poorly defined.

In respect of the mechanism of oxygen sensing, a major issue is whether signalling is related directly, or indirectly , to hypoxic compromise of mitochondrial oxidative phosphorylation. Evidence from human hepatoma

* This research was supported by the National Kidney Research Fund (UK) and the Wellcome Trust (UK). C.C.T. is a Wellcome Overseas Fellow at the Institute of Molecular Medicine, Oxford

Pagel/Weiss/Jelkmann (Eds.)
Pathophysiology and Pharmacology
of Erythropoietin

cell lines (Fandrey et al. 1991) indicates that this is not the case, but the effects of inhibition of oxidative phosphorylation have not been rigorously examined in normal kidney.

We have optimized the conditions of perfusion of isolated rat kidneys to provide an appropriate model in which these issues can be addresed. In a first series of experiments, we have examined the interplay of changes in perfusate pAO_2 and flow rate, and the effect of reduced glomerular filtration and transport work on modulation of Epo mRNA. In a second set of kidney perfusions, the action of inhibitors of mitochondrial oxidative phosphorylation were studied.

Methods

Male Wistar rats weighing 350–375 g were used for all experiments, and isolated kidney perfusion was performed using modifications of the method described by Ross et al. (1973). Kidneys were perfused in a recirculation circuit for 120 min, at a constant perfusion flow rate, which was clamped for each experiment, to ensure that renal oxygen delivery remained constant throughout the perfusion. The perfusate consisted of Krebs-Henseleit buffer containing bovine serum albumin (Fraction V, Miles Laboratories, Slough, UK) with glucose (5 mmol/l) and the 20 physiological amino acids as substrates (Epstein et al. 1982), and supplemented with washed human erythrocytes to achieve a final haematocrit of about 10%. The perfusion medium was gassed with the appropriate mixture of 95% O_2–5% CO_2, 95% N_2–5% CO_2 or 95% air–5% CO_2 to achieve the desired perfusate pAO_2 level. Oxygen tensions were measured using an ABL-2 blood gas analyser (Radiometer, Copenhagen, Denmark). The haemoglobin-bound oxygen content (μmol/l) was calculated from the formula: 0.62 × saturation (fractional) × haemoglobin concentration (g/dl). The dissolved oxygen content of the perfusate was calculated according to the formula: O_2 content (μmol/l) = $1.257 \times 10^{-3} \times [pO_2(\text{mmHg}) \times (1\text{–haematocrit})]$ (Christoforides et al. 1969). Renal oxygen consumption was derived from these calculations. As the right kidney was homogenized after perfusion, oxygen consumption was expressed per gram wet weight of the left kidney.

In all experiments, RNA was prepared from contralateral control kidneys excised at the start of perfusion, as well as from the perfused kidney under study. RNA was purified on a caesium chloride gradient and stored as an ethanol precipitate at –80°C until analysis. Epo mRNA was analysed used RNase protection by a [^{32}P]-GTP rat genomic riboprobe containing 132 bp of exon V, and adjoining intron, as described previously (Ratcliffe 1990). RNA concentration was determined by absorbance measurement at 260 nm, and for analysis of Epo mRNA 250 μg total RNA was hybridised at 60°C with 0.5–1 × 10^6 cpm radiolabelled Epo probe. RNase digestion was carried out at 20°C for 30 min. Purified protected species were separated on a denaturing 10% polyacrylamide gel. Each assay was internally controlled by parallel hybridization with a genomic rat ß-actin riboprobe, but to allow for the greater abundance of actin in kidney the actin hybridization was performed on a 50 μl aliquot of sample RNA containing 2.5 μg total RNA. Hybridization was at 60°C with 1–2 × 10^6 cpm radiolabelled actin probe. After RNase digestion, one-tenth of the final product was added to the corresponding Epo sample so that subsequent analysis was performed together. After autoradiography, Epo and actin mRNA levels were quantified by counting the excised bands from the gel in a flat-bed liquid scintillation counter (1205, Beta Plate, Pharmacia-Wallac, Turku, Finland). Actin mRNA did not change during perfusion, and in all experiments, the Epo mRNa signal was normalized to the corresponding actin mRNA signal.

Studies on Renal oxygen Delivery and the Reduction of Glomerular Filtration. Three sets of experiments were performed at constant perfusion flow rates and with perfusate albumin concentrations adjusted to maintain glomerular filtration as follows: group 1, perfusion flow rate 6 ml/min, perfusate albumin concentration 1 g/dl; group 2, 12 ml/min, 3 g/dl; and group 3, 18 ml/min, 3 g/dl. Perfusate pAO_2 was kept constant for each perfusion, but was varied from about 20 to 500 mmHg within each group of experiments. To study the effect of reduction of tubular work on Epo production, glomerular filtration was stopped in a series of 12 experiments, performed at a constant flow rate of 6 ml/min, by using hyperoncotic perfusion medium (11 g/dl bovine serum albumin; (Maack 1980). Perfusate pAO_2 levels were varied in the same range as in group 1.

Studies with Metabolic Inhibitors. As the application of metabolic inhibitors to filtering kidneys causes a marked diuresis, natriuresis and increased vascular resistance (Roos et al. 1974), which could potentially confound the effects of the intended perturbation of cellular metabolism, these experiments were performed in non-filtering kidneys perfused at a flow rate of 12 ml/min. Four well-oxygenated and four hypoxic control perfusions were performed with the perfusion media gassed with 95% O_2 – 5% CO_2 and 95% N_2 – 5% CO_2, respectively.

A stock solution of potassium cyanide (BDH, Poole, UK), dissolved in phosphate buffered saline, was made on the day of use. Similarly, stock solutions of oligomycin and antimycin (Sigma Chemical, Poole, UK) were made by dissolving the compounds in absolute ethanol. A series of experiments with the inhibitors of oxidative phosphorylation were performed with the perfusate gassed with 95% O_2–5% CO_2. After 15 min of perfusion, arterial and venous samples were taken for measurements of oxygen tension, and the agent was added to the perfusion medium to achieve the appropriate dose. Arterial and venous oxygen measurements were repeated after a further 15 min and 75 min. The following doses of inhibitor were studied: cyanide 10 μ*M*, 100 μ*M*, 300 μ*M* and 1 m*M*; antimycin 0.01 μ*M*, 0.1 μ*M*, 0.5 μ*M* and 1 μ*M*; and oligomycin 0.01 μ*M*, 0.1 μ*M* and 1 μ*M*. A further set of three experiments was performed with kidneys perfused under hypoxic conditions, gassed with 95% $N_2-5\%$ CO_2, and exposed to 1 m*M* potassium cyanide, 0.5 μ*M* antimycin or 1 μ*M* oligomycin, respectively. A submaximal dose of antimycin was used because the highest dose of 1 μ*M* caused vasoconstriction in the well-oxygenated experiments.

Student's *t* test was used to assess the statistical significance of differences between the means of variables.

Results

Stability of Isolated Kidneys Perfused Under Different Conditions. The functional parameters of perfused kidneys in groups 1–3 and of the non-filtering preparations, are shown in Table 1. Within each experiment, perfusion pressure and renal oxygen consumption did not alter significantly after the first few minutes of perfusion. Mean glomerular filtration rates also remained relatively unchanged for experiments at 6 and 12 ml/min but decreased progressively during perfusion at 18 ml/min from 1.2±0.2 ml/min^{-1} at 30 min to 0.5±0.2 ml min^{-1} at 115 min. The fractional excretion of sodium (FeNA) was significantly increased in experiments performed under hypoxic conditions, as has been previously reported (Ross 1974). In contrast, in the kidneys perfused at pAO_2 levels greater than 150 mmHg the mean FeNa was less than 10% in all three groups.

Effect of Perfusion flow rate and pAO_2 on Epo mRNA. Figure 1a illustrates the effect of alterations in perfusion flow rate and pAO_2 on renal Epo mRNA. At each flow rate, oxygen-regulated Epo mRNA production was evident. Epo mRNA levels increased approximately 30-fold when perfusate pAO_2 was decreased from 500 mmHg to 30 mmHg at perfusion flow rates of 12 and 18 ml/min. In addition, changes in perfusion flow rate had a striking effect on Epo mRNA production. At perfusate arterial oxygen tensions of greater than 300 mmHg, Epo mRNA levels in kidneys perfused at 12 and 18 ml/min were slightly lower than the levels in corresponding contralateral, non-perfused kidneys (mean of 0.07 versus 0.11), but the quantity of Epo mRNA in kidneys perfused at 6 ml/min. was approximately 25-fold greater. This relationship was maintained under hypoxic perfusion conditions, with mRNA levels at perfusate pAO_2 of about 150 mmHg, being approximately ten-fold greater in experiments at 6 ml/min compared with those at 12 and 18 ml/min. When the total oxygen delivery to the perfused kidneys was considered (Fig. 1b), an inverse relationship was observed between renal oxygen supply and Epo mRNA levels, regardless of the perfusion flow rate.

Effects on Epo mRNA of Hyperoncotic Perfusion. The use of hyperoncotic perfusion medium stopped glomerular filtration, as evidenced by the lack of urine production and the maintenance of stable perfusate ^{14}C inulin levels throughout the experiment. In perfusions at 6 ml/min this was not associated

Table 1. Functional parameters of isolated rat kidneys after 30, 60, 90, and 115 min of perfusion, under different conditions

	Group 1 (q 6 ml/min; n=12)				Group 2 (q 12 ml/min; n=11)				Group 3 (q 18 ml/min; n=8)				Non-filtering perfusions (q 6 ml/min; n=12)			
Time (min)	30	60	90	115	30	60	90	115	30	60	90	115	30	60	90	115
Perfusion pressure (mmHG)	48±8	45±7	42±8	43±6	58±14	52±9	52±8	57±12	79±15	71±13	72±13	80±28	39±6	38±6	40±7	41±7
GFR (ml min^{-1} g^{-1})	0.5±0.3	0.5±0.2	0.5±0.1	0.6±0.3	0.6±0.3	0.7±0.4	0.6±0.3	0.6±0.3	1.2±0.2	0.9±0.1	0.7±0.2	0.5±0.2				
FeNa (%) $pA0_2$>150 mmHg	10±5	6±5	5±4	8±6	3±1	2±1	2±2	2±1	9±3	4±1	4±1	6±2				
$pA0_2$>150 mmHg					18±14	12±10	12±14	10±9	27±8	21±11	21±12	24±13				
QO_2 (μmol min^{-1} g^{-1})	3.2±0.6		3.4±0.8		4.1±0.8		3.9±0.8		5.0±2.2		4.8±1.6		2.9±0.5		2.9±0.9	

GFR, Glomerular Filtration rate;
q, perfusion flow rate;
n, numer of experiments in the group;
QO_2, renal oxygen consumption per gram of left kidney weight.

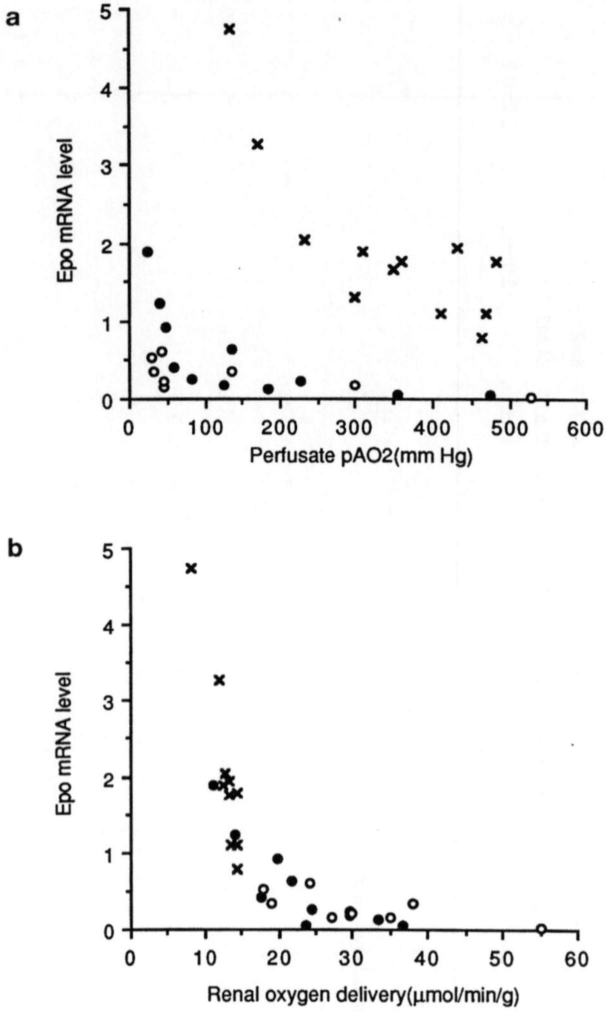

Fig. 1. a The effect of perfusate pAO_2 on Epo mRNA levels in kidneys perfused at 6 ml/min (×), 12 ml/min (●) and 18 ml/min (○). **b** The relationship between Epo mRNA and renal oxygen delivery in kidneys perfused at 6 ml/min (×), 12 ml/min (●) and 18 ml/min (○)

with a statistically significant reduction in renal oxygen consumption (3.0±0.6 compared with 3.3±0.6 µmol min^{-1} g^{-1} in group 1; $p=0.2$). Despite this, Epo mRNA levels in the non-filtering kidneys were decreased about four-fold over a wide range of perfusate pAO_2 (Fig. 2). When only well-oxygenated experiments with perfusate pAO_2 levels greater than 300 mmHg were considered, Epo mRNA levels in the non-filtering kidneys were again significantly lower than the corresponding filtering kidneys in group 1, the values being 0.4±0.2 versus 1.5±0.4 respectively ($P<0.01$).

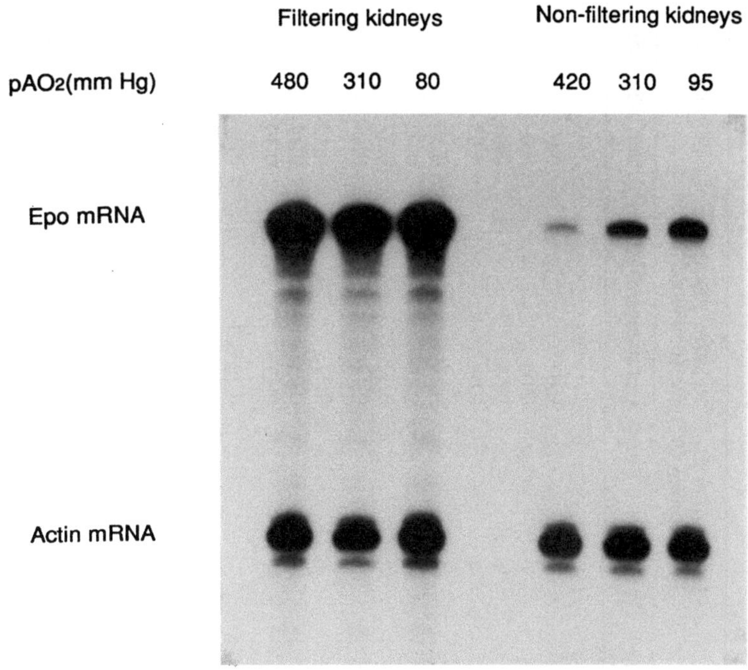

Fig. 2. Autoradiograph of an RNase protection assay of Epo and actin mRNA in three representative kidneys from group 1, and three non-filtering kidneys perfused at the same flow rate and under similar conditions of oxygenation but with hyperoncotic perfusate to stop glomerular filtration

Renal oxygen consumption was, however, not significantly different in this sub-group of non filtering and filtering kidneys (3.0±0.7 versus 3.5±0.5 μmol min^{-1} g^{-1}; $p=0.06$).

Effect of Inhibitors of Oxidative Phosphorylation. To provide appropriate conditions for the study of metabolic inhibitors, non-filtering kidneys perfused at 12 ml/min were used. To demonstrate the potential for modulation of Epo mRNA under these conditions, four perfusions were performed at a pAO_2 of 443±67 mmHg, and four were carried out with a pAO_2 of 32±4 mmHg. Epo mRNA levels were approximately 15-fold higher in the hypoxically perfused kidneys. In kidneys perfused under similar conditions to the well-oxygenated control group, the administration of increasing concentrations of cyanide, oligomycin and antimycin resulted in progressive reductions in renal oxygen consumption (Table 2). Despite this, no concentration of any inhibitor increased Epo mRNA. While Epo mRNA levels after perfusion with cyanide were slightly lower than those with antimycin or oligomycin, this was also true of the non-perfused contralateral kidneys for the cyanide group, and all values for Epo mRNA fell within the range obtained in the well-oxygenated control experiments. Thus metabolic

Table 2. Effect of cyanide, antimycin and oligomycin on EPO mRNA in perfused kidneys

Agent	Dose (μM)	pAO_2 (mmHg)	O_2 consumption (μmol min^{-1} g^{-1})			Decrease in O_2 consumption (%)	Relative abundance of EPO mRNA ($\times 10^{-3}$)[b]	
			15 min	30 min	Mean[a]		Contralateral kidney	Perfused kidney
Well-Oxygenated perfusions								
Cyanide	10	400	4.8	4.7		0	0.07	0.07
	100	390	3.1	3.2		0	0.06	0.03
	300	260	3.8	2.8		26	0.02	0.04
	1000	500	5.8	1.5		74	0.08	0.03
Antimycin	0.01	540	4.3	4.2		0	0.14	0.10
	0.1	540	4.2	3.7		12	0.16	0.02
	0.5	560	4.7	3.2		32	0.12	0.17
	1.0	530	4.7	1.8		62	0.18	0.13
Oligomycin	0.01	490	4.3	4.4		0	0.07	0.12
	0.1	510	5.4	4.8		11	0.05	0.14
	1.0	480	4.9	2.4		51	0.11	0.10
Controls[c]		443±67			3.9±0.6		0.1±0.05	0.07±0.05
Hypoxic perfusions								
Cyanide	1000	40	3.5	0		100	0.05	0.27
Antimycin	0.5	30	4.6	1.6		65	0.10	1.40
Oligomycin	1.0	30	4.6	1.5		67	0.08	0.41
Controls		32±4			3.8±0.6		0.09±0.03	1.15±0.23

[a] Mean renal oxygen consumption (±1 SD) measured after 30 and 90 min of perfusion for control experiments.
[b] Relative to actin mRNA.
[c] Data for controls are mean ±1 SD for four experiments each.

inhibitors did not mimic the effects of hypoxia on Epo mRNA production. In the hypoxic perfusions with large doses of cyanide (1 mM), oligomycin (1 μM) and antimycin (0.5 μM), Epo mRNA levels were higher than the values obtained fo the well-oxygenated controls, demonstrating that the potential for increased Epo production was retained even with the doses of inhibitor used, although the Epo mRNA response to hypoxia in the presence of cyanide and of oligomycin was less than in control kidneys perfused at comparable oxygen tensions.

Discussion

The regulation of Epo production is principally through modulation of its mRNA (Schuster et al. 1989), thus providing an avenue for highly specific assessments of Epo synthesis through the measurement of mRNA levels. As the Epo response to stimulation first manifests as an increase in mRNA, followed by a rise in hormone concentrations (Schuster et al. 1989), an advantage of measuring Epo mRNA is that a shorter duration of experimental observation is required before significant changes in Epo production are detectable. This is of particular importance in in vitro models such as the isolated perfused kidney in which progressive deterioration of the preparation occurs with increasing duration of perfusion (Maack 1980). In addition, the possibility of non-specific Epo hormone release arising from renal cell damage would not confound an assay system based on mRNA measurement. In this study, the use of a sensitive RNase protection technique has demonstrated large, oxygen-dependent increases in Epo mRNA levels in isolated kidneys perfused for 2 h with erythrocyte-supplemented medium.

The experiments in this study provide further evidence that renal oxygen delivery is the dominant factor in the regulation of Epo mRNA production. Epo mRNA levels increased significantly when perfusate pAO_2 was reduced, an observation which is in concurrence with several reports of similar changes in Epo hormone production in isolated perfused rat kidneys (Pagel et al. 1990; Scholz et al. 1991). In addition, however, a striking effect of perfusion flow rate on Epo mRNA production was also evident. This effect is most likely to be due to the differences in renal oxygen delivery arising from the different perfusion flow rates and, since flow rate may vary in kidneys perfused at constant pressure, constitutes a potential pitfall in the use of isolated kidney perfusions for the study of Epo regulation. Our demonstration of flow-dependent changes in Epo mRNA in isolated perfused kidneys is not necessarily in conflict with recent in vivo data demonstrating that reduction in renal blood flow to very low levels by renal artery constriction is a poor stimulus for Epo production in comparison with anaemia (Pagel et al. 1988). Firstly, the range of flow rates under study was different, and, secondly, intrarenal haemodynamic regulation is unlikely to be well preserved in the isolated kidney model.

As tubular sodium reabsorption is believed to be the primary determinant of renal oxygen consumption (Kiil et al. 1961), we attempted to reduce tubular oxygen consumption by stopping glomerular filtration with hyperoncotic perfusion. This, however, failed to susbstantially reduce total renal oxygen consumption. Silva et al. (1980) have shown that renal gluconeogenesis is stimulated when tubular sodium reabsorption is depressed in isolated kidneys of steroid treated rats, suggesting that the two processes may compete for the energy available to renal tubules, and it is possible that similar changes in non-transport work accounted for continued oxygen consumption in our non-filtering kidneys. Although the elimination of glomerular filtration did not greatly alter overall oxygen consumption, Epo mRNA levels were, nevertheless, reduced by about four-fold compared with corresponding filtering kidneys. Since we failed to show a major change in oxygen consumption, the mechanisms underlying this reduction are unclear. One possibility is that glomerular filtration, or tubular transport, itself is necessary for Epo regulation by a mechanism wholly independent of oxygen consumption. This appears unlikely as non-filtering kidneys perfused at a flow rate of 12 ml/min were still capable of a 15-fold increase in Epo mRNA under hypoxic conditions. Another explanation is that, while total oxygen consumption was unchanged in the non-filtering kidneys, regional alterations in oxygen consumption may have occurred in critical areas within the kidney. Alternatively, hyperoncotic perfusate has a higher viscosity, and this may have altered local oxygen consumption delivery (Thorling and Erslev 1968) or have other effects within the kidney. Eckardt et al. (1989c) have demonstrated that the administration of acetazolamide in vivo led to a significant decrease in the Epo hormone response to hypoxia, while diuretic agents acting on more distal nephron segments had no effect. They have suggested that this action of acetazolamide is likely to be mediated by the reduction in oxygen demand of the proximal tubule. While our data do not allow us to make an assessment of this, it is possible that such a mechanism may also be responsible for the findings noted in non-filtering kidneys. However, further definition of the processes involved is severely hampered by the current uncertainties about the nature of the Epo-producing cell in the kidney, and whether other populations of renal cells might be involved in oxygen-sensing.

The administration of three agents which act at different points in oxidative phosphorylation failed to stimulate Epo mRNA over a range of doses which resulted in varying degrees of reduction in renal oxygen consumption. These results are in agreement with the early observations of Necas and Thorling (1972) who found that infusions of cyanide into rats and rabbits did not increase serum Epo concentrations and did not block the Epo hormone response to hypoxia. They are also consistent with recent findings in the hepatoma cell line HepG2, in which the administration of cyanide had no effect on Epo production (Fandrey et al. 1991). We conclude that oxygen-dependent expression of the Epo gene in kidney does not appear to be effected by changes in oxidative phosphorylation.

Acknowledgements. We thank Ms. Kay Yeates and Ms. June Schultz for excellent technical assistance.

References

Adamson JW, Finch CA (1975) Hemoglobin function, oxygen affinity and erythropoietin. Annu Rev Physiol 37:351–369

Bondurant MC, Koury MJ (1986) Anemia induces accumulation of erythropoietin mRNA in the kidney and liver. Mol Cell Biol 6(7):2731–2733

Christoforides C, Laasberg LH, Hedley-Whyte J (1969) Effect of temperature on solubility of O_2 in human plasma. J Appl Physiol 26(1):56–60

Cotes PM (1982) Immunoreactive erythropoietin in serum. Br J Haematol 50:427–438

Eckhardt KU, Kurtz A, Scholz H, Bauer C (1989a) Regulation of erythropoietin formation in vivo. Contrib Nephrol 76:33

Eckhardt K, Boutellier U, Kurtz A, Schopen M, Koller EA, Bauer C (1989b) Rate of erythropoietin formation in humans in response to acute hypobaric hypoxia. J Appl Physiol 66:1785–1788

Eckhardt K, Kurtz A, Bauer C (1989c) Regulation of erythropoietin production is related to proximal tubular function. Am J Physiol 256 (Renal Fluid Electrolyte Physiol 25): F942–947

Epstein FH, Brosnan JT, Tange JD, Ross BD (1982) Improved function with amino acids in the isolated perfused kidney. Am J Physiol 243 (Renal Fluid Electrolyte Physiol 12): F284–292

Erslev AJ, Caro J, Besarab A (1985) Why the kidney? Nephron 41:213–216

Erslev AJ, Wilson J, Caro J (1987) Erythropoietin titers in anemic, nonuremic patients. J Lab Clin Med 109:429–433

Fandrey J, Jelkmann W, Siegers CP (1991) Control of the production of erythropoietin in hepatoma cell cultures (Hep G2). Funktionsanal Biol Syst (in press)

Fried W, Plazak L, Jacobson LO, Goldwasser E (1957) Studies on erythropoiesis. III. Factors controlling erythropoietin production. Proc Soc Exp Biol Med 94:237

Jacobson LO, Goldwasser E, Fried W, Plazak L (1957) Role of the kidney in erythropoiesis. Nature 179:633

Kiil F, Aukland K, Refsum HE (1961) Renal sodium transport and oxygen consumption. Am J Physiol 201:51–516

Krantz SB, Jacobson LO (1970) Erythropoietin and the regulation of erythropoiesis. University of Chicago Press, Chicago, pp17–23

Maack T (1980) Physiological evaluation of the isolated perfused rat kidney. Am J Physiol 238 (Renal Fluid Electrolyte Physiol): F71–78

Necas E, Thorling EB (1972) Unresponsiveness of erythropoietin-producing cells to cyanide. Am J Physiol 222(5):1187–1190

Pagel H, Jelkmann W, Weiss C (1988) A comparison of the effects of renal artery constrictin and anemia on the production of erythropoietin. Pflugers Arch 413(1):62–66

Pagel H, Jelkmann W, Weiss C (1990) Erythropoietin production in the isolated perfused kidney. Biomed Biochim Acta 49:S271–274

Ratcliffe PJ, Jones RW, Phillips RE, Nicholls LG, Bell JI (1990) Oxygen-dependent modulation of erythropoietin mRNA levels in isolated rat kidneys studied by RNAase protection. J Exp Med 172:657–660

Ross BD, Epstein FH, Leaf A (1973) Sodium reabsorption in the perfused rat kidney. Am J Physiol 225:1191–1200

Ross BD, Leaf A, Silvia P, Epstein FH (1974) Na-K-ATPase in sodium transport by the perfused rat kidney. Am J Physiol 226(3):624–629

Scholz H, Schurek H-J, Eckardt K-U, Kurtz A, Bauer C (1991) Oxygen-dependent erythropoietin production by the isolated perfused rat kidney. Pflugers Arch 417:1–6

Schuster SJ, Badiavas EV, Costa-Giomi P, Weinmann R, Erslev AJ, Caro J (1989) Stimulation of erythropoietin gene transcription during hypoxia and cobalt exposure. Blood 73(1):13–16

Silva, P., Ross, BD, Spokes K (1980) Competition between sodium reabsorption and gluconeogenesis in kidneys of steroid-treated rats. Am J Physiol 238 (Renal Fluid Electrolyte Physiol): F290–295

Thorling EB, Erslev AJ (1968) The tissue tension of oxygen and its relation to hematocrit and erythrocytosis. Blood 31:322–343
Tan and P.J. Ratcliffe,

Analysis of Erythropoietin Gene Expression by Means of Competitive Polymerase Chain Reaction*

J.K. FANDREY and H.F. BUNN

Introduction

For a better understanding of the oxygen-sensing mechanism(s) that control erythropoietin (Epo) production, further insights into the regulation of Epo gene expression are necessary. In the hepatoma cell lines HepG2 and Hep3B, Epo is produced in response to the main physiological stimulus, hypoxia (Goldberg et al. 1987). The ability to accurately quantitate the levels of Epo mRNA in these hepatoma cells should provide useful new information about regulation of Epo gene expression. Conventional methods of mRNA analysis such as Northern blot and "dot blot" hybridization and even nuclease protection permit only crude quantitation of mRNA. The sensitivity of these methods is often too low to detect Epo mRNA in uninduced cells or in a small number of cells.

Because of its high sensitivity the polymerase chain reaction (PCR) has been widely used to amplify cDNA copies of low abundance mRNA (Chelly et al. 1988; Frohmann et al. 1988). However, the yield of PCR products fails to provide accurate measurement of the amount of mRNA present in the starting material. The main constraint in obtaining quantitative data is inherent in the amplification process. Because of the exponential increase in products with each cycle of amplification even very small differences in reaction conditions can affect the efficiency of amplification. This can dramatically alter product yield independently of the amount of template input. To circumvent this problem we have adapted competitive PCR (Gilliland et al. 1990) for the quantitation of Epo mRNA from Hep3B cells.

Figure 1 depicts the reactions involved in the competitive PCR system. Epo mRNA is reverse transcribed from total RNA into wild-type Epo cDNA and is coamplified with a competitive template in the PCR. The competitor is a mutant Epo (mutDNA) prepared by site-directed mutagenesis of genomic Epo DNA (Perrin and Gilliland 1990). MutDNA differs from the wild-type Epo DNA by an alteration of only three base pairs (bp) which ablates an *Acc*I site, creates a *Hin*dIII site a few base pairs upstream and does not alter coding (Fig. 2, upper half). The same pair of primers amplifies cDNA and mutDNA

* This research was supported by a grant from the Deutsche Forschungsgemeinschaft (Fa 225/1-1).

Pagel/Weiss/Jelkmann (Eds.)
Pathophysiology and Pharmacology
of Erythropoietin

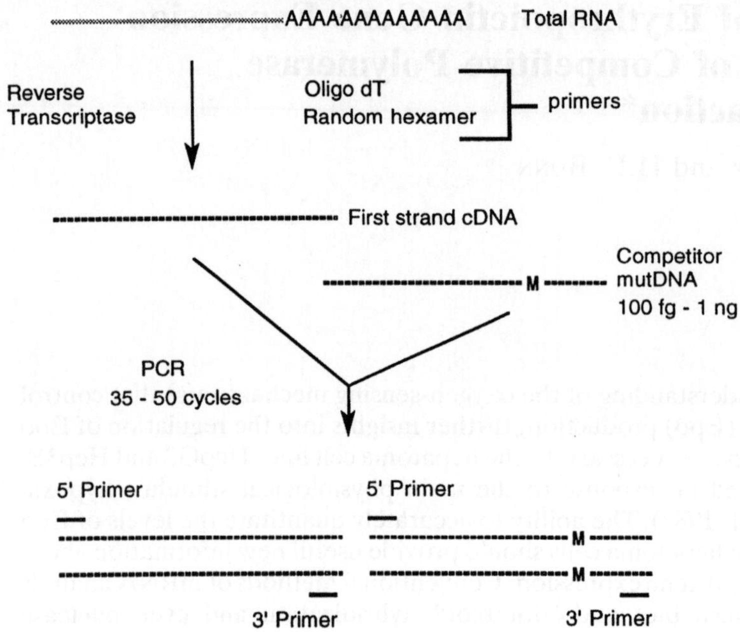

Fig. 1. Flow diagram of competitive PCR for reverse transcribed RNA

fragments. Because the cDNA and the mutDNA compete for available PCR substrates, the efficiency of amplification is equal, and consequently the ratio of products remains constant through the amplification process. By titrating an unknown amount of cDNA template against a dilution series of known concentrations of mutDNA it is possible to reliably and reproducibly quantitate the amount of cDNA (Gilliland et al. 1990). At mutDNA concentration where mutDNA and cDNA products are in equivalence the starting concentration of the mutDNA and the cDNA have been the same.

Identification of the amplified products is achieved by digesting the PCR products with restriction enzymes. Whereas only the cDNA-derived products are cut with *Acc*I, only the mutDNA products are digested by *Hin*dIII.

This method enabled us to define the differences in the amount of mRNA in unstimulated, normoxic cells compared to cells exposed to low oxygen tensions. The high sensitivity of PCR allowed us to detect Epo mRNA in normoxic Hep3B cells which constitutively expressed the Epo gene. Moreover, we could define a dose-response relationship between different oxygen concentrations and corresponding levels of Epo mRNA in Hep3B cells.

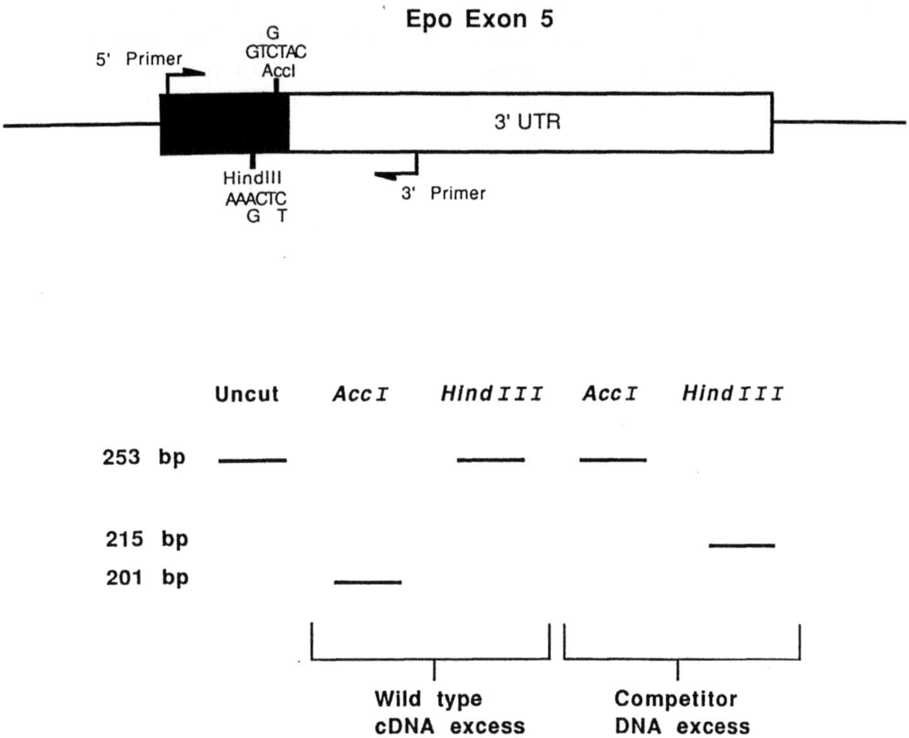

Fig. 2. *Above*, schematic drawing of the mutant competitor DNA (mutDNA). Only exon 5 of the Epo gene is depicted with the binding sites of the upstream 5' primer and the reverse 3' primer for the PCR. The *Acc*I restriction site of the wild-type cDNA and the *Hin*dIII site of the mutDNA are shown with the respective alterations in the nucleotide sequence. *Closed box*, the coding sequence of exon 5; *open box*, the 3'untranslated region (*3'UTR*). *Below*, banding pattern of digested PCR products for either wild-type cDNA excess or competitor mutDNA excess

Methods

All experiments in this study were performed with Hep3B cells which produce Epo in a physiologically controlled manner (Goldberg et al. 1987). Hep3B cells respond to hypoxia with an increase both in Epo gene transcription and in Epo mRNA stability. This results in roughly a 50- to 100-fold increase in Epo mRNA levels in hypoxia-stimulated cells compared with normoxic control cells, as detected by Northern blot analysis (Goldberg et al. 1991). Hep3B cells were grown to confluency in alpha minimal essential medium with 10% iron-enriched calf serum and penicillin (100 U/ml) and streptomycin (100 μg/ml) in 25-cm^2 tissue culture flasks. For experiments the flasks were fitted with rubber stoppers providing gas inlet and outlet through 18 G needles.

Experimental gases containing 21% to 1% O_2, 5% CO_2 and the balance gas N_2 were hydrated at 37°C and circulated at a constant flow rate through the tissue culture flasks. The flasks were immersed in a 37°C constant-temperature water bath and shaken at a frequency of 25 times per minute. The pO_2 was monitored by analyzing the outflow gas in an automated blood gas analyzer (Intrumentation Laboratory).

At the end of the experiments the media were collected and Epo protein determined by radioimmunoassay, as described (Goldberg et al. 1987). The cells were washed with sterile phosphate-buffered saline and lysed with 4 *M* guanidinium isothiocyanate with 0.1 *M* ß-mercaptoethanol. Total RNA was isolated by CsCl centrifugation, as described (Chirgwin et al. 1979), redissolved in water and the concentration determined by absorption at 260 nm.

Reverse Transcription of RNA into cDNA

A quantity of 350 ng – 1 µg total RNA from either hypoxic or normoxic Hep3B cells was reverse transcribed into first-strand cDNA. Primers for the avian myoblastosis virus reverse transcriptase (AMV-Super RT, Molecular Genetic Resources; 4 U/reaction) were either oligo $dT_{(15)}$ or random hexamers (Boehringer, Mannheim, FRG) used at a concentration of 4 p*M*. After initial denaturation of the RNA at 68°C to eliminate possible secondary structure the reaction was allowed to proceed for 1 h at 42°C. The transcription reaction was terminated by boiling the samples for 10 min. The cDNA stocks were kept at –20°C until quantitation by competitive PCR.

In all experiments the presence of possible contaminating genomic DNA from the cells, cDNA from previous PCRs or plasmid DNA was checked by control reactions. Amplification was carried out on samples in which (a) reverse transcriptase was omitted from the reverse transcription reaction mixture, or (b) no RNA was added to the reaction mixture.

Competitive PCR

PCR was performed in 1 × PCR buffer (50 m*M* Tris-HCl, pH 8.3; 50 m*M* KCl; 1.5 m*M* $MgCl_2$; 0.01% wt/vol gelatin), 200 µ*M* of each dNTP, 30 p*M* of each 5' and 3' primer and 5 U/ml *Taq* polymerase (Perkin-Elmer, Cetus) in a final volume of 100 µl. To each tube 1 µl cDNA of unknown concentration and 1 µl of a dilution series containing known amounts of the mutant competitor were added. PCR was then carried out for 35–50 cycles after an initial denaturation for 3 min at 94°C. The amplification profile of each cycle consisted of denaturation at 94°C for 1 min, primer annealing at 58°C for 1 min 30 s, and elongation at 72°C for 3 min.

A competitor mutDNA (Fig. 2, upper half) was prepared from full-length Epo genomic DNA by a PCR-based method (Perrin and Gilliland 1990).

These mutations make amplified PCR products from either cDNA or mutDNA origin distinguishable because they are cut with either *Acc*I or *Hin*dIII. The cut DNA fragments were subsequently separated on a 2% agarose gel, stained with ethidium bromide (0.5 μg/μl) and visualized under UV light.

A schematic drawing of the bands on such a gel is shown in Fig. 2 (lower half). The uncut PCR product appears as a band of 253 bp reflecting the amplified DNA fragment spanned by the pair of primers. After digestion with restriction enzymes a typical banding pattern emerges, depending on the relative amounts of the two different DNAs amplified in the PCR. If wild-type cDNA in the PCR reaction is in excess, it is preferentially amplified. Accordingly, most of the product is cut with *Acc*I. On the ethidium bromide stained gel a 201 bp fragment appears in the lane loaded with *Acc*I-cut DNA whereas the full-length 253 bp fragment remains uncut by *Hin*dIII. With an input excess of competitor mutDNA, however, the opposite pattern is observed: the 253 bp full-length fragment in the *Acc*I lane and the cut 215 bp fragment in the *Hin*dIII lane. If the PCR products are equally cut with *Acc*I and *Hin*dIII, the amount of input cDNA and mutDNA must have been the same.

Results

Figure 3 shows a respresentative experiment in which Hep3B cells were exposed either to hypoxia (1% O_2) or normoxic (21% O_2) for 24 h. Epo protein levels as determined by radioimmunoassay (Goldberg et al. 1987) were 464 mU/ml medium for 1% O_2 and below the limit of detectability (less than 5 mU/ml) for 21%. RNA was isolated. A quantity of 350 ng total RNA was used in the RT reaction with oligo $dT_{(15)}$ as primer. After 35 cycles of PCR

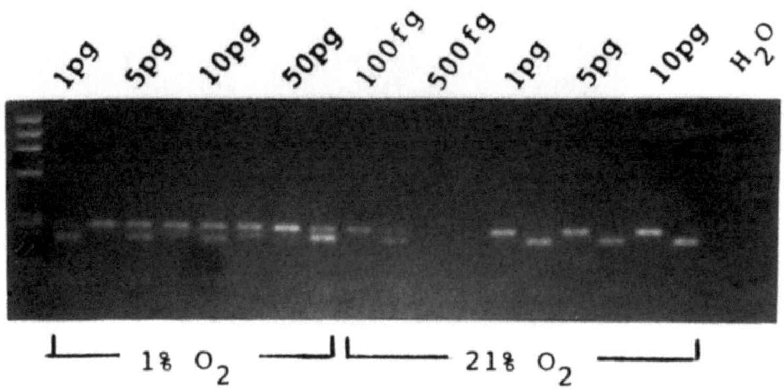

Fig. 3. Competitive PCR of cDNA from either hypoxic (*lanes 2 – 9 from the left*) or normoxic (*lanes 10 – 19*) Hep3B cells following a 24-h incubation. *Lane 1*, molecular weight markers; *lane 20*, negative control (H_2O). At the 500-fg point no amplification was achieved

the DNA was analyzed by restriction enzyme digestion and run on a 2% agarose gel as described above. For 1% O_2 the equilibrium between cDNA and mutDNA was approximately at the 10 pg point of the competitor dilution series (lanes 6 and 7 from the left). For both lanes the product was cut to the same degree by *Acc*I (lane 6) and *Hin*dIII (lane 7). The small size difference of the cut fragments can readily be distinguished, reflecting the *Acc*I fragment of 201 bp compared with the *Hin*dIII fragment of 215 bp. For 21% O_2 the equilibrium is below the lowest concentration of the mutDNA dilution series (100 fg), indicating that the input of cDNA was less than 100 fg. Nevertheless some uncut product is seen in the *Hin*dIII lane (lane 11), providing evidence for the presence of some Epo mRNA even at 21% O_2 in Hep3B cells. From these two concentrations the ratio of Epo mRNA at 1% O_2 versus 21% O_2 is greater than 100, reflecting a more then 100-fold difference in the amount of mRNA in hypoxic versus normoxic Hep3B cells. At the 500 fg point no amplification of any of the DNA by PCR was achieved, demonstrating variability in efficiency of amplification in different reactions mixed and the need for competitive templates for quantitation. This randomly occurring problem of PCR has no implications in this experiment because the 100 fg point shows the same banding pattern as the 1 pg point.

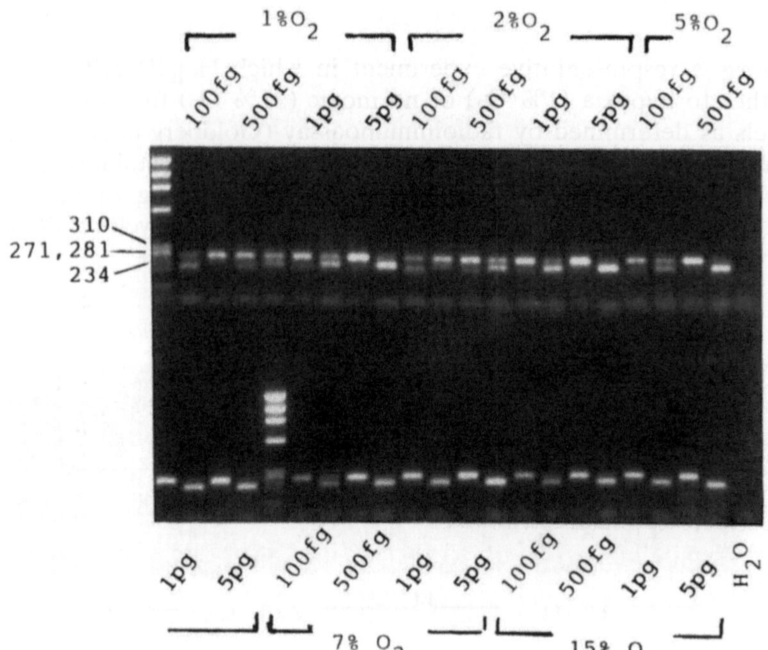

Fig. 4. Competitive PCR of cDNA from Hep3B cells exposed to different oxygen tensions for 6 h. *Lane 1, upper panel from the left*, and *lane 5, lower panel*, molecular weight markers; *lane 22, lower panel*, depicts the negative control (H_2O)

In a different set of experiments Hep3B cells were exposed to increasing O_2 concentrations for 6 h. Here 1 μg total RNA and random nucleohexamer primers were used in the RT reaction. Figure 4 shows the digested PCR fragments after 50 cycles of amplification. A gradual shift of the equilibrium concentrations of mutDNA and cDNA to lower mutDNA concentrations is seen with increasing O_2 concentrations. At 1% O_2 the amount of cDNA is about 500 fg (Epo protein 35 mU/ml), at 2% O_2 it is 250 fg (between 100 fg and 500 fg; Epo protein 30 mU/ml), at 5% O_2 100 fg (Epo protein 18 mU/ml), and at 7% O_2 less than 100 fg (Epo protein 5 mU/ml). Although the 100 fg value is at the lower end of the mutDNA titration curve, the amount of uncut DNA in the *Hin*dIII lane for 100 fg at 7% O_2 (lane 7, lower panel of Fig. 4) seems to be more than in the respective lane at 15% O_2 (lane 15). This indicates a further decrease in the amount of Epo mRNA at 15% O_2 compared to 7% O_2. The Epo protein levels at 15% O_2 were below the limit of detectability of the radioimmunoassay (less than 5 mU/ml). Figure 5 summarizes these results plotting Epo cDNA levels and protein levels versus the different O_2 concentrations. Note that the absolute levels of Epo cDNA as well as protein are considerably higher than in the second set of experiments. The cells here were exposed to 21%/1% O_2 for 24 h whereas the experiment with different oxygen concentrations lasted for only 6 h. Differences between the used subcultures of Hep3B cells may also account for this variation in Epo gene expression, as reported for HepG2 cells (Fandrey

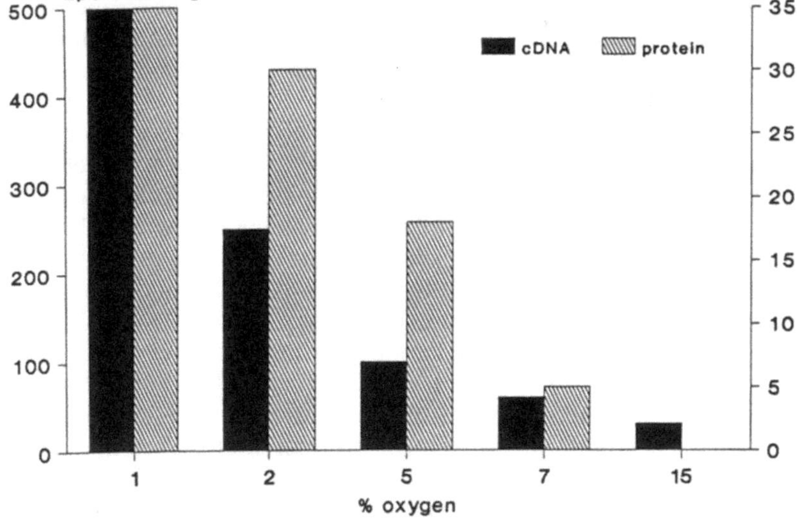

Fig. 5. Epo cDNA and Epo protein are plotted versus the different oxygen concentrations. Data are taken from the gel shown in Fig. 4. The values for 7% and 15% oxygen are estimates, as no equilibrium was achieved below 100 fg

et al. 1990). Therefore for a set of experiments with different oxygen tensions only cells subcultured from one "mother" culture were used.

Discussion

The ability to accurately quantitate Epo mRNA should help to understand better Epo gene expression and how it is influenced by changes in the Epo-producing cells' ambient pO_2. We have adapted the sensitive competitive PCR for quantitation of Epo mRNA and have used this method to address two questions: (a) What is the difference in the amount of Epo mRNA in normoxic versus hypoxic Hep3B cells? (b) How do the Epo mRNA levels change under different oxygen concentrations in the culture medium?

Epo protein production in Hep3B cells exposed to hypoxia can be stimulated more than 90-fold (less than 5 mU/ml versus 464 mU/ml). This increase in protein production is preceded by a more than 100-fold increase in the amount of Epo mRNA in hypoxic cells compared with the normoxic control. These results confirm studies by Goldberg et al. (1991) who estimated a 50- to 100-fold stimulation by analyzing Northern blots of Epo mRNA from Hep3B cells. This increase in mRNA levels in the cells is achieved by an increase in gene transcription rate as well as in mRNA stability under hypoxic conditions (Goldberg et al. 1991).

A major advantage of competitive PCR over other methods is sensitivity. We were able to detect signals from 350 ng total RNA from hypoxic as well as normoxic Hep3B cells. Furthermore, the dilution series of the competitor mutDNA can be extended to 10 fg, which is reliably and reproducibly amplified in this method (data not shown). Competitve PCR does not require radioactive compounds and avoids the problem of nonlinearity associated with autoradiography.

The constitutive expression of the Epo gene under normoxic conditions in Hep3B cells raises several questions. First, one must be certain that the oxygen concentration in the medium surrounding the cells is actually 21% or 160 mmHg. Although we did not measure the pO_2 in the cell layer, the setup of the experiment (high gas flow through the tissue culture flasks and continuous shaking of the medium) should provide equilibration of the fluid with the ambient gas. Oxygen tension lower than 100 mmHg is appropriate for physiological studies since Epo-producing cells in vivo are never exposed to such a high pO_2. Second, if there is constitutive Epo gene expression, could it be caused by the malignant origin of Hep3B cells, or does it reflect the physiological steady-state low-level expression of Epo in the kidney in vivo (Koury et al. 1989)?

If a heme-protein is involved in the oxygen-sensing process, and if the increase in Epo gene expression is dependent on a conformational change upon binding of oxygen to the heme moiety (Goldberg et al. 1988), does Epo gene expression under normoxic conditions reflect a very low oxygen affinity of this heme protein? Would this be advantageous for fine tuning of the oxygen sensor in the kidney in regions with a pO_2 of 25–40 mmHg?

Further characterization of the oxygen-sensing mechanism in Hep3B cells is provided by our experiments in which the cells were exposed to different oxygen concentrations. The gradual decrease in Epo cDNA levels with increasing oxygen concentrations was informative. Koury et al. (1989) have shown by in situ hybridization of sections of mouse kidney that Epo-producing cells respond to increasing degress of anemia with a thresholdlike all-or-none behavior. The number of Epo mRNA molecules per cells remained constant as the hematocrit was lowered, but new cells expressing Epo mRNA were recruited (Koury et al. 1989). Very recently, however, Koury et al. (1991) reported in situ hybridization studies on livers of mice transgenic for the human Epo gene. Lowering the hematocrit in these mice by repeated bleeding caused in increase in the number of mRNA copies per cell and not the recruitment of new Epo expressing cells (Koury et al. 1991). The majority of these cells were in the areas surrounding the central veins and could be recognized as hepatocytes. These findings are consistent with our results in Hep3B cells. As Hep3B cells are of hepatic origin, they might be a suitable model for studying Epo gene expression in the liver. However, it is questionable whether Epo-producing cells in the kidney respond in the same way when exposed to hypoxia.

Semenza et al. (1989, 1990) have reported that upstream of the Epo gene regulatory sequences may be responsible for tissue-specific expression either in the liver or kidney. These regulatory regions could interact differently with oxygen-sensing mechanisms and subsequent transduction into an elevation of Epo mRNA levels under conditions of lowered oxygen tension. To address these questions a kidney cell line in which Epo is produced in a physiologically controlled manner would be desirable.

Acknowledgements. The authors thank Dr. Deborah Galson for her help with the computer graphics and Dr. Gary Gilliland for his critical review of the manuscript.

References

Chelly J, Kaplan J-C, Maire P, Gautron S, Kahn A (1988) Transcription of the dystrophin gene in muscle and non-muscle tissues. Nature 333:858–860

Chirgwin JM, Pryzbyla AE, MacDonald RJ, Rutter WJ (1979) Isolation of biologically active ribonucleic acid from sources enriched in ribonuclease. Biochemistry 18:5294–5299

Fandrey J, Seydel FP, Siegers C-P, Jelkmann W (1990) Role of cytochrome P_{450} in the control of the production of erythropoietin. Life Sci 47:127–134

Frohman MA, Dush MK, Martin GR (1988) Rapid production of full-length cDNA from rare transcripts: amplification using a single gene-specific oligonucleotide primer. Proc Natl Acad Sci USA 85:8998–9002

Gilliland G, Perrin S, Blanchard K, Bunn HF (1990) Analysis of cytokine mRNA and DNA: detection and quantification by competitive polymerase chain reaction. Proc Natl Acad Sci USA 87:2725–2729

Goldberg MA, Glass CA, Cunningham JM, Bunn HF (1987) The regulated expression of erythropoietin by two human hepatoma cell lines. Proc Natl Acad Sci USA 84:7972–7976

Goldberg MA, Dunning SP, Bunn HF (1988) Regulation of the erythropoietin gene: evidence that the oxygen sensor is a heme protein. Science 242:1412–1416

Goldberg MA, Gaut CC, Bunn HF (1991) Erythropoietin mRNA levels are governed by both the rate of gene transcription and posttranscriptional events. Blood 77:271–277

Koury ST, Koury MJ, Bondurant MC, Caro J, Graber SE (1989) Quantitation of erythropoietin-producing cells in kidneys of mice by in situ hybridization: correlation with hematocrit, renal erythropoietin mRNA and serum erythropoietin concentration. Blood 74:645–651

Koury ST, Bondurant MC, Koury MJ, Semenza GL (1991) Localization of cells producing erythropoietin in murine liver by in situ hybridization. Blood 77:2497–2503

Perrin S, Gilliland G (1990) Site-specific mutagenesis using asymmetric polymerase chain reaction and a single mutant primer. Nucleic Acids Res 18:7433–7438

Semenza GL, Traystman MD, Gearhart JD, Antonarakis SE (1989) Polycythemia in transgenic mice expressing the human erythropoietin gene. Proc Natl Acad Sci USA 86:2301–2305

Semenza GL, Dureza RC, Traystman MD, Gearhart JD, Antonarakis SE (1990) Human erythropoietin gene expression in transgenic mice: multiple transcription initiation sites and cis-acting regulatory elements. Mol Cell Biol 10:930–938

The Developmental Biology of Erythropoiesis and Erythropoietin: Is Erythropoietin a True Hormone or a Paracrine Factor?

I.N. Rich and G. Noé

Introduction

We usually take it for granted that erythropoietin is a true hormone – that is, it is produced from an organ and carried through the circulation to exert its effect on a distant tissue. This exocrine production is, in fact, a function of the kidney and the distant tissue is the bone marrow. Under many circumstances, erythropoietin cannot be considered a true hormone because it can be produced in the same organ in which erythropoiesis is occurring. In other words, its production is paracrine. This may appear to contradict what we have learned since the 1950s and what students read in text books. Indeed, for mammals and man in particular, there is certainly a strong argument for erythropoietin being a hormone. But let us first go back in time and consider the phylogeny of erythropoiesis before coming to any forgone conclusions.

The Kidney As an Erythropoietic Organ in Early Vertebrates

Table 1 shows the phylogeny of erythropoiesis in vertebrates. The important point to notice is that in the fish orders *Agnatha* and *Osteichyes* and especially the suborder *Teleosti* or "bony fish" (e.g., salmon and trout), hemopoiesis occurs in the kidney. It is not known whether fish have the equivalent of erythropoietin. Considering the highly conserved nature of erythropoietin, it may be assumed that an erythropoietin-like molecule exists and functions in fish. Is erythropoiesis in fish controlled in a paracrine manner in the kidney? This question remains open, but a precedence for paracrine production does appear to exist.

In the embryo of the amphibian, erythropoiesis occurs in the yolk sac. This has recently been elegantly demonstrated using whole mount in situ hybridization (Hemmati-Brivanlou et al. 1990). In the larval stage of the amphibian order *Anura*, erythropoiesis occurs in the kidney and liver, while in the other two orders of amphibia (*Urodela* and *Apoda*), the spleen is the only source of erythrocytes in both the larval and adult stages of development (Table 1).

Pagel/Weiss/Jelkmann (Eds.)
Pathophysiology and Pharmacology
of Erythropoietin

Table 1. Phylogeny of erythropoiesis in vertebrates

Phylum	Order	Stage	Erythropoietic organs
Pisces	Agnatha		Pronephros kidney
	Chrondrich-thyes		Splenic red pulp, blood (?)
	Osteichyes		Pronephric (head) kidney, opisthonephric (trunk) kidney, spleen
Amphibia	Variety	Early	Ventral blood islands
	Anura	Larva	kidney, liver (spleen?)
		Adult	Liver, spleen, bone marrow
	Urodela		Spleen (liver?)
	Apoda		Spleen
Repitilia		Early	Yolk sac
		Fetal	Liver, spleen
		Adult	Bone marrow (extramedullary: spleen, liver)
Avia		Embryo	Yolk sac, loose mesenchyme
		Adult	Spleen, bone marrow
Mammalia		Embryo	Yolk sac
		Fetal	Liver, spleen
		Neonatal	Liver, spleen, bone marrow
		Adult	Bone marrow (extramedullary: spleen, liver)

However, amphibian hemopoietic development is particularly interesting since these are the first vertebrates in which hemopoiesis can occur in the bone marrow. However, as mentioned above, not all members of amphibia show this ability. In fact only certain members of the order *Anura* demonstrate this. In Xenopus, for example, the bone marrow is poorly developed. The reason for the transfer of hemopoiesis to the bone marrow is not completely clear, but may be due to the change to an air-breathing existence and the need to protect the proliferating cells from the radiation hazard on land. It should be pointed out that in this order, adult erythropoiesis occurs in the spleen and liver also, and all organs have apparently equal importance. In reptiles, the bone marrow is the primary source of erythroid cells in the adult with extramedullary erythropoiesis occurring in the spleen and liver.

The Control of Erythropoiesis in Early Vertebrates

Little is known about the control of erythropoiesis in fish, amphibians and reptiles. However, the stimuli affecting erythropoiesis appear to be similar to those in the mammal. Thus, one can stimulate erythropoiesis by starvation, bleeding, hypoxia and phenylhydrazine treatment. Zanjani et al. (Zanjani et al. 1969) postulated that a plasma factor similar to erythropoietin was present in anemic fish, and Weinberg et al. (Weinberg et al. 1976) showed that high concentrations of native human erythropoietin could stimulate erythropoiesis in starved fish. Some frogs, however, are insensitive to the effect of human erythropoietin.

The Liver and Spleen As Transient Erythropoietic Organs

The liver and spleen play a role in erythropoiesis in nearly all vertebrates. In the early vertebrates, this can be a permanent function. However, later, the liver and spleen are only stopping-off points on the way to the final destination, the bone marrow. That the liver and spleen possess the necessary microenvironment to support erythropoiesis appears obvious, but why that should be the case is not clear. From the developmental point of view, the liver is the only erythropoietic production site that is not derived from the mesoderm (the liver is derived from the endoderm, Fig. 1). However, nearly all organs derived from the endoderm have secondary inclusions of mesodermal origin. When extramedullary erythropoiesis occurs, the spleen (derived from a thickening in the dorsal mesentery of the stomach, the thickening being part of the splanchnic mesoderm plate, Fig. 1) is the first organ to

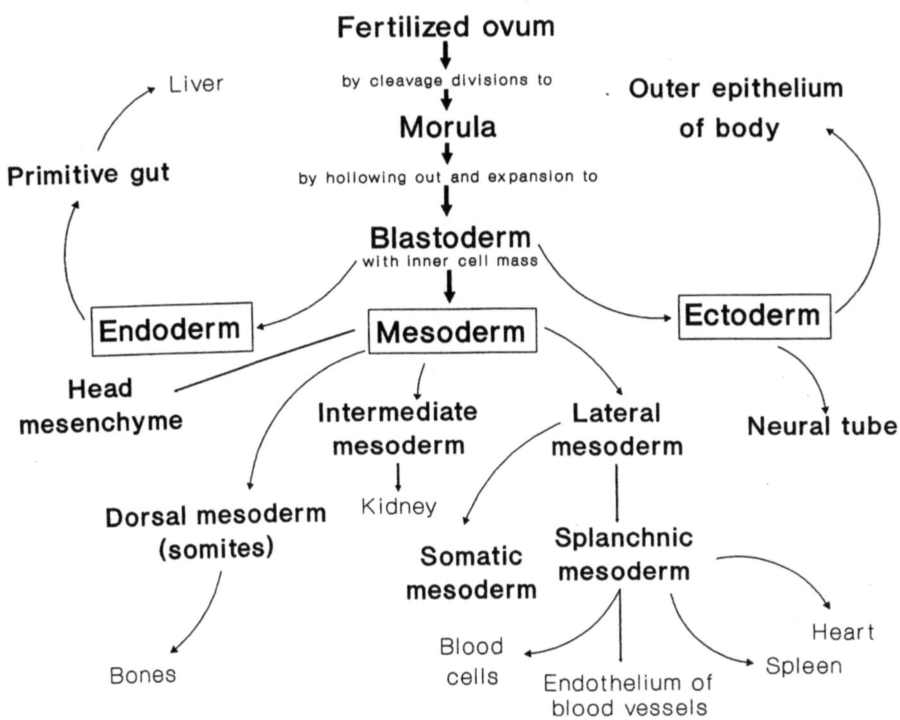

Fig. 1. The origin and developmental biology of blood cells and the hemopoietic and associated organs

reinitiate this function. Only under extreme perturbation is hemopoiesis observed in the liver.

Is the mesodermal origin in some way important? It would appear to be. Erythropoiesis is observed in the mammalian fetal liver associated with macrophages and form blood islands (a central macrophage surrounded by developing erythroblasts). There is no association with any other cell of non-mesodermal origin. It is the mouse fetal liver macrophage that is the source of erythropoietin during early hepatic erythropoiesis (Zucali et al. 1975). This is a prime example of paracrine regulation.

Where Is Erythropoietin Produced in the Kidney?

As recently discussed in detail (Rich 1991), it appears that all groups working in the field agree that erythropoietin is produced in the renal cortex, but not in the glomerular apparatus. Some believe that the peritubular cell is the source (Koury et al. 1988; Lacombe et al. 1988); others believe the tubular (probably the proximal tubule) is responsible for erythropoietin production (Maxwell et al. 1990; Rich et al. 1990). The present controversy is based on the interpretation of in situ hybridization and immunohistochemical photographs.

There is no classification for a peritubular cell. Developmental biology does not help us either. If, as suggested, the peritubular cell is a capillary endothelial cell, then both this and the tubular cell are of mesodermal origin (Fig. 1). We cannot imply that the erythropoietin-producing cell in the kidney is derived from another germ line. We have to look elsewhere.

We suggest that this problem is not one of biology, but rather of methodology, although at first site a flaw is not easy to find. The subjectivity of the in situ hybridization is based on photographs, and careful examination of these brings to light an interesting phenomenon.

Much of the evidence demonstrating the peritubular cells as the site of erythropoietin production shows radioactive labeling concentrated over the nucleus of cells lying external to the tubular cells and not, as would be expected, in the cytoplasm, where the mRNA should be found (Koury et al. 1988; Lacombe et al. 1988; Bruneval et al. 1989; Krantz 1991). Furthermore, "trails" of exposed silver grains can be seen emanating away from the central core of grains lying over the nuclei of these cells. This phenomenon is not observed in the in situ hybridization studies of Maxwell et al. (Maxwell et al. 1990), and certainly not in mouse (Fig. 2) or human kidney tissue (results not shown) using non-radioactive labeling (Rich et al. 1990).

We suggest that this discrepancy is due to spontaneous grain activation during autoradiography. This artifact, called "positive chemography," is caused by the tissue itself or materials used (Quade et al. 1970; Seelig and Head 1984). We, therefore, suggest that until studies have been performed controlling for chemography, the case for the peritubular cell being the site of

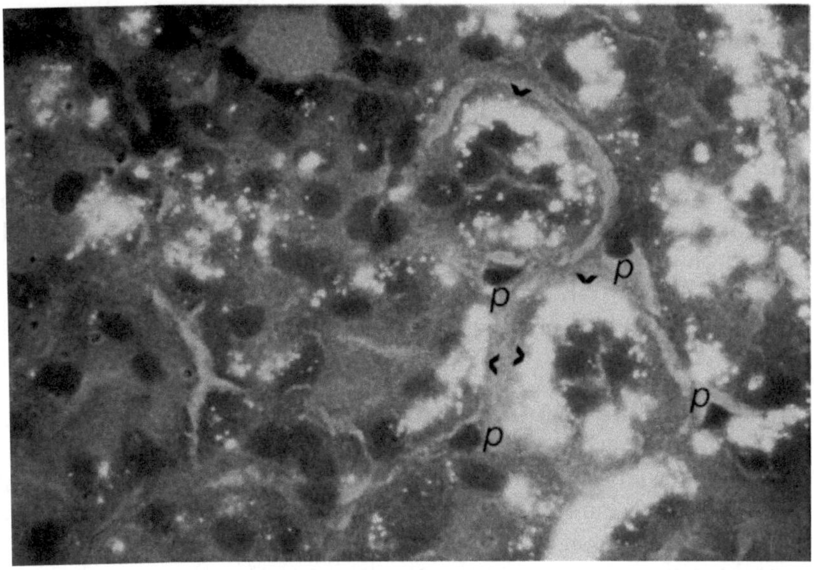

Fig. 2. Nonradioisotopic in situ hybridization performed on hypoxic mouse kidney. A biotinylated genomic erythropoietin probe was used for hybridization. Streptavidin gold (5 nm) and reflection-contrast microscopy was used as signal detection system. The gold is seen as white accumulated light scatter. The *arrows* show particularly dense areas where the gold signal is bound to the hybridized probe. These areas are in tubules. Note that the signal is *not* concentrated over nuclei, but in the cytoplasm. Small cells external to the tubules (p) designate the peritubular cells. These cells do not show any signal whatsoever

erythropoietin production stands on very shaky foundations and is not as conclusive as recently suggested by Krantz (1991).

Finally, taking into account the anatomy of the primitive vertebrate kidney which consists only of ducts (tubules), one could postulate that control of erythropoiesis is paracrine because it can only occur in these nephric ducts. An extratubular intervention is not present or required.

Is Erythropoietin Production a Remnant Kidney Function?

Let us now return to the problem at hand. With the evolution of land animals and transfer of erythropoiesis to the bone marrow, we see an apparent divergence of functional regulation. In the mammal, under normal steady-state conditions, erythroid cells are produced in the bone marrow and erythropoietin is produced in the kidney. At least, this is what we have taken for granted since Jacobson et al. (Jacobson et al. 1957) showed that erythropoietin was produced in the kidney. The fact of the matter is, there is still very little conclusive data on whether erythropoietin is produced in the kidney under normal, steady-state conditions (Rich 1991), and our own results using the polymerase chain reaction have not shown any gene expression under these conditions (results to be published elsewhere). There is no question, however, that under hypoxic or anemic conditions the kidney is responsible for the increase in erythropoietin production (see Rich 1991 for review).

Is it possible that this kidney function is really a relic from ancient times? In other words, is it possible that erythropoietin in the kidney is really a remnant, but nevertheless a sensitive remnant, seen in fish and the larva of some amphibians, remaining in higher vertebrates and used under emergency conditions? We would like to suggest that this is indeed the case, and that normal erythropoietin regulation is under paracrine control.

The Role of the Blood Island in the Adult

We have already mentioned that in the early mammalian fetal liver, erythropoiesis occurs in blood islands. In fact, active erythropoiesis nearly always occurs in blood islands. Our own studies on the role of the macrophage in erythropoiesis have led to the conclusion that in the normal murine bone marrow, a subpopulation of macrophages expresses the gene and synthesizes mature erythropoietin (Rich 1991). This macrophage function occurs only in blood islands, and under erythropoietic stress conditions there is an increase in the number of macrophages expressing the erythropoietin gene and a corresponding increase in the the number of blood islands present (Vogt et al. 1990). That a discrepancy between erythropoietin produced by the bone marrow macrophage and the kidney could be due to different oxygen sensor set points has been discussed in detail elsewhere (Rich 1991). Thus, during erythropoietic ontogeny in the mammal, there is a paracrine production of erythropoietin in at least two organs, the fetal liver and the bone marrow.

During extramedullary erythropoiesis, a partial repetition of phyletic history occurs, with erythropoiesis returning to the spleen. To date, however, erythropoietin gene expression has not been detected in the spleen, even though one may expect it. The reason for this is not clear, but certainly warrants further investigation. Nevertheless, histological examination shows that active splenic erythropoiesis is seen to occur in erythropoietic functional units or blood islands.

Blood Islands in the Mouse Embryo

The induction of mesoderm at about day 7 of gestation followed by the production of the amniotic cavity in the mouse embryo later results in cells from this germ layer migrating onto the inner surface of the visceral endoderm. These migrating cells form the first hemopoietic tissue and the blood islands of the visceral yolk sac. We do not know the differentiation potential of the first hemopoietic stem cells. However, we do know that

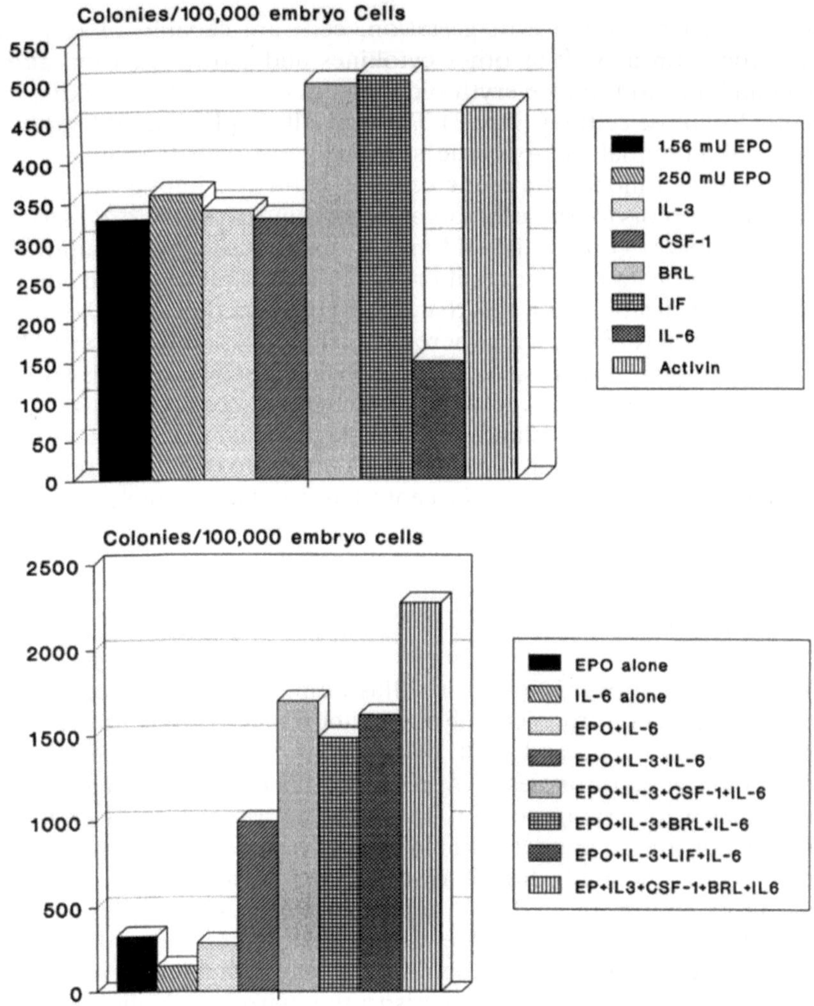

Background colony number = 150

Fig. 3. Effect of erythropoietin (*EPO*) and other growth factors on the stimulation of 8-day embryo cells. Cells were obtained from disaggregated embryos and plated at a concentration of 2.5×10^4 cells/ml in methylcellulose containing 30% fetal calf serum, $1 \times 10^{-4} M$ α-thioglycerol, and $1 \times 10^{-12} M$ mouse transferrin. The cultures were incubated for 3 days under 3.5% O_2 and 5% CO_2. Factors plated were recombinant interleukin-3 (*IL-3*), macrophage colony-stimulating factor (*CSF-1*), *BRL* supernatant (a source of stem cell factor, SCF), leukemia inhibitory factor (*LIF*), interleukin-6 (*IL-6*), and activin. The types of colonies obtained were either pure erythroid, pure macrophage, or erythroid macrophage mixed between days 7.5 and 8 of gestation, cells are capable of responding to erythropoietin as well as other cytokines and growth factors, alone or in combination, to produce erythroid colonies or macrophage-like colonies in methylcellulose cultures (Fig. 3). These results imply that, at least in culture, the multipotential hemopoietic stem cells are capable of giving rise to erythroid and monocytopoietic cells. It is further implied that these cells possess receptors for these factors and that some of the factors may be present in the embryo. Using 8-day-old mouse embryos, we have found, using the polymerase chain reaction, that erythropoietin is present, and whole mount in situ hybridization indicates that the erythropoietin may be totally located in the visceral yolk sac, i. e., where the first blood islands are located, and not in the embryo proper (results to be published elsewhere). This latter finding would imply paracrine regulation of embryonic erythropoiesis.

Whether the blood islands of the embryo are similar in structure to those in the fetal liver and adult bone marrow remains to be seen. If this is the case, the next question is whether the central macrophage of the embryonic blood islands produces erythropoietin.

Conclusions

We have put forward arguments illustrating the precedence of paracrine production of erythropoietin during vertebrate development. At least in the mouse embryo, fetus and adult, there is considerable evidence of a paracrine regulation of erythropoiesis. We consider this to be the primary production mechanism for erythropoietin. The intimate association of erythropoietin production and responder cells means that only minute amounts of erythropoietin are required. Together with the low number of erythropoietin receptors present on erythroid precursor cells, this provides an extremely specific mode of regulation. We also suggest that kidney-produced erythropoietin occurs under stress situations and may be a remnant function caused by the divergence of erythropoiesis to the marrow in higher vertebrates.

Acknowledgements. This work was supported by the German Red Cross and the Deutsche Sonderforschungsbereich SFB 322, Project A8.

References

Bruneval P, Da Silva JL, Lacombe C et al. (1989) Erythropoietin synthesis in the anemic mouse kidney as observed by morphological techniques. In: Jelkmann W, Gross AJ (eds) Erythropoietin. Springer, Berlin Heidelberg New York, pp26–35

Hemmati-Brivanlou A, Frank D, Bolce ME, Brown BD, Sive HL, Harland RM (1990) Localization of specific mRNAs in Xenopus embryos by whole-mount in situ hybridization. Development 110:325–330

Jacobsen LO, Goldwasser E, Fried W, Plzak L (1957) The role of the kidney in erythropoiesis. Nature 179:633–634

Koury ST, Bondurant MC, Koury MJ (1988) Localization of erythropoietin synthesizing cells in murine kidneys by in situ hybridization. Blood 7:524–527
Krantz S (1991) Erythropoietin. Blood 77:419–434
Kurtz A, Eckardt KU, Neumann R, Kaissling B, Le Hir M, Bauer C (1989) Site of erythropoietin formation. Contrib Nephrol 76:14–23
Lacombe C, Da Silva J-L, Bruneval P, Fournier J-G, Wendling F, Casadevall N, Camilleri J-P, Bariety J, Varet B, Tambourin P (1988) Peritubular cells are the site of erythropoietin synthesis in the murine hypoxic kidney. J Clin Invest 81:620–623
Maxwell AP, Lappin TRJ, Johnston CF, Bridges JM, McGeown MG (1990) Erythropoietin production in kidney tubular cells. Br J Haematol 74:535–539
Quade R, Halbhuber K-J, Stibenz J (1970) Die Verwendung von Zaponlackmembranen als inertes Intermedium zur Verhinderung positiv-chemographischer Effekte in Emulsionen lichtoptischer Autoradiogramme. Acta Histochem (Jena) 38:270–277
Rich IN (1991) Erythropoietin production – a personal view. Exp Hematol 19:985–990
Rich IN, Noé G, Vogt C, Lemke C (1990) Erythropoietin is produced in the tubules of the renal cortex (Abstr) Blood 76 (Suppl 1):162a
Seelig LL, Head JR (1984) Spontaneous autoradiographic grain activation associated with mast cells in methacrylate plastic embedded tissue sections. Stain Technol 59:43–47
Vogt C, Noé G, Rich IN (1990) Normal steady-state haemopoiesis assay of erythropoietin by ELISA, localization of erythropoietin in the kidney and expression of haemopoietic regulator molecules in the bone marrow. In: Sachs L, Abraham NG, Wiedermann CJ, Levine AS, Konwalinka G (eds) Molecular biology of haematopoiesis. Intercept, Andover, pp 401–412
Weinberg SR, LoBue J, Siegel CD, Gordon AS (1976) Hematopoiesis of the kissing gourami (Helostoma temmincki). Effects of starvation, bleeding, and plasma-stimulation on its erythropoiesis. Can J Zool 54:1115–1127
Zanjani ED, Yu ML, Permutter A, Gordon AS (1969) Humoral factors influencing erythropoiesis in the fish (blue gourami. Trichogaster trichopterus). Blood 33:573–581
Zucali JR, Stevens V, Mirand EA (1975) In vitro production of erythropoietin by mouse fetal liver. Blood 46:85–90

Insulin-like Growth Factor 1 and Erythropoiesis During Growth

A. Bechensteen and S. Halvorsen

Introduction

Our research in recent years has dealt with the regulation of erythropoiesis during growth. Results indicate that there are factors in addition to erythropoietin which have stimulatory effects on erythropoiesis in this period. Our hypothesis is that one or more factors stimulating growth in general also stimulate erythropoiesis. Insulin-like growth factor (IGF) 1 is a candidate for this function. The IGFs are a group of serum proteins that have mitogenic properties and insulin-like metabolic activity. IGF-1 is a liver-derived peptide with low molecular mass (7649 Da) bound in the circulation to larger binding proteins (40 000 and 150 000 Da).

There have recently been several reports about the erythropoietic stimulatory effect of IGF-1, both in vitro and in vivo. Kurtz et al. [1] found that IGF-1 stimulates colony-forming units–erythroid (CFU-E) from murine fetal liver and human bone marrow cells in a similar way as erythropoietin (Epo) does. Later several works have confirmed this stimulatory in vitro effect of IGF-1 upon erythropoietic cells [2–4]. Kurtz et al. [5] showed that the activity in fetal bovine serum that stimulates CFU-E in fetal mouse livers is IGF-1. Kurtz et al. [6] examined the correlation between the erythropoietic activity, Epo and IGF-1 in the rapidly growing rat. IGF-1, rather than Epo, was clearly correlated with iron incorporation into red blood cells. Most in vivo studies with IGF-1 are conducted in animals without intact production of growth hormone (GH; hypophysectomized rats, Snell dwarf mice) using treatment that more resembles a substitutional therapy. In these studies IGF-1 injection or infusion normalizes the body and organ weight of the IGF-1 treated animals. Kurtz et al. [7] also showed that the decreased erythropoiesis following hypophysectomy was reversed after 6 day's treatment with IGF-1, that reticulocyte count and iron incorporation increased compared to the nontreated animals, but that the hematocrit level did not change during the IGF-1 treatment time. Philipps et al. [8] injected IGF-1 into neonatal rats with intact hypophysis and GH production. Increased body weight, tail length, organ weight (liver, brain, heart, and testes) were observed; erythropoiesis measured by maximal CFU-E response also showed an increase after IGF-1 treatment, but hematocrit and reticulocyte count did not differ from those in the control group.

Pagel/Weiss/Jelkmann (Eds.)
Pathophysiology and Pharmacology
of Erythropoietin

IGF-1 clearly has stimulatory effects upon erythroid cells in vitro. The physiological role of IGF-1, rather than Epo, as the important stimulator of erythropoiesis in the neonatal or rapidly growing age is not fully established, nor is the effect upon normal erythropoiesis in vivo. In this study we investigated the effect of human recombinant IGF-1 upon murine erythropoiesis in vivo. We also studied the effect of IGF-1 upon murine erythropoiesis in vitro.

Materials and Methods

AKR/O and WLO mice of both sexes were studied. The AKR/O strain has been bred in our laboratory since 1965 and was originally derived from the AKR/J strain (Arnesen, personal communication) [9]. The WLO strain is described in detail elsewhere [10]. The litter size was five or six pups in the in vivo studies. Each litter contained both treatment groups and controls. In the in vitro studies the litter size varied from 4 to 12.

In the in vivo studies the mice were divided into three groups. In group A human recombinant IGF-1 (a gift from Kabi Vitrum, Stockholm) was given in two equal daily doses from day 6 to day 20. The doses were 500 and 2000 μg/kg per day administered subcutaneously at 0900 and 2100 hours. The volume was initially 0.05 ml, increasing to 0.1 ml on day 9. IGF-1 was diluted in a buffer (NaCl phosphate buffer, pH 6), and this buffer also served as placebo. In groups B and C IGF-1 was given during a shorter period of time (days 17–20), with shorter intervals (every 6 h: at 0800, 1400, 2000, and 0200 hours), and the daily doses were thus divided in four equal doses. The volume was 0.05 ml in each injection. The doses were 500 μg/kg per day and 1500 μg/kg per day in groups B and C, respectively. All mice were weighed daily before the injections were given.

On day 20 the animals were exsanguinated by cervical vessel transection. Autopsy was performed and organs (liver, spleen, and thymus) were inspected and weighed. Packed cell volume (PCV), white blood cell count, and reticulocyte count were determined in each animal – PCV using a microhematocrit method and white blood cell count by direct count. For the reticulocyte count 1000 red blood cells were counted. Plasma samples from each animal were stored at –20°C until thawed for analysis. Plasma samples from each individual animal or pooled from two animals were used in measuring the plasma erythropoietin level by enzyme-linked immunosorbent assay [11].

Since IGF-1 has insulin-like properties, AKR/O mice 10- to 20-days-old were tested for hypoglycemia due to IGF-1 injections. Doses similar to those used above (i.e. 250 and 1000 μg/kg) were injected subcutaneously, and blood glucose was measured initially and after 15, 30, 60, and 75 min and every 30 min until 180 min after injection. Blood glucose was measured using a hemoglucose test. Buffer was used as placebo.

In about half of the experimental animals bone marrow and spleen cells were cultured for CFU-E determinations. The spleens were homogenized in alpha medium containing 20% fetal bovine serum, the femora flushed with alpha medium, and the cells prepared and cultured as described for neonatal liver cells [12]. The counts are expressed as the number of colonies per 0.5×10^5 mononuclear cells plated.

For the in vitro study CFU-E from murine liver, bone marrow, and spleen was examined in both AKR/O and WLO mice at different ages. In addition to the methods described above in the CFU-E investigations, in which dose-response curves of Epo were included each time, IGF-1 at different concentrations (1, 10, 100, 1000, and 10 000 ng/ml) was added to the cultures to obtain a dose-response curve. IGF-1 was diluted in NaCl phosphate buffer, which also served as a control.

The statistical evaluation was performed using the nonparametric Wilcoxon–van Eltern test. The level of significance was chosen as $p=0.05$; median values and 95% confidence intervals are reported here.

Results

Mice treated with IGF-1 (500 µg/kg per day) on days 6–20 (group A, AKR/O mice) showed a significant increase in weight compared to controls and compared to the high-dose group (2000 µg/kg per day; $p<0.05$). Figure 1 shows the weight curve of these mice. Absolute and relative spleen weights in both treatment groups also differed significantly compared to controls; the differences in thymus weight were not significant (Table 1). In groups B and C (WL/O mice), where treatment time was 60 h, no statistically significant differences in body or organ weight were observed.

Glucose measurements after IGF-1 injection showed a clear fall in blood glucose values 30–150 min after subcutaneous administration; significant

Table 1. Weight results obtained in AKR/O mice treated with IGF-1 on days 6–20 median values and 95% confidence intervals

	Control (n=10)	IGF-1; 500 µg (n=10)	IGF-1; 2000 µg (n=10)
Weight gain	5.0 (4.4–5.2)	5.1[a] (5.0–5.6)	5.0 (4.8–5.4)
Spleen weight	32 (21–40)	43[a] (36–49)	41[a] (40–49)
Spleen weight: % bw	4.0 (2.7–5.1)	5.1[a] (4.4–5.5)	4.8[a] (4.6–5.5)
Thymus weight	55 (44–66)	59 (48–75)	59 (53–68)

[a] $p<0.05$ (Wilcoxan–van Eltern test)

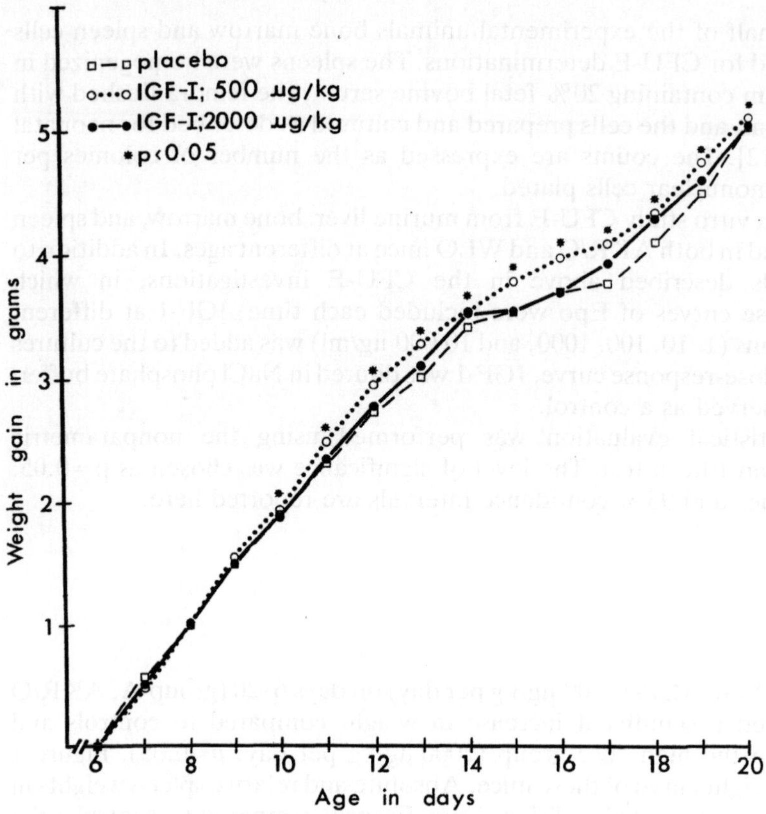

Fig. 1. Daily weight gain from day 6 to day 20

Table 2. Hematological values in AKR/O mice treated with IGF-1 on days 6–20: median values and 95% confidence intervals. No significant ($p<0.05$) differences

	Control (n=10)	IGF-1; 500 µg (n=10)	IGF-1; 2000 µg (n=10)
PCV	38 (36–39)	38 (36–40)	39 (37–40)
WBC	2.8 (1.3–3.1)	2.0 (1.4–2.7)	2.3 (1.7–3.6)
Reticulocytes (%)	7.0 (3.9–10.7)	6.7 (3.3–8.0)	7.4 (3.3–9.3)
CFU-E, bone marrow	150	157	171
CFU-E, spleen	101	111	84

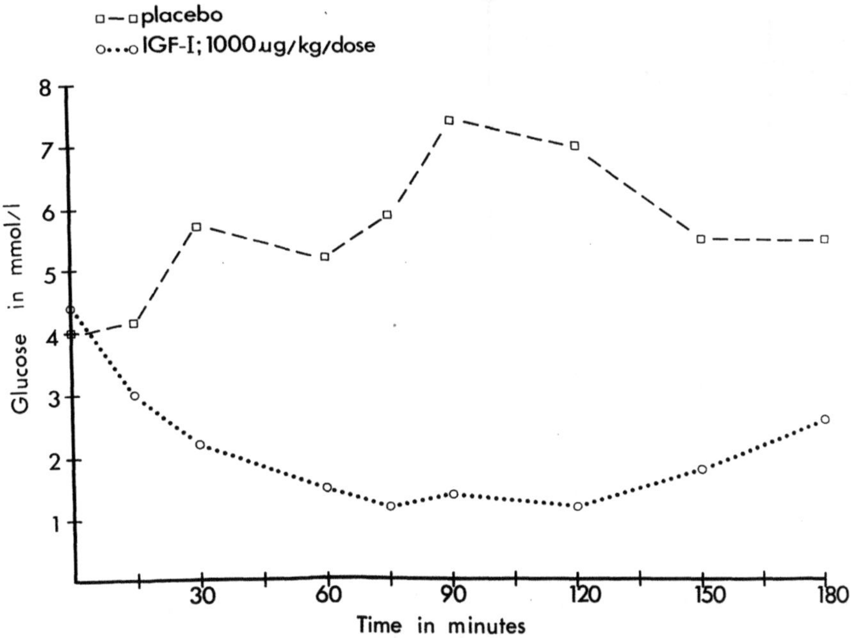

Fig. 2. Glucose levels following subcutaneous injection of IGF-1 (1000 μg/kg)

Table 3. Hematological results in WL/O mice treated with IGF-1 on days 17–20, injections every 6th h: median values and 95% confidence intervals. No significant ($p<0.05$) differences

	Group B		Group C	
	Control (n=12)	IGF-1: 500 μg (n=12)	Control (n=6)	IGF-1: 1500 μg (n=6)
PCV	34 (27–37)	34 (28–40)	27 (24–30)	27 (23–31)
WBC	2.5 (1.9–3.9)	2.9 (2.6–3.8)	3.0 (2.7–4.3)	3.6 (3.0–4.1)
Reticulocytes (%)	13.7 (9.7–16.6)	14.2 (11.4–17.6)	17.2 (15.2–18.0)	17.7 (14.2–22.9)
CFU-E, bone marrow	207 (20–416)	266 (117–344)	30 (7–129)	56 (16–126)

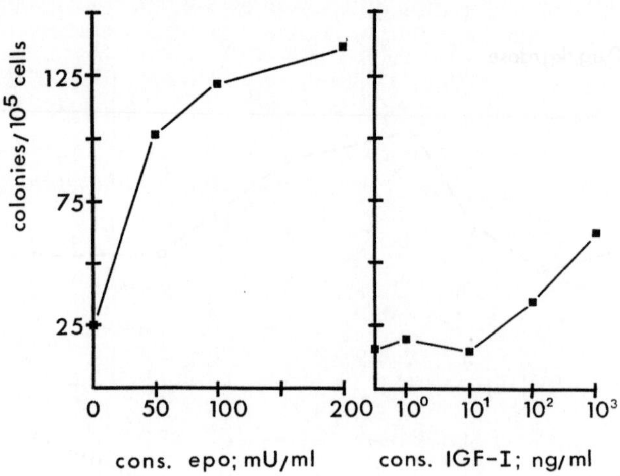

Fig. 3. Dose-response curves for Epo and IGF-1 in an erythropoietic cell culture from murine bone marrow

differences were observed at 15, 30, 60, 75, 90, 120, 150, and 180 min after injection ($p<0.05$, Fig. 2). This occurred only when the doses were 1000 μg/kg, and no similar effect was observed when the dose was 250 μg/kg. There were no significant differences in the hematological values either in the groups in which injections were given on days 6–20 or in the groups in which the duration of treatment was only 60 h (Tables 2 and 3). Epo analyses showed a clear correlation between PCV and erythropoietin values, and no differences between treatment groups were seen.

In the in vitro studies, increasing doses of IGF-1 led to clear dose-response increases in CFU-E. With the culture technique IGF-1 stimulated CFU-E, but to a lesser extent than Epo (Fig. 3). No Epo was added to these cultures. The maximal response to IGF-1 was obtained at a concentration of 1000 ng/ml in the cultures, and a tenfold increase in IGF-1 concentration did not stimulate the erythroid cells further.

Discussion

This study confirms the in vitro stimulatory effect of IGF-1 upon erythropoietic cells. The concentration of IGF-1 required for a maximal CFU-E response was 1000 ng/ml. A tenfold higher concentration does not increase the response further.

The in vivo study shows that IGF-1 has growth-inducing properties in young mice with intact hypophysis. Doses of 2000 μg/kg per day, however, did not induce increased growth. The hypoglycemia caused by such doses of IGF-1 [13] might interfere with growth and thus explain this result. According to previous results [14, 15], IGF-1 induces growth in specific organs. In our study the spleen weights showed significantly higher values in both the IGF-1 treated groups compared to controls. Thymus and liver weights did not differ between groups. In the other studies (B and C) in which IGF-1 was given on days 17–20 we found no differences in overall body weight or organ weight.

In contrast to Philipps et al. [8] we did not find an erythropoietic response to IGF-1 in vivo, either when IGF-1 was administered on days 6–20, or when IGF-1 was injected over a period of 60 h. This applies both to changes in erythroid precursors in bone marrow and spleen and to changes in reticulocyte count and PCV. The mode of administration of IGF-1 in this study was subcutaneous, and IGF-1 was injected two to four times a day. In other studies IGF-1 was infused [7, 13, 14, 15], which perhaps provides a more physiological method of administration.

Philipps et al. [8] found no changes in reticulocyte count or PCV and only a minor but significant change in CFU-E response. The study by Kurtz et al. [7] in which hypophysectomized rats were substituted with IGF-1 did not show normalization of PCV values but only an increase in iron incorporation and reticulocyte count. There are thus partly insufficient and partly conflicting data concerning the role of IGF-1 in regulation of erythropoiesis, and further studies are necessary.

In conclusion, this in vitro study confirms earlier findings about the IGF-1 stimulatory effect upon erythroid precursors in a serum-containing system. In our in vivo study we found no signs of IGF-1 induced stimulation of erythroid activity in mice with intact hypophysis. Further studies to define the IGF-1 role as a stimulatory agent in erythropoiesis are necessary.

References

1. Kurtz A, Jelkmann W, Bauer C (1982) A new candidate for the regulation of erythropoiesis, insulin-like growth factor I. FEBS L 149(1):105–108
2. Clausters M, Chatelain P, Sultan C (1987) Insulin-like growth factor I stimulates human erythroid colony formation in vitro. J Clin Endocrinol Metab 65(1):78–82
3. Akahana K, Tojo A, Urabe A, Takaku F (1987) Pure erythropoietic colony and burst formations in serum-free culture and their enhancement by insulin-like growth factor I. Exp Hematol 15:797–802
4. Sawada K, Krantz SB, Dessypris EN, Sawyer ST (1989) Human colony-forming units erythroid do not require accessory cells, but do require direct interaction with insulin-like growth factor I and/or insulin for erythroid development. J Clin Invest 83:1701–1709
5. Kurtz A, Hartl W, Jelkmann W, Zapf J, Bauer C (1985) Activity in fetal bovine serum that stimulates erythroid colony formation in fetal mouse livers is insulin-like growth factor I. J Clin Invest 76:1643–1648
6. Kurtz A, Matter R, Eckardt K, Zapf J (1990) Erythropoiesis, serum erythropoietin, and serum IGF-1 in rats during accelerated growth. Acta Endocrinol (Copenh)122/3:323–328
7. Kurtz A, Zapf J, Eckardt K, Clemons G, Froesch R, Bauer C (1988) Insulin-like growth factor I stimulates erythropoiesis in hypophysectomized rats. Proc Natl Acad Sci USA 85:7825–7829

8. Philipps A, Persson B, Hall K, Lake M, Skottner A, Sanengen T, Sara V (1988) The effects of biosynthetic insulin-like growth factor-I supplementation on somatic growth, maturation, and erythropoiesis on the neonatal rat. Pediatr Res 23:298–305
9. Hellebostad M, Sanengen T, Halvorsen S (1990) Variations in erythropoiesis throughout a lifetime. Blut 61:358–363
10. Kreyberg L (1952) The origin and the development of the "white label" mouse strain. Br J Cancer 6:140–147
11. Goto M, Murakami A, Akai K, Kawanishi G, Ueda M, Chiba H, Sasaki R (1989) Characterisation and use of monoclonal antibodies directed against human erythropoietin that recognize different antigenic determinants. Blood 74:1415
12. Hågå P, Falkanger B (1979) In vitro assay for erythropoietin: erythroid colony formation in methyl cellulose used for the measurement of erythropoietin in plasma. Blood 53:1172–1181
13. Guler H, Binz K, Eigenmann E, Jaggi S, Zimmermann D, Zapf J, Froesch R (1989) Small stature and insulin-like growth factors: prolonged treatment of mini-poodles with recombinant human insulin-like growth factor I. Acta Endocrinol (Copenh) 121:456–464
14. Skottner A, Clark RG, Fryklund L, Robinson ICAF (1988) Growth responses in a mutant dwarf rat to human growth hormone and recombinant human insulin-like growth factor I. Endocrinology 124(5):2519–2526
15. Guler H, Zapf J, Scheiwiller E, Froesch R (1988) Recombinant human insulin-like growth factor I stimulates growth and has distinct effects on organ size in hypophysectomized rats. Proc Natl Acad Sci USA 85:4889–4893

III. Blood Levels in Health and Disease

Clinical Applicability of the Determination of Erythropoietin

M. Wolff and W. Jelkmann

Introduction

The cloning and expression of the human erythropoietin (Epo) gene in mammalian cells has enabled the production of large quantities of recombinant human Epo (rHuEpo). The availability of rHuEpo for antibody production has initiated the use of radioimmunoassays (RIA) and more recently of enzyme-linked immunosorbent assays (ELISA) for sensitive and accurate determination of Epo in serum. As earlier work was based generally on in vivo bioassay, the question arises as to the correlation between Epo values obtained by different assay methods. Therefore, we compare first the results of Epo measurements by ELISA, RIA, and traditional bioassay in polycythemic mice.

Successful Epo replacement therapy in chronic renal failure has provoked a vast number of studies concerning endogenous Epo levels in other disorders associated with insufficient erythropoiesis. The demonstration of an absolute or relative Epo deficiency is regarded as a prerequisite for the beneficial and economic administration of rHuEpo in nonrenal anemia [15]. In addition, determination of serum Epo may provide a diagnostic tool in distinct hemopoietic disorders. The second part of this report deals with the applicability of the assay of Epo in clinical laboratory routine and with the interpretation of the assay results.

Comparison of Assay Methods

Material and Methods

To compare ELISA with the in vivo bioassay, 50 frozen serum samples from anemic patients with leukemia, aplastic anemia, or ulcerative colitis were reexamined. Epo bioactivity in the sera was measured earlier [29] by bioassay in exhypoxic polycythemic mice as described in [23]. Groups of five polycythemic mice per sample on 2 consecutive days were injected subcutaneously with either 0.5 ml serum or standard human urinary Epo calibrated against the International Reference Preparation B (IRP). On the following day, 0.1 μCi ^{59}Fe-labeled citrate was given intravenously. After 48 h the mice

Pagel/Weiss/Jelkmann (Eds.)
Pathophysiology and Pharmacology
of Erythropoietin

were bled, and the radioactivity of their blood was determined. Epo concentrations were calculated from log dose–log response curves achieved with standard Epo solutions. The lower detection limit of the test is at the best 25 U/l [23].

To compare the ELISA with the RIA, 80 frozen serum samples were reexamined, which were analyzed previously for Epo by RIA [26]. The RIA system included ^{125}I-labeled rHuEpo, rabbit anti-serum raised against rHuEpo and human urinary Epo calibrated against IRP by bioassay. A second antibody technique was applied to separate free and antibody-bound ^{125}I-labeled rHuEpo. The data were evaluated by means of a log-logit plot. The lower detection limit of the RIA is about 2.5 U/l. Referring to a range of 20–50 U/l, the intra-assay coefficient of variation is under 7%, and the inter-assay coefficient of variation under 25%.

ELISA was carried out with a commercial test kit (Medac, Hamburg, FRG). Based on the "sandwich" principle, Epo was first bound to a monoclonal anti-rHuEpo antibody that was attached to the wells of a microtiter plate. After 2 h incubation, a second alkaline phosphatase-conjugated monoclonal antibody was added for 2 h. After repeated rinsing and development of the enzyme reaction with nitrophenyl phosphate, the extinction was read at 405 nm (EAR 400 AT; SLT Instruments, Overath, FRG). The Epo concentration was calculated from a linear calibration curve obtained with the Epo standard enclosed in the kit.

Results

Figure 1 shows that there was a fairly good agreement between the results obtained by ELISA and bioassay. A similar good correlation was seen when the ELISA data were plotted against those measured by RIA (r=0.944, n=80, logarithmic plot; Epo concentration 25–10 000 U/l; data not shown). The mean serum Epo concentration in 64 normal subjects was 14.3 U/l (geometric mean, 95% confidence limit 5.8–34.9 U/l). There was neither a significant difference between the mean value in males (14.2 U/l; n=44) and females (14.5 U/l; n=20) nor a dependency on the age of the blood donors (Fig. 2). ELISA had a low detection limit (2.0 U/l). The variation coefficients, which were determined in the concentration range of 6–90 U/l, were under 11% (intra-assay) and under 28% (inter-assay).

Discussion

For determination of Epo in human serum, bioassays have been replaced almost entirely by immunological methods. Besides being laborious and time consuming, the bioassay in polycythemic mice has poor sensitivity and precision and requires many animals [22]. The in vivo bioassay still serves to calibrate Epo standard preparations against IRP and to quantify the activity

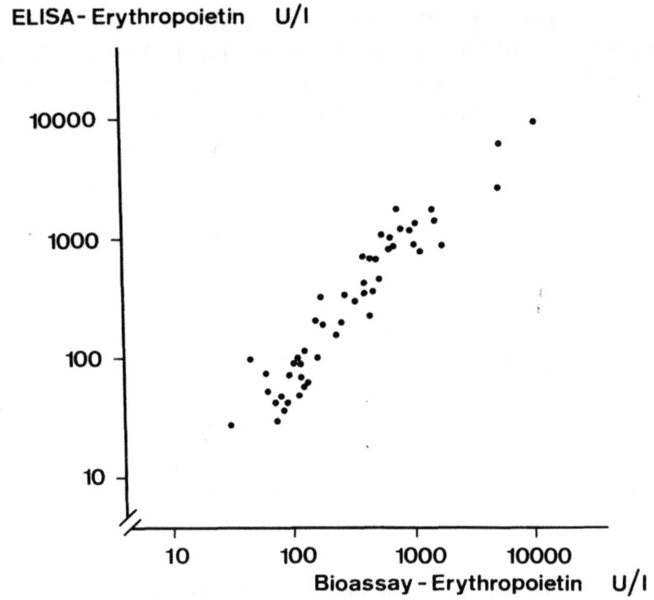

Fig. 1. Correlation between immunoreactive (ELISA) and bioactive (polycythemic mouse assay) erythropoietin in 50 human sera with Epo levels above 25 U/l (lower detection limit of the bioassay). Regression statistics: log [Epo ELISA] = 1.07 log [Epo Bioassay] – 0.28; r=0.952. (From [24])

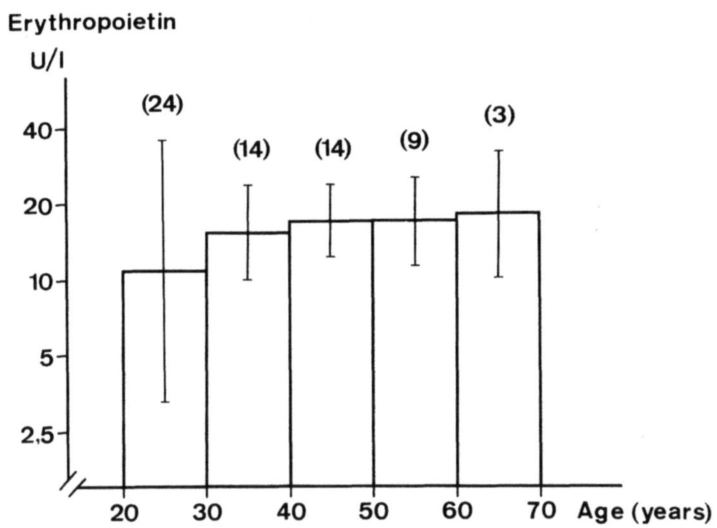

Fig. 2. Serum immunoreactive erythropoietin (ELISA) in relation to age. Data are the geometric mean with calculated 95% range (mean ± 2 SD). Number of subjects are given in parentheses. (From [24])

of rHuEpo preparations for therapeutic use. In vitro bioassay of Epo in cell cultures should be avoided because of its lack of specificity [22, 32]. The World Health Organization has recently initiated a multicenter study to establish a recombinant DNA-derived analogue of human Epo as the new international Epo standard (approximately 130 000 U/mg glycoprotein).

In addition to the radioimmunological determination of Epo, which is widely approved [7, 8, 14, 17, 26, 31, 32], ELISA methods have been developed recently. While RIAs can be carried out with polyclonal antisera, ELISAs require the use of monoclonal antibodies. These have been raised both against human urinary Epo [44] and rHuEpo [19]. ELISAs are preferable for several reasons. First, they are rapid (results were obtained within 5 h with the present kit). Second, exposition to radiation is avoided. A low Epo detection limit (2 U/l) and a sufficient precision in regard to the wide range of Epo levels occurring in man (<2–10 000 U/l) make ELISAs convenient for both basic research purposes and clinical studies [24]. So far, only few cases have been reported of discrepancies between bioactive and immunoreactive Epo in human serum [11].

Instructions for Clinical Application

The stability of Epo permits storage of the serum samples at room temperature for 14 days without loss of immunological activity [13]. The time of day for blood sampling for Epo measurements should be standardized because there is some diurnal variation of the Epo level in blood [36, 43]. Our

Table 1. Disorders associated with relatively low or high serum Epo levels

	Disorder	Suggested mechanism	References (selected)
Low Epo level	Chronic renal failure		6, 35
	Acute renal failure	Injury to Epo-producing cells	38
	Nephrotoxic chemotherapy		34
	Chronic inflammatory disease Infections Noninfections	Inhibition of Epo production by the monokines interleukin-1	2,21,27,41
	Malignoma	and tumor necrosis factor-α	37
	Renal transplant rejection		3
	Postinjury anemia		33
High Epo level	Induction of chemotherapy	Unknown (abolished feedback	4
	Bone marrow irradiation	inhibition by proliferating	1
	Aplastic anemia and other forms of bone marrow hypoplasia	erythrocytic progenitors, lowered utilization of Epo?)	26, 39
	Postrenal transplant erythrocytosis	Autonomous Epo production	45
	Paraneoplasia		28,30,42

own preliminary measurements in seven normal subjects revealed the Epo mean level to be 40% higher at midnight than in the morning. Epo values from different laboratories cannot be compared readily. Normal values should be established for each assay technique, as the results may vary considerably [22]. Epo levels appear to be independent of the sex and the age of the blood donors [9, 25]. During pregnancy serum Epo concentrations increase about fourfold [9]. Epo levels must be related to the patient's hemoglobin concentration, as in anemia Epo production increases exponentially [9, 16, 26]. This inverse relationship may be altered under pathophysiological conditions leading to an absolute or relative Epo deficiency or to inappropriately high Epo levels. Several examples of this kind are listed in Table 1.

Erythropoietin Determination as a Diagnostic Tool

In respect to the increasing number and costs of diagnostic procedures, rational and individualized testing must replace excessive, unreflected laboratory work [40]. A useful laboratory test should lead to conclusions which may significantly alter a diagnostic or therapeutic decision [20]. A diagnostic test serves basically a discriminatory function, separating patients into groups that should be managed in different ways. Often, however, there is an overlap of results among patients with and without a specific disease. This is also the crucial point in evaluating clinical Epo measurement. The validity of Epo assay results in terms of specificity and sensitivity for a distinct disorder should be assessed. In addition, the clinical usefulness must be considered by estimating predictive values and posttest probabilities under specific clinical conditions [20]. Unfortunately, there is still a lack of data concerning these test parameters for Epo measurements. Hence, clinical application and evaluation of Epo assays should be performed in controlled clinical studies whenever possible.

Test criteria such as validity and predictive value of cause depend heavily on the proposed clinical purpose of the test. Thus, a clear statement of the clinical objective should precede ordering and evaluating serum Epo measurements. Suggested clinical functions can be referred to as diagnosis, staging, screening, and monitoring.

The diagnostic impact of Epo measurements, i.e., including or excluding a suspected diagnosis, is still a matter of controversy. Possible indications for measurements of serum Epo include: (a) differential diagnosis of erythrocytoses, (b) follow-up of paraneoplastic Epo production, (c) differential diagnosis of anemias, and (d) election for rHuEpo therapy of patients with nonrenal anemia. The recently developed sensitive assays for estimation of immunoreactive Epo may contribute to the diagnosis of otherwise unclear forms of polycythemia or anemia. In polycythemia vera Epo production is expected to be suppressed, whereas secondary polycythemia due to respiratory, cardiac, and hemoglobin abnormalities or in chronic mountain sickness should be associated with elevated Epo formation. There are, indeed, several reports demonstrating that the mean Epo levels in patients with polycythemia vera and with secondary polycythemia are significantly different from those in normal subjects [17, 31]. However, Cotes et al. [10] have pointed out that this difference is detectable only when large groups of patients are compared. Unfortunately, knowledge of the Epo concentration in individual patients is of low predictive value.

Erythrocytosis occurring intermittently after renal transplantation is in part due to excessive Epo production by the native kidneys [45]. If persistently high levels of circulating Epo are confirmed, bilateral nephrectomy may be considered as an alternative for repeated phlebotomies.

Extremely high Epo levels, in the order of 1000 to 10 000 U/l have been observed in aplastic anemia [16, 39] and other forms of bone marrow hypoplasia [26]. In contrast, Epo production is inadequately low for the degree of anemia in patients with chronic renal failure or with inflammatory and malignant diseases (Table 1). As shown in Fig. 3, there is considerable individual variation in Epo levels among subjects with the same hemoglobin concentration. This might be attributed to the fact that Epo production is only one regulatory principle in response to anemia. Intraerythrocytic and circulatory adaptations also compensate for the decrease in arterial oxygen

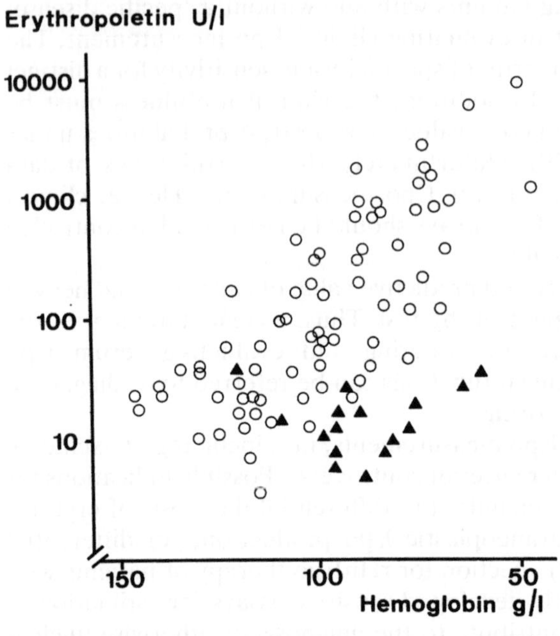

Fig. 3. Semilogarithmic presentation of the relationship between the concentration of serum immunoreactive erythropoietin (ELISA) and blood hemoglobin in anemic patients with normal kidney function (*open circles*, 73 subjects) or chronic renal failure (*closed triangles*, 15 subjects)

content. In addition, the Epo response to anemia may be blunted by circulating cytokines such as tumor necrosis factor-α and interleukin-1 [27] or by pharmacologic agents.

If anemia becomes symptomatic, and therapy with rHuEpo is considered, the basal Epo level should be measured. As recently suggested by Erslev [15], anemia with an Epo concentration of less than 100 U/l can be expected to respond well to doses of rHuEpo which are commonly used in renal failure. Epo levels above 500 U/l reflect a relative refractoriness toward the hormone. Nevertheless, in some cases extremely high doses of rHuEpo were effective even in patients with bone marrow aplasia [12]. Thus, the rationale for the arbitrary borderline of 500 U/l Epo for a successful treatment with rHuEpo results from a fairly economic point of view.

The decision whether anemia should be treated either with blood transfusions or with rHuEpo depends on physical findings and symptoms rather than on Epo measurements. Hence, the suggestion of von Bormann et al. [5] to employ serum Epo estimations as screening parameter for a deficient whole-body oxygen supply seems not to be warranted.

Finally, the potential value of Epo as a tumor marker is to be considered. Ectopic and autonomous Epo production as a paraneoplastic phenomenon has been noted mainly in renal cell carcinoma, Wilms' tumor, hepatocellular carcinoma, and cerebellar hemangioblastoma [28]. Erythrocytosis can result from paraneoplastic Epo production although it is often missed due to chronic bleeding or malignancy-associated anemia. Apart from a number of case reports, increased Epo activity has been reported in 63% of renal [42] and in 23% of hepatocellular carcinomata [30]. In some cases serum Epo levels dropped after surgical resection but increased again with recurrance. Although some tumors are able to secrete excessive quantities of the hormone, there is no threshold difference in Epo levels that allows for separating benign from malignant disorders. Thus, the principal role for Epo measurements in Epo-producing tumors is in monitoring the response to treatment: complete surgical resection should normalize paraneoplastically elevated Epo levels, whereas residual, recurrent, or metastatic tumor growth should be evidenced by progressively increasing Epo in serum.

References

1. Abedi MR, Bäckman L, Boström L, Lindbäck B, Ringden O (1990) Markedly increased serum erythropoietin levels following conditioning for allogeneic bone marrow transplantation. Bone Marrow Transplant 6:121–126
2. Baer AN, Dessypris EN, Goldwasser E, Krantz SB (1987) Blunted erythropoietin response to anaemia in rheumatoid arthritis. Br J Haematol 66:559–564
3. Besarab A, Caro J, Jarrell BE, Francos G, Erslev AJ (1987) Dynamics of erythropoiesis following renal transplantation. Kidney Int 32:526–536
4. Birgegard G, Wide L, Simonsson B (1989) Marked erythropoietin increase before fall in Hb after treatment with cytostatic drugs suggests mechanism other than anaemia for stimulation. Br J Haematol 72:462–466

5. von Bormann B, Weidler B, Schwanen N, Rotthey K, Hempelmann G (1990) Perioperative Anämie und Erythropoese. Chirurg 61:124–128
6. Chandra M, Clemons GK, McVicar M (1988) Relation of serum erythropoietin levels to renal excretory function: evidence for lowered set point for erythropoietin production in chronic renal failure. J Pediatr 113:1015–1921
7. Cohen RA, Clemons GK, Ebbe S (1985) Correlation between bioassay and radioimmunoassay for erythropoietin in human serum and urine concentrates. Proc Soc Exp Biol Med 179:296–299
8. Cotes PM (1982) Immunoreactive erythropoietin in serum. Br J Haematol 50:427–438
9. Cotes PM (1989) Physiological studies of erythropoietin in plasma. In: Jelkmann W, Gross AJ (eds) Erythropoietin. Springer, Berlin Heidelberg New York, pp 57–79
10. Cotes PM, Dore CJ, Yin JAL, Lewis SM, Messinerzy M, Pearson TC, Reid C (1986) Determination of serum immunoreactive erythropoietin in the investigation of erythrocytosis. N Engl J Med 315:283–287
11. Cotes PM, Tam RC, Reed P, Hellebostad M (1989) An immunological cross-reactant of erythropoietin in serum which may invalidate EPO radioimmunoassay. Br J Haematol 73:265–268
12. DaCosta NA, Hultin MB (1991) Effective therapy of human immunodeficiency virus-associated anemia with recombinant human erythropoietin despite high endogenous erythropoietin. Am J Hematol 36:71–72
13. Eckardt K-U, Kurtz A, Hirth P, Scigalla P, Wieczorek L, Bauer C (1988) Evaluation of the stability of human erythropoietin in samples for radioimmunoassay. Klin Wochenschr 66:241–245
14. Egrie JC, Cotes PM, Lane J, Gaines Das RE, Tam RC (1987) Development of radioimmunoassays for human erythropoietin using recombinant erythropoietin as tracer and immunogen. J Immunol Methods 99:235–241
15. Erslev AJ (1991) Erythropoietin. N Engl J Med 324:1339–1344
16. Erslev AJ, Wilson J, Caro J (1987) Erythropoietin titers in anemic, nonuremic patients. J Lab Clin Med 109:429–433
17. Garcia JF, Ebbe SN, Hollander L, Cutting HO, Miller ME, Cronkite EP (1982) Radioimmunoassay of erythropoietin: circulating levels in normal and polycythemic human beings. J Lab Clin Med 99:624–635
18. Garcia JF, Sherwood J, Goldwasser E (1979) Radioimmunoassay of erythropoietin. Blood Cells 5:405–419
19. Goto M, Murakami A, Akai K, Kawanishi G, Ueda M, Chiba H, Sasaki R (1989) Characterization and use of monoclonal antibodies directed against human erythropoietin that recognize different antigenic determinants. Blood 74:1415–1423
20. Griner PF, Mayewski RJ, Mushlin AI et al. (1981) Selection and interpretation of tests and procedures: principles and applications. Ann Intern Med 94:557–600
21. Hochberg MC, Arnold CM, Hogans BB, Spivak JL (1988) Serum immunoreactive erythropoietin in rheumatoid arthritis: impaired response to anemia. Arthritis Rheum 31:1318–1321
22. Jelkmann W (1986) Renal erythropoietin. Properties and production. Rev Physiol Biochem Pharmacol 104:139–215
23. Jelkmann W, Bauer C (1981) Demonstration of high levels of erythropoietin in rat kidneys following hypoxic hypoxia. Pflugers Arch 392:34–39
24. Jelkmann W, Piccardi K (1991) Erythropoietinbestimmungen im Serum. Klin Lab 37:229–234
25. Jelkmann W, Wiedemann G (1989) Lack of sex dependence of the serum level of immunoreactive erythropoietin in chronic anemia. Klin Wochenschr 67:1218
26. Jelkmann W, Wiedemann G (1990) Serum erythropoietin level: relationships to blood hemoglobin concentration and erythrocytic activity of the bone marrow. Klin Wochenschr 68:403–407

27. Jelkmann W, Wolff M, Fandrey J (1990) Modulation of the production of erythropoietin by cytokines. In vitro studies and their clinical implications. Contrib Nephrol 87:68–77
28. Johannsen H, Gross AJ, Jelkmann W (1989) Erythropoietin production in malignancy. In: Jelkmann W, Gross AJ (eds) Erythropoietin. Springer, Berlin Heidelberg New York, pp 80–91
29. Johannsen H, Jelkmann W, Wiedemann G, Otte M, Wagner T (1989) Erythropoietin/haemoglobin relationship in leukaemia and ulcerative colitis. Eur J Haematol 43:201–206
30. Kew MC, Fisher JW (1986) Serum erythropoietin concentrations in patients with hepatocellular carcinoma. Cancer 58:2485–2488
31. Koeffler HP, Goldwasser E (1981) Erythropoietin radioimmunoassay in evaluating patients with polycythemia. Ann Intern Med 94:44–47
32. Lappin TRJ, Elder E, Taylor T, McMullin MF, Bridges JM (1988) Comparison of the mouse spleen cell assay and a radioimmunoassay for the measurement of serum erythropoietin. Br J Haematol 70:117–120
33. Levine EA, Rosen AL, Sehgal LR, Gould SA, Egrie JC, Moss GS (1991) Erythropoietin deficiency after coronary artery bypass procedures. Ann Thorac Surg 51:764–766
34. Lin AC, Goldwasser E, Bernard EM, Chapman SW (1990) Amphotericin B blunts erythropoietin response to anemia. J Infect Dis 161:348–351
35. McGonigle RJS, Wallin JD, Shadduck RK, Fisher JW (1984) Erythropoietin deficiency and inhibition of erythropoiesis in renal insufficiency. Kidney Int 25:437–444
36. Miller ME, Garcia JF, Cohen RA, Cronkite EP, Moccia G, Acevedo J (1981) Diurnal levels of immunoreactive erythropoietin in normal subjects and subjects with chronic lung disease. Br J Haematol 19:189–200
37. Miller CB, Jones RJ, Piantadosi MD, Abeloff JL, Spivak JL (1990) Decreased erythropoietin response in patients with the anemia of cancer. N Engl J Med 322:1689–1692
38. Nielsen OJ, Thaysen JH (1990) Erythropoietin deficiency in acute tubular necrosis. J Intern Med 227:373–380
39. Pavlovic-Kentera V, Milenkovic P, Ruvidic R, Jovanovic V, Biljanovic-Paunovic L (1979) Erythropoietin in aplastic anemia. Blut 39:345–350
40. Schechter MT, Sheps SB, Diagnostic testing revisited: pathways through uncertainty. Can Med Assoc J 132:755–760
41. Spivak JL, Barnes DC, Fuchs E, Quinn TC (1989) Serum immunoreactive erythropoietin in HIV-infected patients. JAMA 261:3104–3107
42. Sufrin G, Mirand EA, Moore RH, Chu TM, Murphy GP (1977) Hormones in renal cancer. J Urol 117:433–438
43. Wide L, Bengtsson C, Birgegard G (1989) Circadian rhythm of erythropoietin in human serum. Br J Haematol 72:85–90
44. Wognum AW, Landdorp PM, Eaves AC, Krystal G (1989) An enzyme linked immunosorbent assay for erythropoietin using monoclonal antibodies, tetrameric immune complexes, and substrate amplification. Blood 74:622–628
45. Wolff M, Jelkmann W (1991) Erythropoiesis and erythropoietin levels in renal transplant recipients. Klin Wochenschr 69:53–58

A Sensitive ELISA for Erythropoietin and Its Application in Various Diseases

G. Noé, I. Rich, and B. Kubanek

Introduction

Radioimmunoassays (RIA) (Sherwood and Goldwasser 1979; Cotes 1982; Egrie et al. 1987) have shown many advantages concerning specificity, accuracy, speed, and sensitivity over in vivo and in vitro bioassays (Cotes and Bangham 1961; Goldwasser et al. 1975; Rich and Kubanek 1976; Krystal 1983). We decided to develop another immunological assay, namely an enzyme-linked immunosorbent assay (ELISA), which has a primary advantage over RIAs in the avoidance of radioactivity. All aspects of the ELISA have been established, so that we can present a well-functioning assay with which it is possible to reliably measure the serum erythropoietin (EPO) concentration in various diseases.

ELISA Requirements

The requirements for a reliable and sensitive ELISA are:

1. *Anti-EPO antibodies:*
   The anti-EPO antibodies must have a high affinity and they must be purified. We decided to produce polyclonal antibodies to EPO because, in general, these are easier to obtain than monoclonal antibodies, although, once obtained high-affinity polyclonal antibodies have to be purified.
2. *Pure EPO:*
   Pure EPO is required for the purification of the antibodies. Furthermore pure EPO is recommended, although not absolutely necessary, for immunization and as a standard in the test. Without recombinant EPO, development of the ELISA would not have been possible.

Production and Purification of Antibody

Rabbits were immunized with pure recombinant human EPO (Merckle, Ulm FRG) in the microgram range (25–50 μg per immunization). Usually after the first or second booster, the rabbits responded with a reasonable antibody titer.

Pagel/Weiss/Jelkmann (Eds.)
Pathophysiology and Pharmacology
of Erythropoietin

Once the antiserum had been obtained the next step was to purify the anti-EPO antibodies. We developed a three-step purification procedure of the antiserum employing affinity chromatography on

- Protein A Sepharose
- Sepharose coupled with recombinant EPO (Merckle)
- Sepharose coupled with EPO-free human serum

The last step turned out to be very important to eliminate false positive reactions when measuring serum samples.

Development of an EPO ELISA

The quantification of the antigen molecule by ELISA is usually performed by sandwiching the antigen between two antibody layers. Based on this principle many different variations were examined. The principle of the present ELISA is shown diagrammatically in Fig. 1, in which the EPO molecule is bound by two layers of the purified anti-EPO antibody. The upper anti-EPO antibody, however, is conjugated to biotin. The biotin is then bound by anti-biotin conjugated to the enzyme alkaline phosphatase. The last step is the addition of the substrate, development of the enzyme-substrate reaction, and finally the photometric measurement of the yellow product at 405 nm.

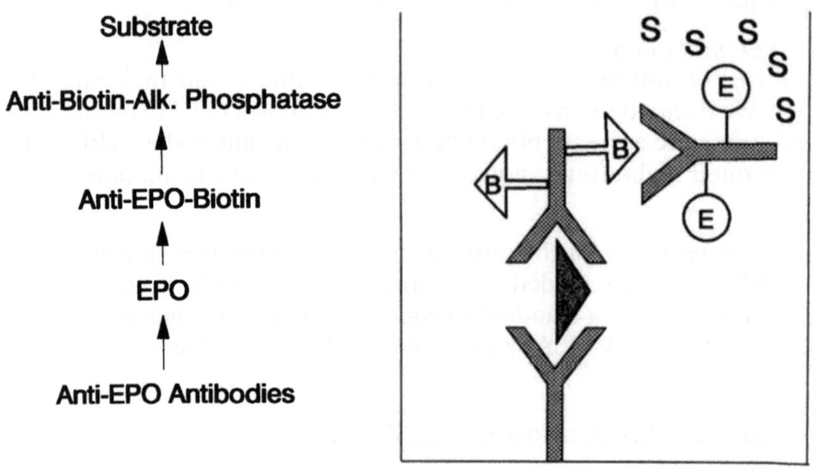

Fig. 1. Schematic representation of the EPO-ELISA. *B*, biotin; *E*, enzyme (alkaline phosphatase); *S*, substrate

Assay Optimization

The following parameters were investigated to optimize the assay: assay buffer, concentration of the antibodies, reaction volume, and incubation time.

A typical standard curve of the optimized assay measures the EPO concentration range between 0.3 mU/ml and 10 mU/ml (Fig. 2). The minimal detectable dose determined as three standard deviations above the zero standard is 0.3–0.5 mU/ml corresponding to 1.2–2 mU/ml of EPO in serum. The EPO standard (Merckle, Ulm, FRG) was validated against the Second International Reference Preparation.

Validation of the ELISA

Specificity

We tested various blood components, such as albumin, IgG, transferrin, lactoferrin, hemoglobin in physiological concentrations, and various cytokines (interleukin-1, IL-1; IL-2; IL-3; IL-6; granulocyte macrophage colony-stimulating factor, GM-CSF; tumor necrosis factor, TNF) in the range

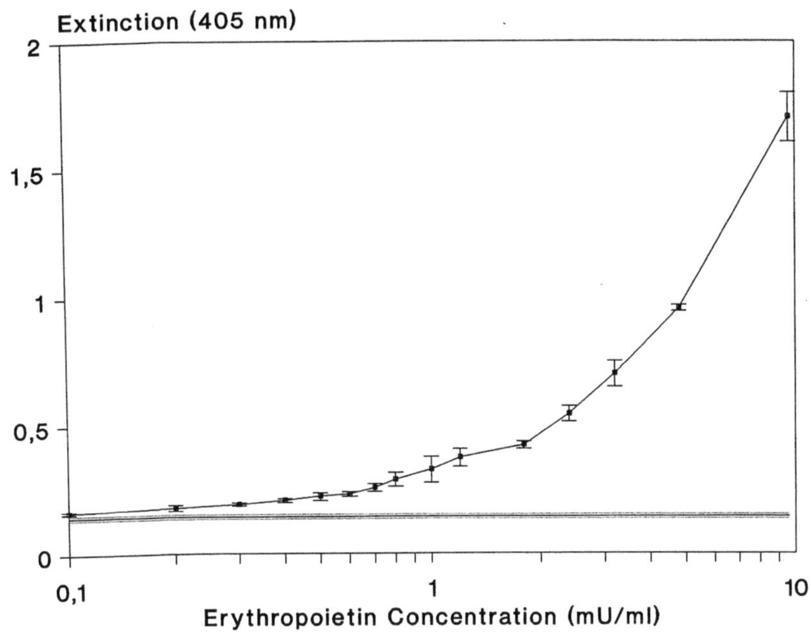

Fig. 2. EPO dose response curve with pure recombinant EPO (mean ± SD)

between 10 and 1000 U/ml. In contrast to EPO or whole serum, none of these components were bound by the anti-EPO antibody.

Recovery of Exogenous EPO in Human Serum

In these experiments 25% serum was spiked with known concentrations of recombinant EPO and the measured EPO concentration was compared with the expected EPO concentration. The recovery of added EPO was between 90% and 100%. Thus, the assay is free of factors which could inhibit the binding of EPO to its antibody.

Dilution Experiments

Regardless of the dilution of the sample used, the same end concentration must be obtained. This means that the EPO in serum must behave as the EPO used for the standard curve. This could also be demonstrated.

Correlation of the ELISA with an RIA

There is a linear relationship between the ELISA and a commercially available RIA (Incstar, Stillwater, Minns. USA) with a regression coefficient of r=0.91.

Measurement of Serum EPO Samples

First, we investigated the EPO levels in 100 healthy persons (50 men, 50 women). The mean EPO concentration ± standard deviation (SD) was 8.7 ± 3.96 mU/ml. This is in the range which has been reported for some RIAs (Rhyner et al. 1989, Wide et al. 1989; Jelkmann and Wiedemann 1990). We then investigated the EPO levels in various diseases.

Polycythemia Vera and Erythrocytosis

The EPO concentrations in the polycythemia vera group (n=20, hematocrit range 46%–61%) are mainly below the normal range and many are even below the detection limit of our assay, that is, below about 1.2 mU/ml. In contrast the EPO values of erythrocytosis patients (n=9, hematocrit range 49%–55%) are in the normal or above normal range (Fig. 3). The mean ± SD of the polycythemia vera group is 2.6 ± 2.4 mU/ml, which is significantly below ($p<0.05$) that for the normal values. The mean of the erythrocytosis

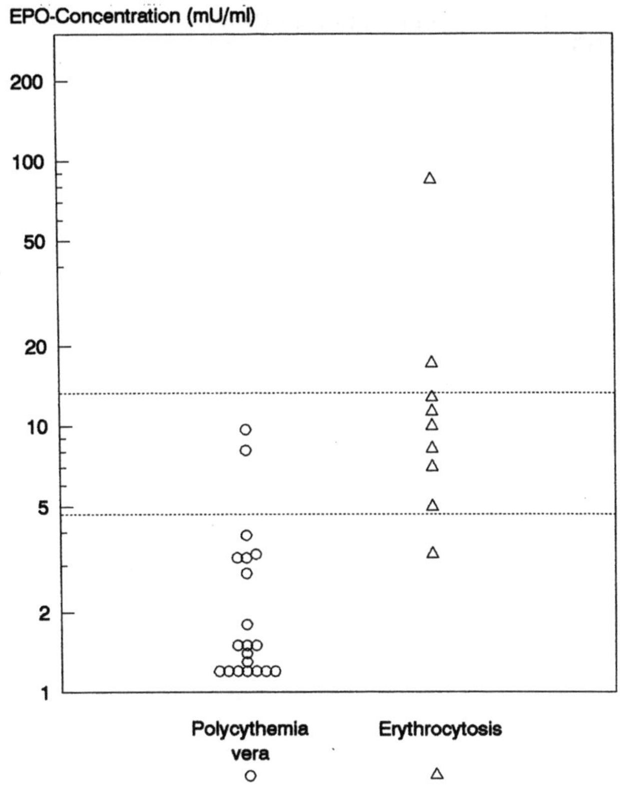

Fig. 3. Comparison of the EPO concentrations in sera from polycythemia vera and erythrocytosis patients. The range between the *dotted line* represents the mean ± SD for normal values

group is 17.8±25.6 mU/ml, which is significantly greater ($p<0.05$) than the mean for normal values. Thus, the ELISA can distinguish between these two polycythemic states.

Chronic Leukemia

Under the title "chronic leukemia" we have grouped together chronic myelogenous leukemia (CML), chronic lymphatic leukemia (CLL), hairy cell leukemia (HCL), myeloproliferative syndrome (MPS), and plasmacytoma (Fig. 4). Plotting the hematocrit against the log EPO concentration, a negative correlation between these two parameters is easily seen; the lower the hematocrit, the higher the EPO concentration. This correlation is defined by the regression line with a regression coefficient of $r=-0.77$.

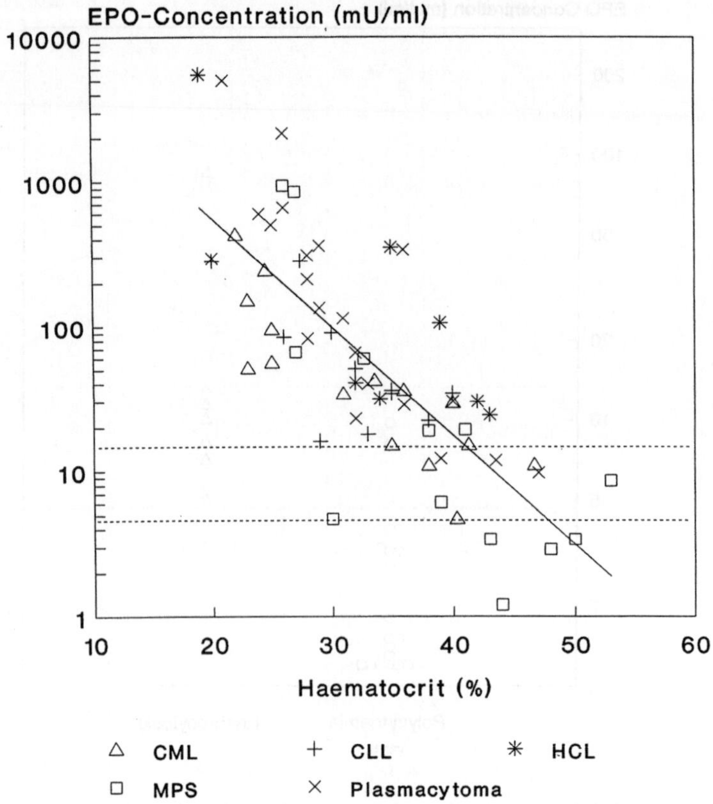

Fig. 4. EPO concentration in relation to hematocrit in chronic leukemia (*CML*, chronic myelogenous leukemia; *CLL*, chronic lympathic leukemia; *HCL*, hairy cell leukemia; *MPS*, myeloproliferative syndrome; and plasmacytoma). The exponential regression line of all values is shown. The range between the *dotted lines* represents the mean ± SD for normal values (n=67; r=−0.77)

Comparison of Renal and Tumor Anemia with Nonrenal anemia

The EPO levels in nonrenal anemia (19 hemolytic anemia, 4 iron-deficiency anemia) demonstrate a clear correlation between hematocrit and EPO ($r=-0.73$) (Fig. 5). In contrast, however, no correlation is seen with the sera of renal anemia patients; although the hematocrit is rather low, the EPO levels are in the normal or subnormal range.

An interesting feature is shown in patients with neoplasia (gastrointestinal carcinoma; carcinoma of prostate, testes, kidney, thyroid gland, and pancreas). Two observations should be pointed out. First, the regression coefficient is only −0.54, and therefore much worse than the correlation demonstrated by the other diseases we investigated. Second, the y-intercept

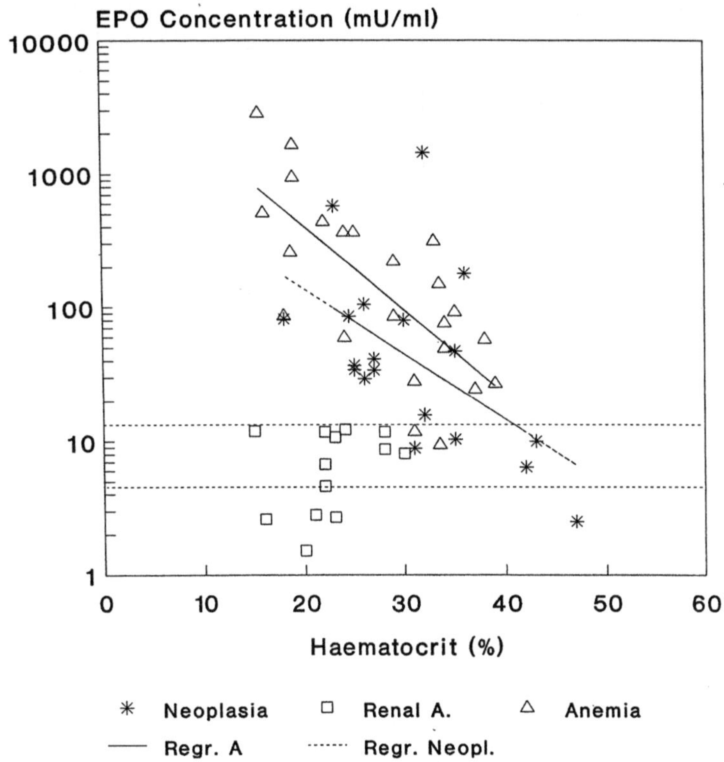

Fig. 5. EPO concentration as a function of hematocrit in patients with neoplasia, renal anemia, and nonrenal anemia. The *solid line* represents the exponential regression of nonrenal anemia values (*Regr. A*), the *broken line* of neoplasia values (*Regr. Neopl.*). The area between the *dotted lines* shows the normal EPO concentration range (mean ± SD) (Anemia n=23; r=−0.73. Neoplasia n=19; r=–0.54)

of the regression line is lower than the y-intercept of the regression line of the other anemias.

Conclusion

We investigated a large variety of diseases and found a negative correlation between hematocrit and EPO; the more severe the anemia the higher the EPO concentration. An exception to this rule was demonstrated in the case of renal anemias where no correlation exists. The tumor anemias could be a candidate for impaired EPO responsiveness as demonstrated by the slightly lower EPO levels in response to the degree of anemia and a poor correlation between EPO and hematocrit. This correlation is not as strict as in the hemolytic and iron-deficiency anemias. These results however are not totally in accordance with those of Miller et al. (1990), who found an inadequate EPO response and no correlation between EPO and hemoglobin in patients with solid tumors.

The assay characteristics and the data of EPO levels in various deseases clearly show that not only RIAs, but also an ELISA such as the one presented here, is well suited to routinely measure the EPO concentration in various diseases.

Acknowledgments. This work was supported by German Red Cross and the Wilhelm Sander Foundation, Munich.

References

Cotes PM (1982) Immunoreactive erythropoietin in serum. I. Evidence for the validity of the assay method and the physiological relevance of estimates. Br J Haematol 50:427–438

Cotes PM, Bangham DR (1961) Bio-assay of erythropoietin in mice made polycythemic by exposure to air at a reduced pressure. Nature 191:1065–1067

Egrie JC, Cotes PM, Lane J, Gaines Das RE, Tam RC (1987) Development of radioimmunoassay for human erythropoietin using recombinant erythropoietin as tracer and immunogen. J Immunol Methods 99:235–241

Goldwasser E, Eliason JF, Sikkema D (1975) Assay for erythropoietin in vitro at the milliunit level. Endocrinology 97:315–323

Jelkmann W, Wiedemann G (1990) Serum erythropoietin level: relationships to blood hemoglobin concentration and erythrocytic activity of the bone marrow. Klin Wochenschr 68:403–407

Krystal G (1983) A simple microassay for erythropoietin based on ^{3}H-thymidine incorporation into spleen cells from phenylhydrazine treated mice. Exp Hematol 11:649–660

Miller C, Jones RJ, Piantadosi S, Abeloff MD, Spival J (1990) Decreased erythropoietin response in patients with anemia of cancer. N Engl J Med 322:1689–1692

Rhyner K, Egli F, Niemöller M, Wieczorek A, Greminger P, Vetter W (1989) Serumerythropoetinwerte bei verschiedenen Krankheitszuständen. Nephron 51 [Suppl 1]:39–46

Rich IN, Kubanek B (1976) Erythroid colony formation in foetal liver and adult bone marrow and spleen from mouse. Blut 33:171–180

Sherwood JB, Goldwasser E (1979) A radioimmunoassay for erythropoietin. Blood 54:885–893

Wide L, Bengtsson C, Birgegard G (1989) Circadian rhythm of erythropoietin in human serum. Br J Haematol 72:85–90

Use of Erythropoietin Radioimmunoassay in Polycythemias

M.H. Schlageter, M.E. Toubert, M.P. Podgorniak, and Y.Y. Najean

Introduction

The recent production of erythropoietin (EPO) by cloning and expression of the human gene in mammalian cells has enabled, not only the treatment of severe anemias due to a defect in EPO production by the kidney, but also the development of sensitive immunoassays (Erslev et al. 1987) which allow large-scale study of EPO levels in human disorders.

The determination of EPO concentration by in vivo or in vitro bioassays has been difficult, long, and expensive (Erslev 1977; Lewis et al. 1989). In vivo bioassays measure the incorporation of ^{59}Fe into polycythemic (post-transfusion or exhypoxic) mice after injection of EPO (or of any sample to be tested). The rate of incorporation is related to the amount of injected EPO. This method of reference is very specific but lacks sensitivity. In vitro bioassays are based on in vitro growth of erythrocytic colonies of normal or erythroleukemic cell lines in the presence of EPO. The number of colonies or the amount of hemoglobin or RNA synthesized is then measured. These in vitro assays are more sensitive but less specific than the in vivo bioassays (Table 1).

The first immunoassays, which used purified EPO from urine of anemic people, showed a lack of specificity because of incomplete purification of the antigen. The use of recombinant EPO as standard and tracer allows the

Table 1. Different methods of measurement of EPO

Type of assay	Measured variable	Sensitivity	Specificity
In vivo biologic assay (polycythemic mice)	Incorporation of ^{59}Fe in erythrocytes	50 U/l	Very good
In vitro biologic assay (semi-solid culture of medullary cells)	Erythroid colony formation Hemoglobin synthesis RNA or protein synthesis	20 U/l	Poor
In vitro immunologic assay	Competitive method: displacement of labeled EPO Sandwich method: antibody labeled with ^{125}iodine or enzyme	<10 U/l	Very good

Pagel/Weiss/Jelkmann (Eds.)
Pathophysiology and Pharmacology of Erythropoietin

development of sensitive and specific methods that are well correlated to the in vivo bioassays. The commercially available kits use either ^{125}I-labeled EPO and polyclonal antibody (RIA), ^{125}I-labeled monoclonal antibody (IRMA), or enzyme-labeled monoclonal antibody (ELISA).

The aim of our work is to study the levels of serum EPO in patients with myeloproliferative disorders and particularly to evaluate the usefulness of EPO determination to discriminate polycythemia vera (PV) from pure erythrocytosis, as the prognosis and the choice of therapy differ for the two diseases. PV is considered a monoclonal malignant disorder and will benefit from phosphorus 32 or hydroxyurea. Erythrocytosis is rather a dysregulation of erythropoiesis or secondary to an excess of EPO production and will benefit from bleedings (Berlin 1975).

Methods

We used the radioimmunoassay (RIA) kit developed by Incstar Corp. Stillwater, MN) and marketed actually in France by Sorin Biomedica (92160 Antony), according to the manufacturer procedure. The standard used in this kit is pure recombinant EPO, which has been calibrated against the WHO 2nd International Reference Preparation of Erythropoietin, HUM Urinary/Bioassay, 67/343. The EPO concentrations are expressed as units per liter (U/l).

Briefly, duplicate 200 µl samples of standard EPO (10–260 U/l) or of unknown serum are preincubated with rabbit specific antibody (100 µl) for 2 h at room temperature. After adding ^{125}I-labeled recombinant EPO and incubating the mixture for 16–24 h at 4°C, the precipitating agent (containing normal rabbit serum, goat anti-rabbit serum and surfactant) is added; 30 min later the tubes are centrifugated and decanted. The amount of radioactivity in the pellets is counted for 1 min in a gamma scintillation counter. The standardization curve is fitted with a spline method, allowing the determination of EPO concentration in unknown samples by intrapolation. Sera with values higher than 260 U/l EPO are diluted and assayed again.

Patients

A total of 66 normal subjects without anemia (hematocrit Ht, between 39% and 49%) were studied; they comprised members of the staff and patients who were studied for documented thrombocytopenic purpura with normal Ht and hemoglobin concentration.

Of the 158 patients with PV, defined according to the PV Study Group (Berlin, 1975) who were studied, 82 were in the active phase of the disease, 57 were in remission (induced by ^{32}P, hydroxyurea, or phlebotomy), and 19 were in the spent phase or demonstrated complications from myelofibrosis with splenic myeloid metaplasia.

Patients studied with pure erythrocytosis defined by an excess of red cell volume in the absence of the criteria of PV numbered 131.

In addition, we studied 79 patients with chronic anemia, but without malignant and renal diseases. For comparison, 14 patients with a chronic renal deficiency, undergoing chronic hemodialysis were studied.

Analytical Performances of the EPO RIA

The analytical performances of the assay have been described (Schlageter et al. 1990). Briefly, the precision profile of EPO RIA which represents the variation coefficient as a function of EPO concentration showed that the EPO concentration is measured with intraassay and interassay variation coefficients lower than 10% between 12 and 150 U/l, and 12 and 100 U/l respectively.

The accuracy of the method was assessed by dilution and recovery tests (Fig. 1). It is important to note that the antibody recognizes natural EPO as well as recombinant EPO.

Results and Discussion

Controls

In 66 normal adults, a log normal distribution of EPO levels was observed, with a mean value of 13.4 U/l and a standard deviation of 2.45 U/l.

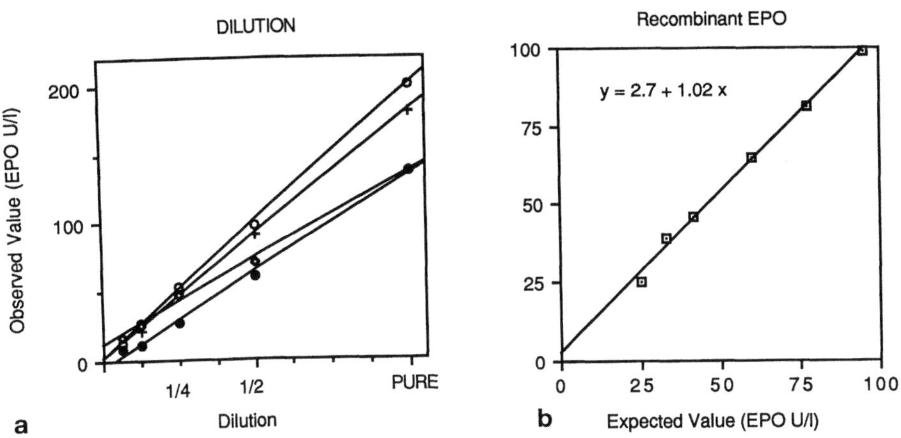

Fig. 1a,b. Accuracy of EPO RIA assessed by dilution test (**a**) whereby three sera were diluted in zero standard (○, ●, +) and one in a serum with a low EPO value (◇), and recovery test (**b**) whereby increasing known amounts of recombinant EPO were added to a serum with a low EPO content. The same type of curve was obtained when a serum with a high EPO value was added

Anemic Patients

In anemic patients, a logarithmic relation is observed between the EPO level and the Ht value (Fig. 2), with a large dispersion of the experimental points around the calculated line. A least-square analysis gave a logarithmic regression curve expressed by the equation:

$$\log \text{EPO} = -0.059\ (\text{Ht}) + 3.45$$

where EPO is expressed in U/l and Ht in percentage. The slope of the curve has a standard error of 0.007 (r=0.69).

This type of curve is useful to determine the response of erythropoietin to the anemia in each patient – whether it is "normal", insufficient, or excessive.

For example and as expected, in anemic patients with severe renal failure, EPO was low whatever the degree of anemia (Fig. 2, open circles).

Patients with Polycythemias

In 57 efficiently treated PV cases, the mean EPO value was found at 8.51 U/l, which is significantly lower than the normal mean value (Table 2). No correlation was observed between EPO value and Ht in the cases with normal Ht (45%–51%). EPO levels were slightly elevated in the cases with Ht from 36% to 44% (r=0.34), but less than in anemic patients with similar Ht values.

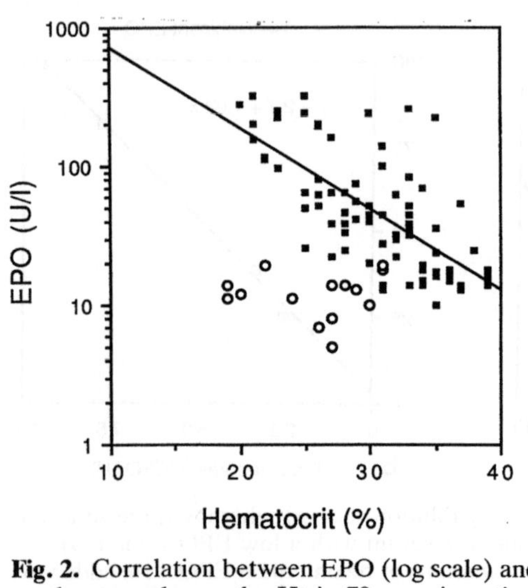

Fig. 2. Correlation between EPO (log scale) and Ht. EPO concentrations were measured on the same day as the Ht in 79 anemic patients. *Open circles* represent the results for patients undergoing dialysis

Table 2. Epo levels in polycythemia vera

	Slightly anemic PV (Ht<45%)	PV in remission (Ht 45%–51%)	PV in the spent phase or myelofibrosis	Active PV (Ht>51%)
No. of cases	26	31	19	82
Mean value ± SD		7.32±3.63		6.59±2.75
Median 95% interval[a]	10 5–15		21 5–110	

[a] Abnormal distribution.

In 82 active PV cases (newly diagnosed or in relapse) with Ht values higher than 51%, no correlation was found between EPO value and Ht. In this group, the mean EPO value was significantly lower than the mean EPO level in the control group, but not than the one observed in the group of PV patients in remission.

In 19 patients studied in the spent phase of PV or with myelofibrosis (Ht 31% to 52%), we found a negative correlation between EPO and Ht, but the EPO values were lower than expected for the degree of anemia.

In the group of patients with erythrocytosis, a large scatter of values was observed:

- In 14 cases with EPO levels higher than 50 U/l (53–650), a primary disorder could be demonstrated (kidney or liver tumor, low oxygen arterial saturation due to lung or heart disease)
- Among the other 117 cases:
  30 could be considered as "secondary" erythrocytosis (chronic bronchitis or heavy smoking): EPO levels were found between 11 and 47 U/l
  68 cases had slightly excessive EPO levels: 17–47 U/l
  34 cases had normal EPO values: 11–16 U/l
  15 cases had low EPO values: 5–10 U/l (with Ht between 42% and 62% and in 13 out of 15 cases, higher than 52)

These results show that the group of PV patients is very homogeneous, with low EPO titers and with no correlation between EPO and Ht values.

In contrast, the group of erythrocytosis patients is heterogeneous, with some secondary cases displaying high EPO values but the majority with slightly excessive, normal, or low values. These results reflect the clinical heterogeneity of erythrocytosis: excess of EPO secretion, abnormal regulation of erythropoiesis, or autonomous proliferation of only the red cell line (Najean et al. 1990).

Kinetic studies of burst-forming unit erythroid (BFU-E) or colony-forming unit erythroid (CFU-E) in the presence of EPO could be useful to classify erythrocytosis with low or normal EPO values, since it has been reported that the abnormal erythroid clone in PV is extremely sensitive to EPO stimulation (Casadevall et al. 1982; Eridani et al. 1987) or completely independent of EPO stimulation (Najman et al. 1987).

To precisely define the sensitivity and the specificity of EPO dosage and to choose a decision concentration, we constructed a Receiver-Operating Characteristic Curve (Fig. 3): for five decision concentrations, the percentage of true positives (PV with EPO value lower than the decision concentration) is represented versus the percentage of false positives (erythrocytosis with EPO value lower than the decision concentration). The area under the curve reflects the performance of the test and the decision concentration should be at the beginning of the "plateau."

For a decision concentration of 13 U/l, the specificity of the test in the diagnosis of PV is 72% (number of cases with EPO higher than 13 U/l in erythrocytosis = true negatives), the sensitivity is 92% (number of cases with EPO lower than 13 U/l in PV patients = true positives) and the positive predictive value is 79.1% (probability of having PV when EPO level is lower than 13 U/l).

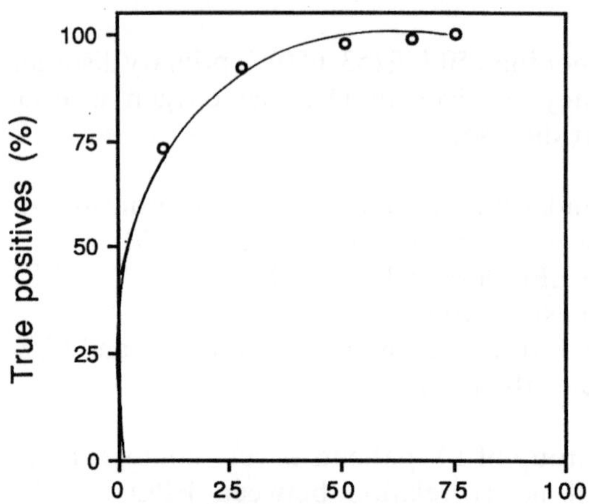

Fig. 3. Receiver-Operating Characteristic Curve (ROC Curve). For five decision concentrations (10, 13, 17, 20, and 31 U/l), the percentage of true positives (proportion of PV with EPO < decision concentration) is represented vs the percentage of false positives (proportion of erythrocytosis with EPO < decision concentration)

References

Berlin NI (1975) Diagnosis and classification of the polycythemias. Semin Hematol 12:5–19

Casadevall N, Vainchenker W, Varet P, Lacombe C, Vinci G, Chapman J, Breton-Gorius J (1982) Erythroid progenitors in polycythemia vera. Blood 59:447–451

Eridani S, Dudley JM, Sawyer BM, Pearson TC (1987) Erythropoietic colonies in a serum-free system: results in primary proliferative polycythaemia and thrombocythaemia. Br J Haematol 67:387–391

Erslev AJ (1977) The clinical usefulness of erythropoietin measurement. In: Fisher JW (ed) Kidney hormones. Academic, New York, pp 571–584

Erslev AJ, Wilson J, Caro J (1987) Erythropoietin titers in anemic, non uremic patients. J Lab Clin Med 109:429–433

Lewis JP, MacKle VC, Faguet GB, Felice AE (1989) Quantitation of erythropoietin stimulatory activity using ^{3}H-Thymidine uptake by K-562 cells. Exp Hematol 17: 102–105

Najean Y, Schlageter MH, Toubert ME, Podgorniak MP (1990) Radioimmunoassay of immunoreactive erythropoietin as a clinical tool for the classification of polycythaemias. Nouv Rev Fr Hematol 32:237–240

Najman A, Baillou C, Drouet X, Krystal G, Isnard F, Gorin NC (1987) Late erythroid progenitors in polycythemia vera are independent of erythropoietin. Blood 70 (Suppl 1):264a

Schlageter MH, Toubert ME, Podgorniak MP, Najean Y (1990) Radioimmunoassay of erythropoietin: analytical performance and clinical use in hematology. Clin Chem 36(10):1731–1735

Effects of Exercise on Plasma Erythropoietin in Natives Living Permanently at High Altitude

W. Schmidt, H. Spielvogel, K.U. Eckardt, A. Quintela, and R. Penaloza

Introduction

The effects of inspiratory hypoxia on the erythropoietin (EPO) system decisively depend on the duration of the exposure to the hypoxic environment. Whereas a stay of some hours in a hypobaric chamber leads to markedly increased serum [EPO] (Eckardt et al. 1989), it could be demonstrated by Milledge and Cotes (1985) that members of a Himalaya expedition had, after a rapid increase in EPO values, presented decreasing EPO values during the second half of their altitude exposure. Furthermore, Winslow et al. (1989) showed low EPO concentrations in the blood of high-altitude natives from Nepal and Chile, which are similar to those of subjects living at sea level. The reasons for the adaptation of the EPO response to chronic hypoxia are not completely understood. It may only in part be attributed to other adaptive processes, e.g., increased hemoglobin concentration [Hb] or lower pH values.

As a physiological stimulus, the higher oxygen demand and the suspected systemic hypoxia during heavy exercise are assumed to increase renal EPO production (Galbo 1983). This hypothesis seemed to be proved in the case of long-lasting skiing competitions, when the erythropoietic activity of the plasma was increased (Vedovato et al. 1988). But all these studies used bioassay, which not only include the effects of erythropoietin, but also of other erythropoiesis-stimulating factors. In recent years it has been demonstrated by Berglund et al. (1988) and our group (Schmidt et al. 1990b, 1991), that short-lasting bouts of maximal exercise and long-lasting bouts of submaximal exercise had no direct influence on the plasma EPO level. Moreover, exercise under acute hypoxic conditions corresponding to 3700 m did not exert any additional effect to the isolated hypoxic influence (Schmidt et al. 1991).

Acknowledgement. This study was financially supported by the Deutsche Forschungsgemeinschaft (DFG; Schm 716/1-2).

The authors wish to thank Dr. Enrique Vargas and Dr. Jean Pierre Dedet in their function as director and co-director of the Instituto Boliviano de Biologia de Altura (IBBA) for their helpful support during the study. The authors are furthermore gratefully indebted to Sra. Martha Aguilar, Sra. Esperanza Caceres, and Sra. Gladys de Contreras for their excellent technical assistance.

Pagel/Weiss/Jelkmann (Eds.)
Pathophysiology and Pharmacology
of Erythropoietin

The aim of this study therefore was to investigate the resting EPO levels of natives living at high altitude and to evaluate whether maximal or submaximal types of exercise have any effect on these subjects. If exercise had no stimulating effects on the EPO system under chronic hypoxic conditions, the hypothesis of a systemic (renal) hypoxia during exercise (Galbo 1983) could be rejected.

Methods

Subjects and Protocol

The experiments were performed with eight subjects in the laboratories of the Bolivian high-altitude institute (IBBA) situated at 3700 m in the city of La Paz.

Seven of the subjects had spent their whole life at the Altiplano between 3700 m and 4100 m. Only one subject was born at 2850 m, but he had lived at 4100 m more than 20 years. All subjects performed a vita maxima test on a cycle ergometer (Monark, Sweden) beginning with 3-min warm-up at 30 w with the work load increased by 15 w every minute until exhaustion. Two weeks after the maximal exercise all subjects performed a 60-min submaximal ergometer test at 60% of their maximal work capacity. Cubital venous blood samples were taken before, during, and up to 48 h after both trials. For determination of the acid base status arterialized samples from an ear lobe were taken before and at the end of both exercise tests.

Measurements

During the tests the expired air was collected in a Douglas bag system. The volume of the bags was determined by a spirometer (Tissaut, France) and the oxygen and CO_2 contents were measured by the oxygen analyser Servomex 570 A (UK) and the Capnograph Mark III (Gould Godart, Netherlands).

With exception of erythropoietin all parameters were determined in the laboratories of IBBA at La Paz. The hormone samples were transported in dry ice via airplane within 24 h from La Paz to the laboratories at Hannover, Germany.

[EPO] was determined in serum of the blood samples by a radioimmunoassay as previously described (Eckardt et al. 1988). Thc assay is based on a rabbit antiserum against pure recombinant human EPO and ^{125}I-labelled recombinant human EPO (Amersham International, Amersham, UK), used as tracer.

The following parameters were measured in heparinized blood (20 mU Liquemin /10 ml):

- [Hb] by the cyanmethemoglobin method (test kit 3317, Merck, FRG)
- Hematocrit (Hct) value by microhematocrit centrifugation at 20 900 g

- Lactic acid concentration (test kit 124 842, Boehringer Mannheim, FRG)
- pH, pO_2, and pCO_2 by the blood micro system BMS 2/ PHM 23 (Radiometer, Denmark)
- Reticulocyte count (expressed as per mill red cells) by counting 2000 erythrocytes of a blood smear stained with brilliant cresyl blue.

For statistical analysis a one-way analysis of variance with repeated measurements and the student's (*t*) test (two-tailed) were used. Linear regression analysis was applied to prove any correlation between the Hct value and [EPO].

Results

Maximal Exercise

The anthropometric data and the performance data of the subjects are given in Table 1. For better comparison, the data of a similar study performed at sea level under normoxic and simulated acute hypoxic conditions, corresponding to those of La Paz, are also presented (Schmidt et al. 1991). Because of lower weight and lower efficiency the maximal workload was clearly decreased in the altitude group, but the maximal oxygen uptake did not show any significant difference.

Table 2 compares the hematological values of the sea level and high altitude groups for resting conditions, as well as the values at the end of maximal and submaximal exercise in the altitude group. At rest [Hb] and Hct values were significantly higher at altitude; the pH of the arterialized blood did not differ and $p0_2$ and pCO_2 showed the well-known lower values in the altitude group. During both types of exercise [Hb] and Hct significantly increased, and pH and pCO_2 decreased. During maximal exercise the fall in pH was more pronounced in the cubital venous blood (from 7.36 to 7.15, $2p<0.001$) and the base excess markedly decreased from -4.0 ± 1.7 mmol/l to -18.5 ± 1.8 mmol/l ($2p<0.001$).

Table 1. Anthropometric and performance data of the test subjects from La Paz (3700 m) and from a similar study performed at sea level under normoxic and hypoxic conditions

	Age (years)	Weight (kg)	Performance (W)	VO_2 max (ml/kg)
Sea level (50 m) (Normoxic conditions)	28.7 ±4.3	76.0 ±9.3	297 ±26	47.0 ±4.6
Acute hypoxia (pO_2 = 92 mmHg)			271 ±19	40.0 ±5.1
High altitude (3700 m)	28.6 ±4.4	61.7 ±5.0	189 ±20	43.6 ±5.4

Table 2. Hematological data and acid base status before and after maximal and submaximal exercise in natives of high altitude. Resting data of the sea level group are also presented

	[Hb] (g/100 ml)	Hct (%)	pH	pO_2 (mmHg)	pCO_2 (mmHg)	[Lac] (mmol/l)	[EPO] (mU/ml)
Sea level (50 m)	15.5 ±0.9	47.1 ±2.5	7.41 ±0.01	93.2 ±3.8	39.3 ±1.6	1.4 ±0.5	18.1 ±1.8
High altitude							
Maximal exercise							
Rest	17.0** ±0.7	53.6** ±2.3	7.38 ±0.01	62.3*** ±5.2	30.2*** ±3.4	1.4 ±0.2	18.5 ±1.9
Exhaustion	18.9*** ±0.5	59.9*** ±2.0	7.25** ±0.02	66.2 ±6.6	21.2*** ±6.8	11.4*** ±1.8	
Submaximal exercise							
Rest	16.0 ±0.7	51.3 ±2.3	7.41 ±0.02	62.5 ±7.5	29.3 ±5.0	1.7 ±0.5	21.1 ±1.6
End (60 min)	16.9** ±0.7	53.5** ±2.3	7.36* ±0.03	67.3 ±6.0	24.9* ±4.4	6.0*** ±1.7	

Significant differences (*t*-test) are indicated for sea level vs high altitude and for exercise vs rest at altitude.
* $2p<0.05$; ** $2p<0.01$; *** $2p<0.001$; lac, lactic acid

The [EPO] at rest did not differ between the sea level and altitude group (Table 2). Five hours after maximal exercise [EPO] significantly decreased ($2p<0.01$) and reached the initial concentration 24 and 48 h thereafter (Fig. 1).

The reticulocyte count significantly increased immediately after exercise and decreased during the following days (Fig. 2).

Submaximal Exercise

The [EPO] after submaximal exercise showed the same time course as after maximal performance (Fig. 3), decreasing significantly after 3 and 5 h and increasing after 48 h ($p<0.01$). As after maximal exercise, the reticulocyte count was also characterized by a rapid increase (from 16.1 ± 2.8‰ to 21.9 ± 4.9‰; $2p<0.01$ and a decrease during the following days.

Discussion

According to Galbo (1983) and others one should expect stimulating effects on EPO production by heavy exercise, which should be due to a decreased renal perfusion (Poortmanns 1984) and the effects of strongly increased catecholamine concentrations. In this study one could assume an enhanced

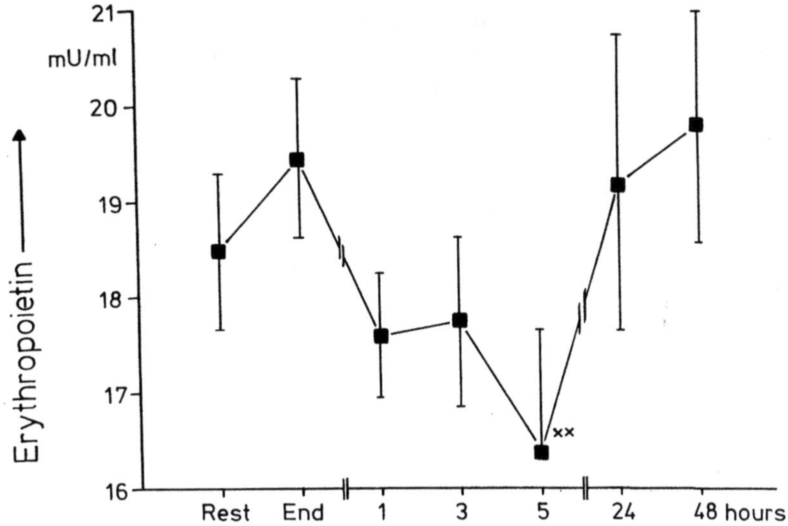

Fig. 1. Erythropoietin concentration before and up to 48 h after maximal exercise under chronic hypoxic conditions (3700 m). ^{xx}, $2p<0.01$

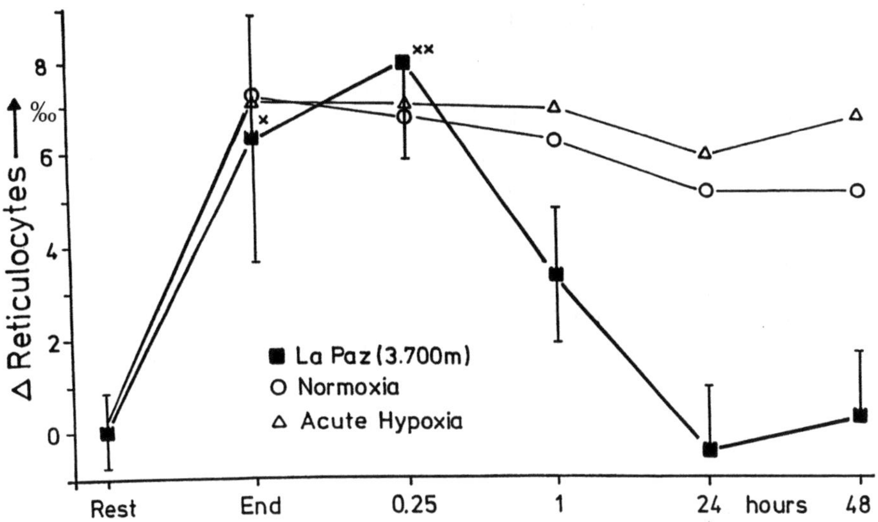

Fig. 2. Reticulocyte count before and after maximal exercise. For better comparison, the values of a similar study at sea level under normoxic and acute hypoxic conditions are indicated. ^{x}, $2p<0.05$; ^{xx}, $2p<0.01$

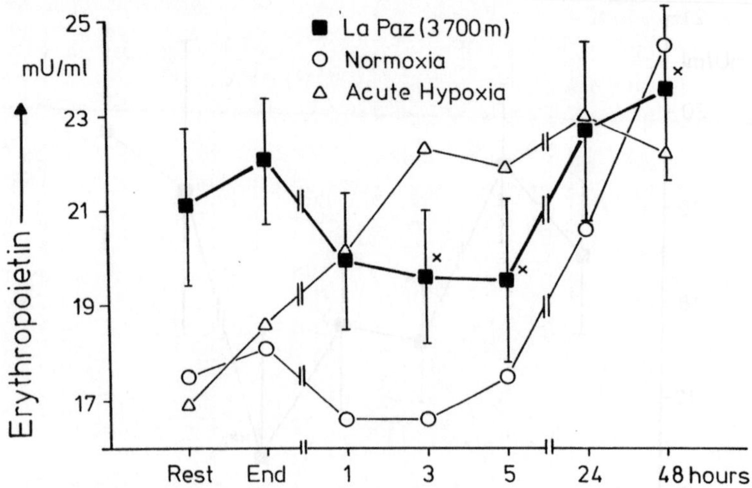

Fig. 3. Erythropoietin concentration before and up to 48 h after submaximal exercise under chronic hypoxic conditions (3700 m). Comparison with similar study at sea level as in Fig. 2. ˣ, $2p < 0.05$

EPO stimulation by the low inspiratory oxygen content at the altitude of 3700 m. The hypothezised [EPO] increase in plasma should be expected about 3 h after the stimulus (Kurtz et al. 1988). But as can be clearly demonstrated, [EPO] decreased during the hours after maximal and submaximal exercises. The tendency to an increase immediately after exercise is caused by hemoconcentration effects.

There are two reasons for these unexpected results:

1. The pH measured in cubital venous blood decreased by about 0.22 units, and the base excess by 15 mmol/l. From a recent study by Eckardt et al. (1990) it is known that a decreasing pH induced by the variation of the pCO_2 or by fixed acids is a potent inhibitor of the renal EPO production even under strong hypoxic conditions.
2. Another cause for the decreasing [EPO] should be the increasing Hct value, which results from a water shift from the intravasal into the interstitial and intracellular water space (Schmidt et al. 1990a). During maximal exercise this water loss explains the increase in Hct by about 6%. Furthermore, Pagel et al. (1988, 1989) showed a very low EPO stimulation when renal perfusion was dramatically reduced, which can be explained by lower renal oxygen demand as a result of lower sodium reabsorption rate.

Immediately after exercise under normoxic conditions there occurs a reticulocytosis, which can be observed for at least 2 or 3 days. The initial part of this reticulocytosis is caused by a rapid expulsion of immature erythrocytes from the bone marrow and is followed by an increased red cell production.

After exercise at high altitude, the rapid increase of the reticulocyte count can be observed but the higher red cell production rate the days after is lacking. One may conclude that the decreased [EPO] protects the human body against unphysiologically high Hct values, which could be induced by regular physical exercise. A pathological reaction to chronic hypoxia is well known in Andean natives as "Monges disease" or "polycythemia" (Monge 1925). It causes illness and even death, and can only be treated by leaving the high-altitude regions.

Compared to sea level exercise, submaximal exercise at altitude is more exhausting as is proved by lower pH and higher lactic acid concentration (Table 2). Therefore, the low [EPO] some hours after submaximal exercise can be explained by the same mechanisms occurring during and after maximal exercise. The low, but significant increase in [EPO] in the days after exercise should be due to hemodilution effects resulting from increasing plasma volume (Schmidt et al. 1990a). The difference in the EPO curves after exercise under acute and chronic hypoxic conditions (Fig. 3) clearly demonstrates the adaptation of the EPO system under chronic hypoxia.

From the relationship between Hct and [EPO] before and some days after submaximal exercise under normoxia (Fig. 4), one would expect lower [EPO] in the blood of high altitude dwellers. But the [EPO] in the blood of high altitude natives does not match the results of sea level subjects, i.e., they posses higher concentrations than one would expect from their Hct values. These findings correspond to those of Winslow et al. (1989), demonstrating higher [EPO] in high altitude natives from Chile, when compared with blood of Sherpas from Nepal with similar Hb concentrations.

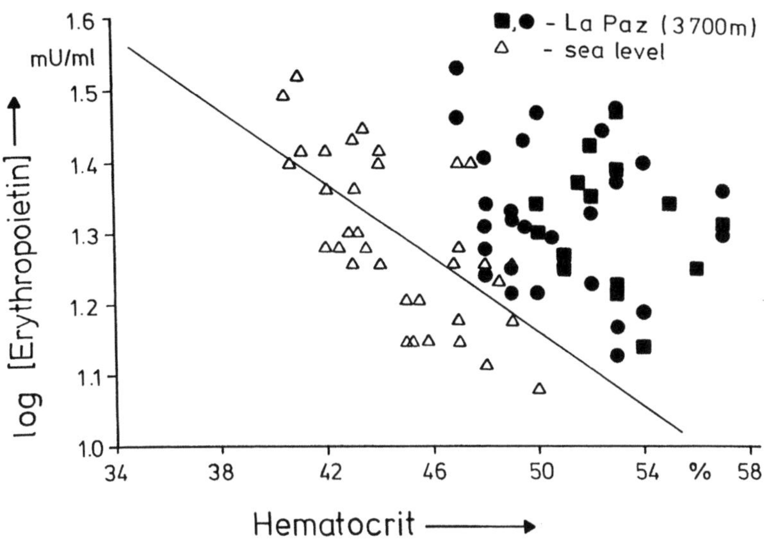

Fig. 4. Relationship between Hct value and [EPO] before and some days after exercise

Conclusion

Our results support the hypothesis that the Andean natives seem to be functionally anemic, reacting under resting conditions with higher [EPO] to the hypoxic environment than one would expect from their Hct values. However, physical exercise does not stimulate the EPO system under chronic hypoxia. Thus, one may conclude that exercise as described in this paper does not induce any systemic hypoxia and does not require adaptation processes of the erythropoietic system.

References

Berglund B, Birgegard G, Hemmingsson P (1988) Serum erythropoietin in cross country skiers. Med Sci Sports Exerc 20:208–209

Eckardt KU, Kurtz A, Hirth P, Scigalla P, Wieczorek L, Bauer C (1988) Evaluation of the stability of human erythropoietin in samples for radioimmunoassay. Klin Wochenschr 66:241–245

Eckardt KU, Bouttellier U, Kurtz A, Schopen M, Koller EA, Bauer C (1989) Rate of erythropoietin formation in humans in response to acute hypobaric hypoxia. J Appl Physiol 66:1785–1788

Eckardt KU, Kurtz A, Bauer C (1990) Triggering of erythropoietin production by hypoxia is inhibited by respiratory and metabolic acidosis. Am J Physiol 258:R678–R683

Galbo J (1983) Hormonal and metabolic adaptation to exercise. Thieme, Stuttgart

Kurtz A, Eckardt KU, Tannahill L, Bauer C (1988) Regulation of erythropoietin production. Contrib Nephrol 66:1–16

Milledge JS, Cotes PM (1985) Serum erythropoietin in humans at high altitude and its relation to plasma renin. J Appl Physiol 59:360–364

Monge MC (1925) Sobre un caso de enfermedad de Vaquez. Communicacion presentada a la Academia Nacional de Medicina, Lima, pp. 1–6

Pagel H, Jelkmann W, Weiss C (1988) A comparison of the effects of renal artery constriction and anemia on the production of erythropoietin. Pflugers Arch 413:62–66

Pagel H, Jelkmann W, Weiss C (1989) O_2 supply to the kidneys and the production of erythropoietin. Respir Physiol 77:111–117

Poortmans JR (1984) Exercise and renal function. Sports Med 1:125–153

Schmidt W, Brabant G, Kröger C, Hilgendorf A, Strauch S (1990a) Atrial natriuretic peptide during and after maximal and submaximal exercise under normoxic and hypoxic conditions. Eur J Appl Physiol 61:398–407

Schmidt W, Maassen N, Heuer D, Heck J, Böning D (1990b) Regulation of plasma volume and erythropoiesis during a ten day lasting cycling race. In: Hermans GPH, Mosterd WL (eds) Sports, medicine and health. Elsevier, Amsterdam, pp 505–512

Schmidt W, Eckardt KU, Hilgendorf A, Strauch S, Bauer C (1991) Effects of maximal and submaximal exercise under normoxic and hypoxic conditions on serum erythropoietin level. Int J Sports Med 12:457–461

Vedovato M, De Paoli Vitali E, Guglielmini C, Casoni I, Ricci G, Masotti M (1988) Erythropoietin in athletes of endurance events. Nephron 48:78–79

Winslow RM, Chapman KW, Gibson C, Samaja M, Monge CC, Goldwasser E, Sherpa M, Blume FD, Santolaya R (1989) Different hematologic responses to hypoxia in Sherpas and Quechua Indians. J Appl Physiol 66:1561–1569

Erythropoietin Concentration in Renal Venous Blood in Patient with Renovascular Hypertension: Is It a Marker of Renal Ischemia?

F. Kokot and A. Więcek

Introduction

Hypoperfusion of the kidney caused by significant unilatral stenosis of the renal artery is a well – known cause of renovascular hypertension (RVH). As already shown by Reissmann (1950) and Erslev (1953) hypoxia is triggering erythropoietin (EPO) release by the kidneys. More recently it has been proven that renal hypoxia is triggering oxygen sensors (probably located in the proximal tubules; Kurtz et al. 1988), which in turn are stimulating EPO synthesis and release by peritubular cells (Lacombe et al. 1988).

Significant unilateral renal artery stenosis (URAS) is the cause of diminished renal perfusion, and consequently of lateralization of both its excretory and endocrine function. Increased plasma renin activity (PRA) (Vaughan et al. 1973) and elevated plasma levels of atrial natriuretic peptide (ANP; – Więcek et al. 1990) in renal venous blood of the "ischemic" kidney are well-established endocrine abnormalities in patients with RVH caused by URAS.

Taking into account the above – mentioned facts the present study aimed to answer the following questions:

1. Does URAS in patients with RVH influence plasma EPO level in renal venous blood of the ischemic kidney?
2. Is plasma EPO levels in renal venous blood related to venous PRA and ANP level of the ischemic kidney?
3. What is the diagnostic value of EPO estimation in renal venous blood in patients with RVH?

Material and Methods

Nineteen patients, 7 female and 12 male, age 28–55 years (mean age 43.8 ± 1.7 years, SEM) were studied. In all these patients RVH caused by significant URAS (≥50%) was diagnosed. The diagnosis of RVH was based on the following two criteria: (a) diminution of the size of the ischemic kidney by more than 3-cm length and more than 1.5-cm width, and (b) lateralization of PRA in renal venous blood (PRA index of the ischemic kidney higher than

Pagel/Weiss/Jelkmann (Eds.)
Pathophysiology and Pharmacology
of Erythropoietin

1.5; Grützmacher et al. 1988). The degree of renal artery stenosis was assessed by renal arteriography using the Seldinger method.

Plasma creatinine level ranged from 82–132 μmol/l and the mean arterial pressure (MAP) from 117–153 mmHg (mean 127 ± 2.5 mmHg). In all examined patients the diastolic blood pressure was > 100 mmHg. Blood samples for the estimation of PRA, ANP, H^+, HCO_3^-, pO_2, pCO_2, and EPO were obtained according to the following protocol. After catheterization of the renal veins by the Seldinger method (using the iliac vein-vena cava inferior route) and the inguinal artery, all patients remained in a supine position for 1 hour. Then blood samples were withdrawn from the renal veins, inguinal artery, and from the vena cava inferior below the orifices of the renal veins. In all blood samples PRA (Kokot and Kuska 1976), ANP (Kokot et al. 1989), and EPO (Wide et al. 1989) were assessed by radioimmunoassay methods while the concentration of H^+, $HCO^-{}_3$, pO_2, and pCO_2 was determined by the micro-Astrup method (Kokot 1969). H^+, $HCO^-{}_3$, pO_2, and pCO_2 was estimated only in eight patients. EPO (obtained from Boehringer, Mannheim) was labelled with ^{125}J using the chloramine method. Sera and standards were incubated with EPO-antibodies (raised in rabbits, final dilution of EPO-antibodies 1:60 000) for 24 h at 4°C. Then ^{125}J-EPO was added and the incubation extended for a further 24 h. Antibody – bound EPO was precipated using anti-gamma globulin antibodies with polyethylene glycoe (PEG). The inter- and intraassay error of our assay was 15% and 9% respectively, while the sensitivity was 1.25 mU/tube. From the values of PRA, ANP and EPO respectively assessed in renal venous blood of the radiographically "normal" (NK) and "ischemic" (IK) kidney and of the vena cava inferior (V) the following ratios were calculated (Vaughan et al. 1973):

$$\frac{IK}{NK}, \frac{IK\text{-}V}{V}, \text{ and } \frac{NK\text{-}V}{V}$$

Results obtained in this study were statistically evaluated using the Student's *t*-test for "unpaired" variables.

Results

As can be seen in Table 1, PRA in renal venous blood of the ischemic kidney was significantly ($p<0.002$) higher than in the contralateral one or in blood of the vena cava inferior ($p<0.02$), but not significantly higher than in arterial blood. The mean $\frac{IK}{NK}$, $\frac{IK\text{-}V}{V}$, and $\frac{NK\text{-}V}{V}$ ratios for PRA were 3.48±0.68, 2.04±0.62, and –0.09±0.1 respectively. According to Vaughan et al. (1973) values of the $\frac{IK\text{-}V}{V}$ ratio ≥ 0.48 and of the $\frac{NK\text{-}V}{V}$ ratio ≤ 0.03 are characteristic for curable RVH.

As shown in Table 1, the plasma level of ANP in the ischemic kidney was statistically higher than in the normal kidney ($p<0.005$) and in blood of the

Table 1. PRA, ANP, and EPO concentration, H^+ and HCO^-_3 levels, and pO_2 and pCO_2 in renal venous blood of the ischemic and normal kidney, and in blood of the vena cava inferior and aorta[a]

	Venous blood of:				Statistical significance of the differences					
	Normal kidney (A)	Ischemic kidney B	Vena cava inferior C	Arterial blood D	A–B	A–C	A–D	B–C	B–D	C–D
PRA (ng/ml per h) $n=19$	2.71 ± 0.48	7.21 ± 1.59	3.43 ± 0.83	4.89 ± 1.05	$p<0.002$	NS	NS	$p<0.02$	NS	NS
ANP (pg/ml) $n=19$	98.00 ± 14.00	161.00 ± 14.00	106.00 ± 17.00	139.00 ± 16.00	$p<0.005$	NS	NS	$p<0.005$	Ns	NS
EPO (mU/ml) $n=19$	16.00 ± 1.40	15.90 ± 1.80	15.80 ± 1.30	17.50 ± 2.00	NS	NS	NS	NS	NS	NS
H^+ (nmol/l) $n=8$	36.75 ± 0.76	36.47 ± 0.81	38.73 ± 0.71	36.19 ± 0.87	NS	$p<0.05$	NS	$p<0.05$	NS	$p<0.05$
HCO^{-3} (mmol/l) $n=8$	26.10 ± 0.84	26.26 0.49	27.35 ± 0.63	26.12 ± 0.44	NS	NS	NS	NS	NS	NS
pO_2 (mmHg) $n=8$	50.49 ± 2.21	52.70 ± 1.62	41.54 ± 1.67	99.06 ± 2.45	NS	$p<0.005$	$p<0.001$	$p<0.001$	$p<0.001$	$p<0.01$
pCO_2 (mmHg) $n=8$	38.96 1.40	38.76 ± 1.24	43.06 ± 1.48	37.61 ± 1.20	NS	$p<0.05$	NS	$p<0.05$	NS	$p<0.01$

NS, statistically not significant
[a] means ± SEM.

vena cava inferior (p<0.005), but was not significantly different from the respective value in arterial blood. The mean $\frac{IK}{NK}$, $\frac{IK\text{-}V}{V}$, and $\frac{NK\text{-}V}{V}$ ratios for ANP were 2.14±0.35, 1.06±0.43, and 0.68±0.69 respectively.

In contrast to PRA and ANP, EPO plasma levels of the ischemic kidney were of the same magnitude as in renal venous blood of the normal kidney, vena cava inferior, and arterial blood. The respective values of the $\frac{IK}{NK}$, $\frac{IK\text{-}V}{V}$, and $\frac{NK\text{-}V}{V}$ ratios for EPO were 1.06±0.1, 0.028±0.09, and 0.018±0.05.

As can be seen in Table 1, blood H^+, HCO^-_3, pO_2, and pCO_2 in renal venous blood of the ischemic kidney did not differ significantly from respective values obtained in venous blood of the normal kidney. It is to be stressed that, pO_2 in renal venous blood both of the ischemic and normal kidney was significantly higher (p<0.05) than in blood of the vena cava inferior.

No significant correlation was found between H^+, HCO^-_3, pO_2, and pCO_2 and all hormonal parameters assessed in this study.

Discussion

Patients with URAS are characterized by higher PRA in renal venous blood of the ischemic kidney than in the normal one (Vaughan et al. 1973). As we have shown in our previous studies, renal venous blood of the ischemic kidneys contains also higher plasma concentrations of ANP than in the nonischemic kidney, showing some parallelism with PRA (Więcek et al. 1990). Since the kidneys synthetize urodilatin (Forssmann et al. 1989), which is the 95–126 terminal fragment of pro-ANP (which probably cross – reacts with ANP antibodies recognizing the 28 – residue amino acid sequence of circulating ANP), we speculated that renal ischemia may trigger renal ANP secretion, and that ANP assessed in renal venous blood may be a marker of renal ischemia. As we have shown in our present study, pO_2 and plasma EPO levels, in contrast to PRA and ANP, showed no lateralization in patients with URAS. These results suggest that different mechanisms regulate renin and ANP secretion or handling than those regulating EPO in the ischemic kidney. The presence of similar pO_2 values both in blood of the ischemic and normal kidney is concordant with our results of EPO estimation in renal venous blood. These results do not agree with those recently reported by Grützmacher et al. (1988). These authors found 1.5 times higher EPO levels in renal venous blood of the ischemic kidney (showing a renal artery stenosis more than 50%) than in the normal kidney. The cause of these discordant results is not clear. It is to be stressed, however, that these authors did not assess pO_2 in renal venous blood. Pagel et al. (1988) have recently compared the magnitude of EPO production in response to either generalized anemia or graded constriction of the renal arteries. They found that EPO levels in anemic animals exceeded a hundred times those found after a 75% reduction of renal blood flow.

The question arises as to why most patients with renal ischemic (= renovascular) hypertension don't show elevated plasma EPO levels and polycythemia?

According to Erslev (1987) the ischemic kidney shows a unique adaptive mechanism which regulates oxygen supply to this organ. As is well known, oxygen is predominantly used in the generation of high energic compounds which are needed in the process of tubular sodium reabsorption. In turn, the tubular load of sodium is dependent upon the magnitude of glomerular filtration rate (GFR). Renal hypoperfusion, caused by renal artery stenosis, by reducing GFR simultaneously diminishes the demand for high energetic compounds and in turn for oxygen. This mechanism may explain why pO_2 in renal venous blood of the ischemic kidney is of a similar magnitude as in renal venous blood of the normally perfused kidney. This hypothesis was recently proved by Eckhardt et al. (1989). They found that acetazolamide, which acts mainly at the proximal tubular site, reduced hypoxia-induced EPO formation in a dose-dependent fashion. In these experiments EPO release correlated with acetazolamide-induced natriuresis.

Summarizing we may conclude, that (a) in contrast to PRA and ANP, URAS in patients with RVH is not followed by lateralization of plasma EPO levels in renal venous blood, and (b) estimation of EPO in renal venous blood is of no diagnostic value in patients with URAS.

References

Eckardt KU, Kurtz A, Bauer C (1989) Erythropoietin formation is related to proximal tubular function. Am J Physiol 256:F942–F947

Erslev A (1953) Humoral regulation of red cell production. Blood 8:349–357

Erslev A (1987) Erythropoietin coming of age. N Engl J Med 316:101–103

Forssmann WG, Feller S, Meyer H, Schultz Knappe P (1989) Morphology of the myoendocrine cardic cell and extra-auricular systems producing cardiac hormones. In: Kaufmann W, Wambach G (eds) Endocrinology of the heart. Springer, Berlin Heidelberg New York, pp 3–26

Grützmacher P, Radtke HW, Stahl RK, Rauber K, Schoeppe W (1988) Renal artery stenosis and erythropoietin. 21st Annual Meeting of the American Society of Nephrology, Oct 11–14 (Abstracts on recombinant erythropoietin, p 20)

Kokot F (1969) Metody badan laboratoryjnych stosowanych w klinice. PZWL, Warsaw

Kokot F, Kuska J (1976) Verhalten der Plasmareninaktivität und des Aldosterons nach Kochsalzbeschränkung und aufrechter Körperlage bei essentieller Hypertonie. Z Inn Med 31:1015–1020

Kokot F, Grzeszczak W, Wieçek A (1989) Water immersion induced alterations of atrial natriuretic factors in patients with non-inflammatory acute renal failure. Nephrol Dial Transplant 4:691–695

Kurtz A, Eckhardt KU, Tannahill L, Bauer C (1988) Regulation of erythropoietin production. Contrib Nephrol 66:1–16

Lacombe C, Da Silva JL, Bruneval P, Fournier JC, Wendling F, Casadevall N, Camilleri JP, Bariety J, Varet B, Tambourin P (1988) Peritubular cells are the site of erythropoietin synthesis in the murine hypoxic kidney. J Clin Invest 81:620–623

Pagel H, Jelkmann W, Weiss C (1988) A comparison of the effects of renal artery constriction and anemia on the production of erythropoietin. Pflugers Arch 412:62–66

Reissmann KR (1950) Studies on the mechanism of erythropoietic stimulation in parabiotic rats during hypoxia. Blood 5:372–380

Vaughan ED, Bühler FR, Laragh JH, Sealey UE, Baer L, Bard BH (1973) Renovascular hypertension: renin measurements to indicate hypersecretion and contralateral suppression, estimate renal plasma flow and score for surgical curability. Am J Med 55:402–414

Wide L, Bengtsson T, Birgegard G (1989) Circadian rhythm of erythropoietin in human serum. Br J Haematol 72:85–90

Wiecek A, Kokot F, Kuczera M, Klin M, Kiersztejn M (1990) Plasma level of atrial natriuretic peptide in renal venous blood-marker of kidney ischemia? Clin Nephrol 34:26–29

Serum Erythropoietin Levels in Von Hippel-Lindau Syndrome

H.P.H. Neumann, W. Jelkmann, H.R. Eggert, P. Schollmeyer, and O.D. Wiestler

Introduction

Von Hippel-Lindau syndrome (HLS) is an autosomal-dominant inherited cancer-prone disorder associated with hemangioblastomas (Hbl) of the CNS, angiomatosis retinae (AR), renal cysts, renal cancer, pancreatic cysts, pancreatic cancer, and pheochromocytoma [10]. The main cause of death in this disease are Hbl of the CNS, followed by renal cancer and pheochromocytoma. AR can lead to unilateral or even bilateral loss of vision [11]. Manifestations of the syndrome usually develop in young adults [8].

As increased production of erythropoietin (EPO) has been reported in association with Hbl of the CNS [13], we investigated whether the serum EPO concentration is an indicator of HLS manifestation, which may facilitate an early diagnosis of affected individuals.

Methods

Assay of Serum EPO and Normal Values

Serum for EPO radioimmunoassay was prepared from venous blood sampled without anticoagulant. The assay was carried out in duplicate using ^{125}I-labelled recombinant human EPO (specific activity 11–33 TBq/mmol; Amersham Buchler, Braunschweig, FRG) and antiserum (1:5000) from a rabbit immunized with human urinary EPO. Human urinary EPO was also used as the standard (10 concentrations in the range 0–250 mU/ml in dilution buffer: phosphate buffered saline, pH 7.4 with bovine serum albumin 0.5 g/l and sodium acid 0.5 g/l, to which bovine gamma globulin 12 g/l was added).

Mixtures of 100 µl antibody solution and 100 µl test serum or EPO standard were incubated at 4°C for 48 h, before 100 µl ^{125}I-EPO (5×10^{-15} mol/l in dilution buffer) was added for another 24 h. Thereafter, the antibody-bound ^{125}I-EPO was precipitated with polyethylene glycol 6000 (160 g/l).

The EPO concentration of the serum samples was calculated from semilogarithmic plots of the standards (radioactivity of the pellet versus log

Pagel/Weiss/Jelkmann (Eds.)
Pathophysiology and Pharmacology of Erythropoietin

EPO concentration). The mean within and between assay coefficients of variation was 7% and 19% in the EPO range 40–50 mU/ml. The detection limit was 5 mU/ml.

Comparative measurements of immunoreactive EPO were peformed on serum samples from 14 normal subjects (5 females, 9 males; age 20–38 years). Their EPO values were essentially normally distributed with a mean of 18.1 ± 7.5 mU/ml (± 1 SD).

Thus, with the assay described, 95.5% of all normal values are in the range 3.1 - 33.1 mU/ml (mean ± 2 SD).

Subjects

Participating in this study were patients with proven positive gene carrier status of HLS. According to the minimal criteria for the diagnosis [9, 10], Hbl or AR in combination with one CNS, ocular, renal, adrenal, or pancreatic lesion were found in two first-degree relatives. Also included were asymptomatic gene carriers without manifestations of HLS. In these cases, a gene carrier status was confirmed by pedigree analysis.

Results

We studied 44 subjects with HLS (21 male, 23 female). The mean age was 38.7 (16–79) years. The following manifestations of HLS were found: Hbl of the CNS in 5, AR in 25, renal cancer in 2, renal cysts in 3, pheochromocytomas in 7 patients; there was a history of surgical treatment for Hbl of the CNS in 4, and for pheochromocytoma in 11 patients. Fifteen subjects presented with multiple lesions. Three individuals were identified as asymptomatic gene carriers by pedigree analysis; our standard screening protocol [11] did not reveal lesions of the HLS.

The results are summarized in Table 1. Serum EPO was elevated (>33.1 mU/ml) in 2 of 5 (40%) patients with Hbl, in 2 of 25 (8%) with AR, in 1 of 7 (14%) with pheochromocytoma, but in none of the patients with renal and pancreatic lesions. No significant correlation was found between elevated EPO values and serum hemoglobin concentrations. One patient with AR and one patient with a history of pheochromocytoma surgery presented with polyglobulia (hemoglobin > 180 g/l), but serum EPO was normal in both cases.

Discussion

The association of polyglobulia and cerebellar hemangioblastoma was first reported by Carpenter et al. in 1943 [2]. Waldman et al. [16] described the production of an erythropoiesis-stimulating factor in Hbl, which was later

Table 1. Serum EPO and hemoglobin (*Hb*) levels in patients with von Hippel-Lindau syndrome

Patient	EPO (mU/ml)	Hb (g/l)
Angiomatosis retinae		
1[a]	15.5	146
2	8	152
3[a]	19	156
4[a]	21	153
5[a]	28	177
6	20.5	136
7[a]	19.5	137
8	32	117
9[a]	*34*	169
10[a]	33	167
11	17	134
12[a]	19	147
13	28	*189*
14	18	138
15[a]	19	175
16	16	136
17[a]	4.5	154
18	19.5	157
19	25	166
20[a]	12.5	169
21	29	
22[a]	27	145
23[a]	*40*	154
24	23.5	153
25[a]	21	146
Gene carriers without manifestations		
26	*37*	141
27	20	153
28	24.5	
Hemangioblastoma (postoperative levels)		
22[a]	27	143
29	*34*	153
30	28	145
31[a]	10.5	164
Hemangioblastoma of the CNS		
1[a]	15.5	146
7[a]	19.5	137
9[a]	*34*	169
12[a]	19	147
23[a]	*40*	154
Pheochromocytoma		
15[a]	19	175
20[a]	12.5	169
32	30.5	157
33	*44*	175
34	14	144
35	21.5	162
36	32	160
Renal cysts		
1[a]	15.5	146
31[a]	10.5	164
37	18.5	163
Renal tumors		
7[a]	19.5	137
38	23	155
Pancreatic cysts		
1[a]	15.5	146
7[a]	19.5	137
39[a]	7	149
Pheochromocytoma (postoperative levels)		
3[a]	19	156
4[a]	21	153
5[a]	28	177
10[a]	33	167
17[a]	4.5	154
25[a]	21	146
40	22	*196*
41	25	141
42	14	149
43	20.5	139
44	24.5	144

Elevated levels are *underlined* (EPO > 33.1 mU/ml; HB > 180 g/l).
[a], Patients with multiple manifestations of the syndrome

identified as EPO [5, 13]. EPO production has been demonstrated in several other neoplasms and tumor-like lesions. Of special interest are the reports of EPO or erythropoiesis-stimulating factors in renal cancer [4, 13], renal cysts [14], and pheochromocytoma [1]. These lesions are among the characteristic manifestations of HLS [10]. In HLS gene carriers, the age of initial manifestations an the pattern of associated lesions are highly variable. We have therefore recently developed a screening program aimed at identifying manifestation of the syndrome in asymptomatic individuals [12]. A genetic locus affected in HLS has recently been mapped to the short arm of chromosome 3 [15]. Commercial linkage tests are, however, not available. Since production of erythropoietin and erythropoiesis-stimulating activities has been reported in several neoplasms, which can develop as manifestations of HLS, we decided to study serum EPO as a potential marker for HLS-affected individuals. The results indicate that EPO is not a suitable marker for the syndrome, neither in asymptomatic nor in symptomatic patients.

None of the five patients with slightly elevated EPO serum concentrations showed signs of polyglobulia. This is in accordance with literature reports which also failed to demonstrate increases in hemoglobin concentrations and red blood cell counts in patients with paraneoplastic EPO production [3, 6, 7]. In two of our patients hemoglobin levels were elevated, but EPO was normal suggesting that increased serum EPO concentration cannot acount for the polyglobulia in these patients.

In summary, the data presented above indicate that serum erythropoietin and hemoglobin are not suitable markers for manifestations of HLS.

References

1. Bradley JE, Young JD Jr, Lentz G (1961) Polycythemia secondary to pheochromocytoma. J Urol 86:1–6
2. Carpenter G, Schwartz H, Walker A (1943) Neurogenic polycythemia. Ann Intern Med 19:470–481
3. Hennessy TG, Stern WE, Herrick SE (1967) Cerebellar hemangioblastoma: erythropoietic activity by radioiron assay. J Nucl Med 8:601–606
4. Hewlett JS, Hoffmann GC, Senhauser DA, Battle JD (1960) Hypernephroma with erythrocythemia: report of a case and assay of the tumor for an erythropoietic-stimulating substance. N Engl J Med 262:1058–1062
5. Jankovic GM, Ristic MS, Pavlovic-Kentera V (1986) Cerebellar hemangioblastoma with erythropoietin in cerebrospinal fluid. Scand J Hematol 36:511–514
6. Jeffreys RV, Napier JAF, Reynolds SH (1982) Erythropoietin levels in posterior fossa hemangioblastoma. J Neurol Neurosurg Psychiatry 45:264–266
7. Kew MC, Fisher JW (1986) Serum erythropoietin concentrations in patients with hepatocellular carcinoma. Cancer 58:2485–2488
8. Lamiell JM, Salazar FG, Hsia YE (1989) Von Hippel-Lindau disease affecting 43 members of a single kindred. Medicine (Baltimore) 68:1–29
9. Melmon KL, Rosen SW (1964) Lindau's disease: review of the literature and study of a large kindred. Am J Med 36:595–617
10. Neumann HPH (1987) Basic criteria for clinical diagnosis and genetic counselling in von Hippel-Lindau syndrome. Vasa 16:220–226

11. Neumann HPH (1987) Prognosis of von Hippel-Lindau syndrome. Vasa 16:309–311
12. Neumann HPH, Eggert HR, Weigel K et al. (1989) Hemangioblastomas of the central nervous system. A ten year study with special reference to the von Hippel-Lindau syndrome. J Neurosurg 70:24–30
13. Rosenlöf K, Fyhrquist F, Grönhagen-Riska C et al. (1985) Erythropoietin and renin substrate in cerebellar hemangioblastoma. Acta Med Scand 218:481–485
14. Rosse WF. Waldman TA, Cohen P (1963) Renal cysts, erythropoietin and polycythemia. Am I Med 34:76–81
15. Seizinger BR, Rouleau GA, Ozelius LJ et al. (1988) Von Hippel-Lindau disease maps to the region of chromosome 3 associated with renal cell carcinoma. Nature 332: 268–269
16. Waldmann TA, Levin EH, Baldwin M (1961) The association of polycythemia with a cerebellar hemangioblastoma. Am I Med 31:318–324

IV. Erythropoietin Replacement in Renal Failure

Iron Metabolism Under Treatment of Renal Anemia with Recombinant Human Erythropoietin

P. GRÜTZMACHER, P. ROTH, J. VLACHOJANNIS, T. TSOBANELIS, and E. WERNER

Introduction

The administration of recombinant human erythropoietin (rEPO) generally corrects renal anemia in a dose-dependent manner. Several reasons for an impaired therapeutic response have been identifed, e.g., absolute and functional iron deficiency, occult gastrointestinal blood loss, infection, and aluminium overload [3, 4, 9, 10]. A decrease in serum ferritin levels is common under rEPO therapy. The exhaustion of the storage iron pool can be preceeded by a functional, i.e., a dynamic iron deficiency where the iron incorporation rate exceeds the iron mobilization and transport capacity. Serum iron levels and total iron-binding capacity show only minor changes, not allowing an adequate monitoring of iron metabolism during rEPO therapy [1, 2]. Development of microcytosis is unusual. Iron supplementation is usually performed either in general or, if ferritin levels fall below a borderline value in the range of 100 μg/ml to prevent iron depletion. Except in patients with severe hemosiderosis, iron supplementation is nearly always necessary to maintain adequate ferritin levels in serum.

Functional iron deficit is indicated by a transferrin saturation below 15%–20% in the presence of normal ferritin levels, resulting in a slower increase rate for the hematocrit. By increasing the transferrin saturation to above 20% by iron administration, the increase rate of the hematocrit can be restored [7]. Functional iron deficiency has been observed in more than 50% of iron-overloaded hemodialysis patients during rEPO therapy [6]. By monitoring red cell zinc protoporphyrin, functional iron deficiency can possibly be detected earlier.

The iron reserve as well as the iron necessary for a desired correction of anemia can also be calculated, using equations which have been proposed by Van Wyck et al. [11].

iron reserve	=	400 empiric constant	x	[log (n) actual ferritin	– log (30)] minimum threshold ferritin
iron demand	=	150	x	[target Hb	– actual Hb]

Pagel/Weiss/Jelkmann (Eds.)
Pathophysiology and Pharmacology of Erythropoietin

Detailed information on iron metabolism under rEPO can be obtained by ferrokinetic and erythrokinetic studies, of which some of our own data shall be discussed in the following.

Ferro- and Erythrokinetic Studies

Twelve patients on maintenance hemodialysis were treated with rEPO (3×40 U/kg body wt. per week, i.v.). Patients were not on iron substitution. Ferro- and erythrokinetic studies with ^{59}Fe-tagged transferrin and ^{51}Cr-labelled red cells were performed before and 4 weeks after the onset of rEPO substitution; each study lasted 6 weeks. Parameters of iron metabolism were obtained by a computerized kinetic analysis based on a multicompartment model (Fig. 1). In addition, the following parameters were studied: bone marrow iron turnover (MIT), red cell iron turnover (RCIT), hemoglobin synthesis rate, tissue iron turnover (TIT), erythrocyte survival rate (^{51}Cr serum half-life), blood volume, and blood loss. The model and analysis employed have been described elsewhere [8].

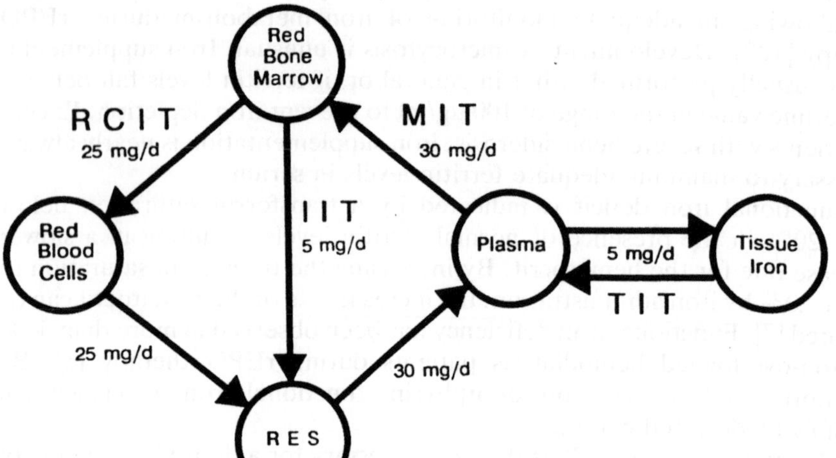

Fig. 1. Compartmental model of iron metabolism and normal daily iron shifts (according to [8]). In the *center* is the plasma compartment, from where iron is transported to either the bone marrow or the tissue storage compartment. Nearly all bone marrow iron appears in the peripheral circulation, incorporated into red cells, reflecting effective erythropoesis. Only a small portion of new red cells is destroyed within the bone marrow and metabolized by the reticuloendothelial system (RES) from where iron is released into the circulation again, i.e., ineffective erythropoesis. *MIT*, marrow iron turnover; *RCIT*, red cell iron turnover; *IIT*, ineffective iron turnover; *TIT*, tissue iron turnover; *PIT*, plasma iron turnover (MIT+TIT)

The erythropoetic activity, indicated by MIT was considerably reduced in hemodialysis patients, but increased by 25% to normal values under rEPO (see Fig. 2). Erythropoetic efficacy, reflected by RCIT was equally reduced to 16 mg/day and increased by the same extent to normal values. Under normal conditions a close functional relation between MIT and RCIT is present, where RCIT covers 70%–100% of MIT. A normal ratio was found in uremic patients prior to rEPO. This ratio was maintained under rEPO administration. This indicates that intramedullary hemolysis is not increased by rEPO in uremic patients.

Hemoglobin synthesis rate was reduced to 4.7 g Hb/day and was normalized to 6.2 g Hb/day by rEPO substitution. The estimated TIT from the plasma to the storage iron pools appeared not to be reduced in uremia; it did not however show any increment under rEPO therapy. The lacking

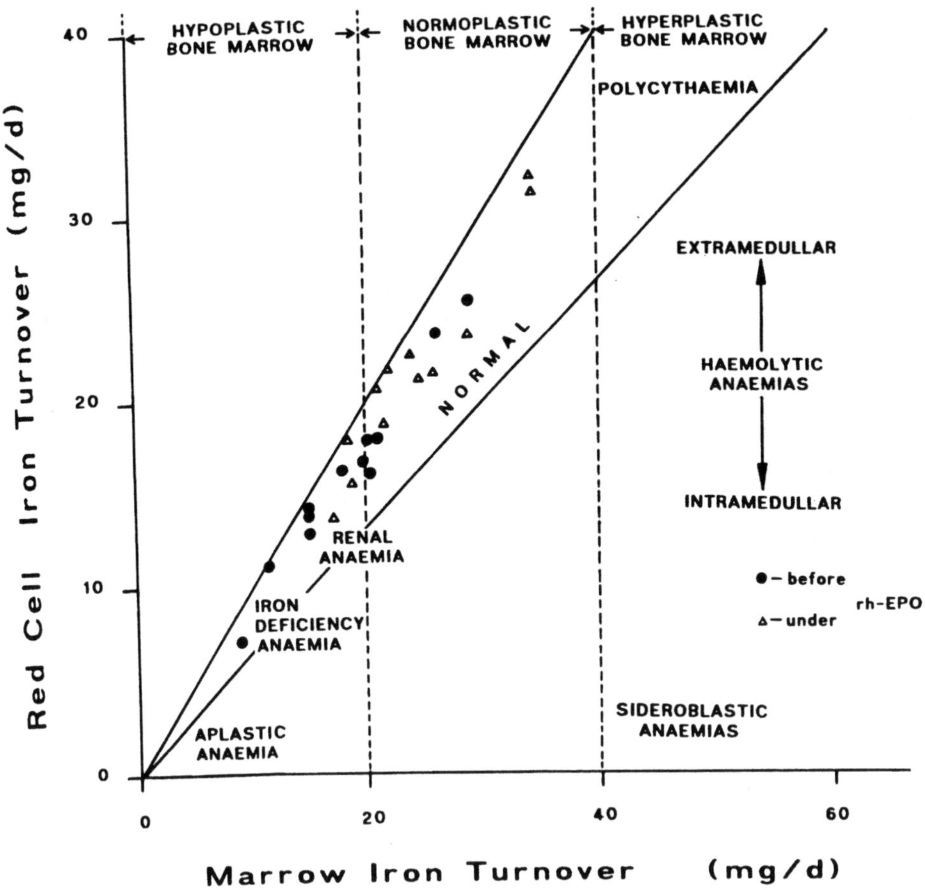

Fig. 2. Functional relationship between red cell iron turnover and bone marrow iron turnover before and under rEPO therapy

Table 1. Influence of rEPO on the characteristies of renal anemia, analyzed by ferro- and erythrokinetic studies

	rEPO therapy		
Parameter	Before 4 weeks[a]	After 4 weeks[a]	p
Hematocrit (vol.%)	22.2± 0.5	27.9± 1.3	<0.01
Hemoglobin (g/ml)	7.1± 0.3	8.8± 0.4	<0.01
Ferritin (µg/l)	1002.0±338.0	429.0±158.0	<0.05
Transferrin saturation (%)	38.8± 6.8	22.2± 3.7	<0.04
Hemoglobin synthesis rate (g/d)	4.6± 0.4	6.2± 0.5	<0.01
Red cell iron turnover (mg/d)	16.0± 1.5	21.6± 1.8	<0.05
Marrow iron turnover (mg/d)	18.2± 1.8	24.3± 1.7	<0.05
Tissue iron turnover (mg/d)	36.8± 9.2	31.6± 8.2	NS
Erythrocyte volume (ml)	862.0± 50.0	1010.0± 62.0	<0.02
Plasma volume (ml)	3255.0±160.0	2987.0±195.0	NS
Blood volume (ml)	4117.0±206.0	3996.0±231.0	NS
Red cell survival (d)	23.6± 0.9	23.9± 1.1	NS

NS, not significant
[a] Sample mean ± SEM.

increase of iron delivery by the storage iron pool during treatment with rEPO is in good accordance with the above-mentioned observations obtained by ferritin and transferrin saturation, which suggest a very high incidence of functional iron defiency under rEPO.

The erythrocyte survival rate was found to be considerably reduced prior to rEPO, confirming previous studies. rEPO therapy had no effect on erythrocyte survival rate, making it possible to estimate that the peripheral hemolysis rate increases by 50% after an increment of hematocrit by 50%, e.g., from 20 vol.% to 30 vol.%. This probably explains the occasional changes in potassium and phosphate observed during rEPO therapy [4, 5].

Daily blood loss was increased in uremia and persisted during rEPO treatment (Table 1).

Clinical Effects of Intensive i.v. Iron Substitution

In a preliminary study on 33 anemic hemodialysis patients the effect of an intensive iron substitution on the therapeutic response to rEPO was investigated; in particular, transferrin iron saturation was monitored weekly. Ferrous gluconate (40 mg) was given i.v. after each dialysis if transferrin

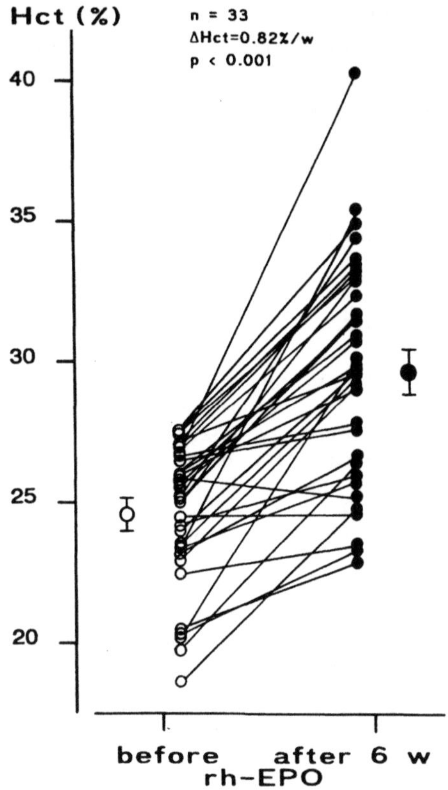

Fig. 3. Short-term effect of intensive i.v. iron substitution, combined with ultra-low dose rEPO in hemodialysis patients Hct, hematocrit

saturation approached 20%. To ensure the successful maintenance of an adequate transferrin saturation, an ultra-low i.v. dose of 20 U rEPO/kg body wt. was given three times weekly. Even under this low dose regimen intensive i.v. iron substitution was necessary in 26 out of 33 patients to maintain a normal transferrin iron saturation. However this combined i.v. iron-rEPO therapy improved renal anemia effectively with a mean increase rate of hematocrit of 0.8 vol.%/week (Fig. 3). This increase rate has been achieved by i.v. rEPO doses in the range of three times 80–100 U/kg body wt. in other studies [4, 9, 10].

Conclusions

Iron metabolism should be closely monitored when using rEPO therapy in patients with renal diseases. Iron needs often exceed body iron stores resulting in absolute iron deficiency. The depletion of body iron stores can be approximately predicted according to formula "hematocrit + 10 vol.% = 500 mg iron." Functional iron deficiency develops very frequently during the correction period, partially dependent on body iron stores and on gastrointestinal blood loss, and can be detected best by monitoring the transferrin iron saturation. As iron deficiency severely compromises a successful correction of renal anemia, rEPO therapy should regulary be accompanied by iron supplementation and should not be started when there is iron deficiency. Intensive iron supplementation allows the use of markedly lower doses of rEPO, resulting in a more economic use of this substance. Simultaneous i.v. iron substitution appears to be of advantage; the mechanisms have to be investigated further.

References

1. Allegra V, Mengozzi G, Vasile A (1991) Iron deficiency in maintenance hemodialysis patients: assessment of diagnosis criteria and of three different iron treatments. Nephron 57:175–182
2. Bergmann M, Grützmacher P, Heuser J, Kaltwasser JP (1990) Iron metabolism under rEPO therapy in patients on maintenance hemodialysis. Int J Artif Organs 13:109–112
3. Donelly SM, Mahmoud AM, Churchill DN (1990) Bioavailability of iron in hemodialysis patients treated with erythropoietin: evidence for the inhibitory role of aluminum. Am Kidney Dis 16:447–451
4. Eschbach JW, Downing MR, Egrie JC, Browne JK, Adamson JW (1989) USA multicenter clinical trial with recombinant human erythropoietin (Amgen). Contrib Nephrol 76:160–165
5. Grützmacher P, Bergmann M, Weinreich T, Nattermann U, Reimers E, Pollok M (1988) Beneficial and adverse effects of correction of anemia by recombinant human erythropoietin in patients on maintenance hemodialysis. Contrib Nephrol 66:104–113
6. Grützmacher P, Ehmer B, Messinger D, Scigalla P (1991) Therapy with recombinant human erythropoietin in hemodialysis patients with transfusion-dependent anemia: report of a European multicenter trial. Nefrologia 11:58–65
7. MacDougall IC, Hutton RD, Cavill J, Coles GA, Williams JD (1989) Poor response to treatment of renal anemia with erythropoietin corrected by iron given intravenously. Br Med J 299:157–158
8. Roth P, Werner E (1986) Entwicklung eines verbesserten Compartmentmodells für den menschlichen Eisenstoffwechsel. In: Poretti G, Schmidt T (eds) Medizinische Physik 1985. Deutsche Gesellschaft für Medizinische Physik, Bern, pp 409–414 (1986)
9. Sobota JT (1989) Recombinant human erythropoietin in patients with anemia due to end-stage renal disease. Contrib Nephrol 76:166–178
10. Suzuki M (1989) Research group for clinical assessment of rhEPO dose-finding double-blind, clinical trial of recombinant human erythropoietin (Chugai) in Japanese patients with end-stage renal disease. Contrib Nephrol 76:179–192
11. Van Wyck DB, Stivelman JC, Ruiz J, Kirlin LF, Katz MA, Ogden DA (1989) Iron status in patients receiving erythropoietin for dialysis-associated anemia. Kidney Int 35:712–716

Erythropoietin Treatment in Children with Renal Anemia

D.E. MÜLLER-WIEFEL, O. AMON, K. SCHÄRER, K. DANIEL, and A. LEDER

Renal anemia is a greater clinical problem in children than it is in adults because it is much more pronounced in this age group due to a more severe bone marrow hypoproliferation [18, 20, 24, 28, 29]. From this point of view there is a special reason to treat renal anemia of children with recombinant human erythropoietin (rhEpo).

Patients and Methods

The first author has treated a total of 77 pediatric patients with rhEpo at the University Children's Hospitals of Hamburg and Heidelberg from November 1987 to April 1991. Of these, 19 were on conservative treatment (CT), 33 on continuous ambulatory peritoneal dialysis (CAPD), 23 on hemodialysis (HD), and 2 had been transplanted. As the effect of rhEpo on HD patients has been published elsewhere [21, 23, 32], we concentrate here on the results with CT before HD and on CAPD, which is an individually tailored procedure of constant blood purification in childhood. During the observation period 15 of 33 CAPD children were successfully grafted.

Ten patients on CT and 15 on CAPD were treated subcutaneously once per week, 9 on CT were treated subcutaneously three times per week, and 18 on CAPD were treated intraperitoneally three times per week. For the patients on CT we used rhEpo provided by Cilag, Sulzbach; to those on CAPD rhEpo manufactured by Behring, Marburg, was administered. The treatment was monitored according to the hemoglobin (Hb) concentrations achieved within 2-week periods. In the case of an insufficient increase below 0.5 g/dl combined with a reticulocytosis below 5% the rhEpo dose was increased by 50%. No change was made in the case of a reticulocytosis above 5% and a Hb change between 0.5 and 1.0 g/dl. In the case of an overshooting increase in Hb above 1.0 g/dl the rhEpo dose was reduced by 50%. The sustaining dose was monitored as follows. After reaching a concentration of Hb above 12.0 g/dl the dose was reduced by 50%, and increased by 50% after a Hb drop below 12.0 g/dl. When increasing above 14.0 g/dl, the rhEpo dose was stopped and restarted below 14.0 g/dl with a dose reduced by 50%. At a concentration between 12.0 and 14.0 g/dl, the rhEpo dose was not changed. CAPD was performed by means of dialysis solutions containing 2.3% glucose (Fresenius, Bad Homburg) with a volume between 500 and 1500 ml (mean

Pagel/Weiss/Jelkmann (Eds.)
Pathophysiology and Pharmacology
of Erythropoietin

1000 ml), an overnight dwell time of 8–12 h and a last bag dwell time on continuous cycling peritoneal dialysis (CCPD) of 12–14 h.

Results

The response dose (defined as an increase in Hb concentration above 2.0 g/dl) was comparable between CT and CAPD, with 135 and 125 U/kg per week and lower than that on HD (Table 1). The response intervals were 7 and 9 weeks, respectively. The sustaining dose was significantly lower on CT, with 125 U/kg per week compared to 175 on CAPD and 200 on HD. The sustaining Hb concentration achieved was highest on CT. There was only one nonresponder to rhEpo treatment under CT- a girl with severe chronic bowel infection due to various fistulas presumably deriving from radiation treatment of a rhabdomyosarcoma.

Conservative Treatment. On CT the response to Epo in terms of the frequency of application is presented on table 2. With a comparable response dose the response interval was much shorter in the case of administration three times per week with 5.3 versus 7.6 weeks. Consequently the correction interval was much shorter, with 8.6 versus 15.6 weeks. Finally, the superiority of administration three times per week is underlined by a sustaining dose of 90 versus 171 U/kg per week. The individual courses of creatinine clearance on

Table 1. Response to Epo treatment

	CT	CAPD	HD
Response dose (U/kg per week)	135	125	275
Response interval (weeks)	7	9	7
Sustaining dose (U/kg per week)	125	175	200
Mean sustaining Hb (g/dl)	12.0	11.4	10.7
Nonresponse	1	0	0

Table 2. Response to Epo on conservative treatment according to the frequency of application

	One application weekly (n=8)	Three applications weekly (n=7)
Response dose (U/kg per week)	140	135
Response interval (weeks)	7.6	5.3
Correction dose (U/kg per week)	172	142
Correction interval (weeks)	15.6	8.6
Sustaining dose (U/kg per week)	171	90

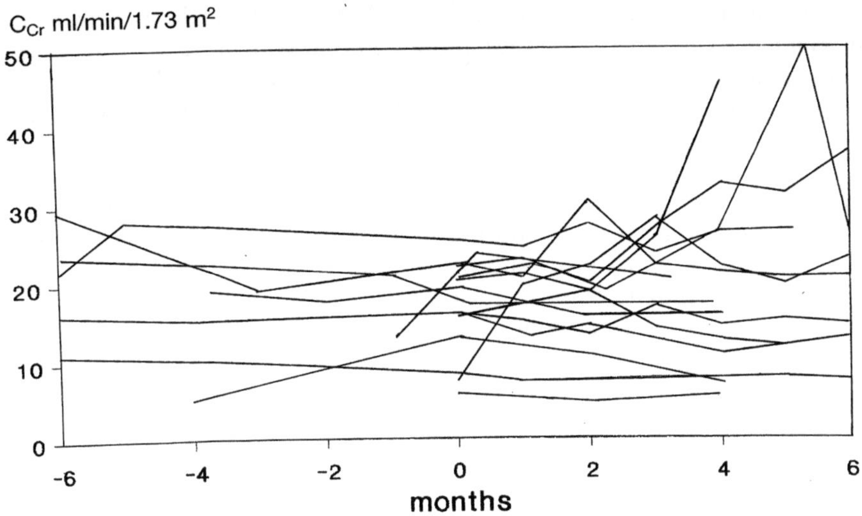

Fig. 1. Course of creatinine clearance (C_{Cr}) in 16 children with preterminal renal failure 6 months before and 6 months after start of rhEpo treatment

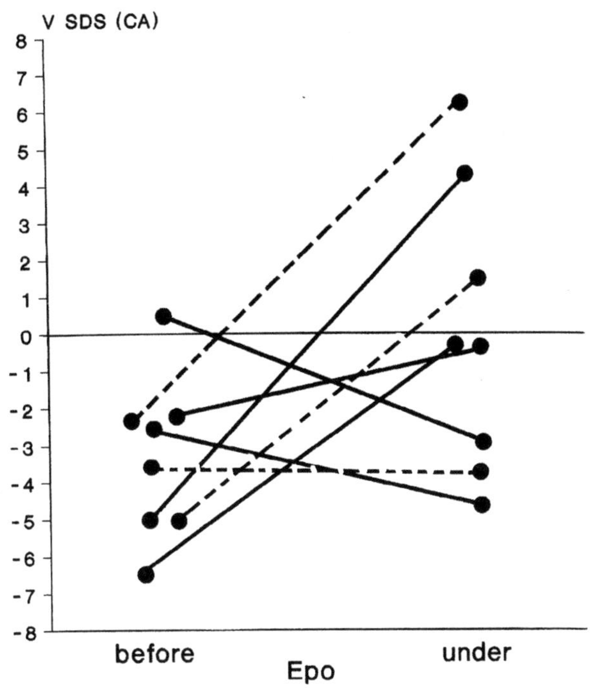

Fig. 2. Growth velocity (*V*) expressed as standard deviation score (*SDS*) for chronological age (*CA*) in eight prepubertal children with preterminal renal failure, comparing a period of 6 months before rhEpo to 6 months under rhEpo treatment. *Broken Lines*, children with an increase of glomerular filtration rate

rhEpo treatment are presented in Fig. 1. Most of the patients demonstrated no change in clearance. In three children there was an increase, but in none an inadequate decrease in renal function. The influence of rhEpo on growth velocity on CT is demonstrated in Fig. 2 with respect to the standard deviation score (SDS) of chronological age. Most of the patients analyzed (n=8) showed an increase in SDS (n=5) and only two patients a decrease in growth velocity (one patient showed no change). Only those patients were analyzed in whom a 6-month period without rhEpo could be compared with a 6-month period on rhEpo. Whereas before rhEpo treatment the mean SDS was around –3, it increased to a positive range on rhEpo treatment. Side effects from rhEpo treatment were lowest on CT, whereas in terminal renal failure either on CAPD or on HD the incidence of hypertension, hypercalemia, hyperphosphatemia and thrombocytosis was rather high (Table 3). In all patients a clear improvement in physical, mental, and psychological well-being was observed by parents and teachers, especially when the Hb concentration exceeded 12 g/dl.

Continuous Ambulatory Peritoneal Dialysis. Examining the rhEpo dosage in terms of the route of administration in children on CAPD, the response dose

Table 3. Side effects under Epo (percentages)

	CT	CAPD	HD
Hypertension (>antihypertensive drugs)	11	33	32
Hyperkalemia (>6.0 mmol/l)	5	39	22
Hyperphosphatemia (>2.5 mmol/l)	0	55	17
Thrombocytosis (>500 000/μl)	0	21	27

Table 4. Epo dosage according to route of administration in 33 children on CAPD

	Median	Range
Starting dose (U/kg per week)		
Subcutaneous	110	50–200
Intraperitoneal	270	75–600
Response dose (U/kg per week)		
Subcutaneous	125	50–250
Intraperitoneal	270	75–600
Response interval (weeks)		
Subcutaneous	9	2– 20
Intraperitoneal	7	3– 10
Sustaining dose (U/kg per week)		
Subcutaneous	175	90–400
Intraperitoneal	350	300–450

was twice as high for the intraperitoneal as for the subcutaneous route (Table 4). After a comparable response interval the same was true for the sustaining dose, which was 350 U/kg per week when administering rhEpo intraperitoneally in contrast to 175 U/kg per week after subcutaneous administration. The clinical effectiveness of an intraperitoneal Epo dose is demonstrated in Fig. 3, with a rapid increase in Hb concentration from 7 to 13 g/dl within a period of 12 weeks by a dosage of 200 U/kg three times/week without a significant change in blood pressure. The incidence of peritonitis per 12 treatment months was 2.0 for intraperitoneal and 1.95 for subcutaneous Epo administration. However, the general frequency of peritonitis was rather high, with only 11 out of 33 patients not achieving peritonitis under Epo. On CAPD treatment no significant change in growth velocity was detected in nine prepubertal children. The growth velocity remained low, at 2 cm per 6 months. During the observation period 15 of 33 CAPD children were successfully grafted. In all five patients with increased cytotoxic antibodies there was a decrease in the percentage panel reactivity from the start of rhEpo treatment to the time of transplantation. Graft failure occurred in only 2 of out of 15 patients (independent of rhEpo treatment).

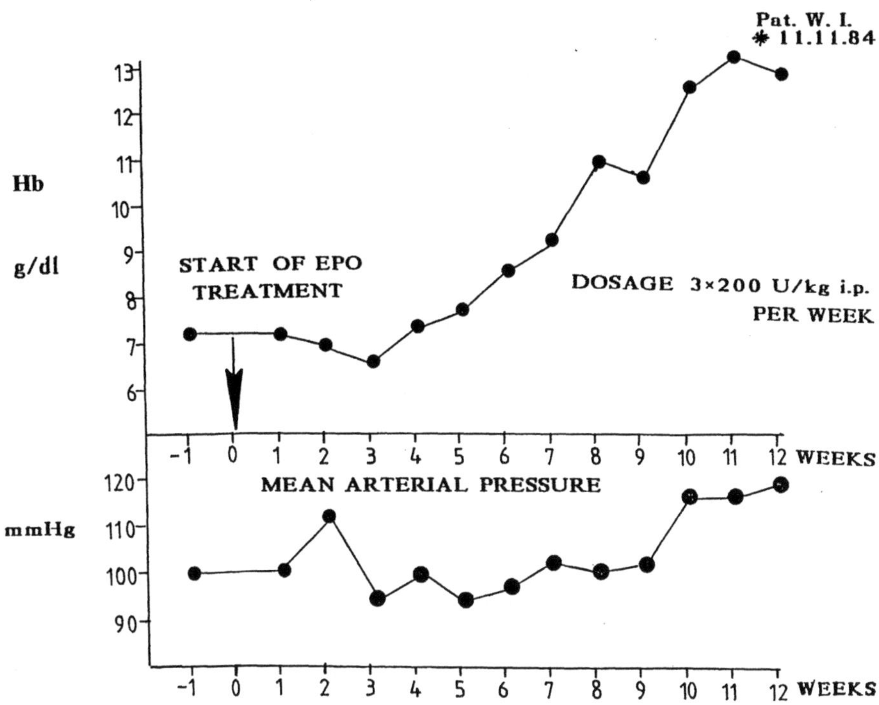

Fig. 3. Clinical effectiveness of intraperitoneal rhEpo administration demonstrated in a 6-year-old girl

Discussion

The experience with rhEpo treatment of renal anemia in the pediatric age group is rather limited up to now, in contrast to its use in adults. Previous observations in children on HD, however, did not reveal an age-dependent effect of rhEpo either from a pharmacokinetic or from a pharmacodynamic point of view. The percentage of side effects under rhEpo was also comparable between children and adults on HD [12, 21, 23, 27, 32].

On the other hand, the use of rhEpo in children with preterminal renal failure on CT has not been reported up to now. With respect to its easy handling and prolonged action we administered rhEpo via the subcutaneous route. However, in childhood the form of injection that is most favored is that which can be avoided. We therefore sought to determine whether one injection is as effective as administration three times per week – providing a constant cumulative dose. With careful randomization regarding the frequency of rhEpo administration we demonstrated that with the same starting dose of 90 U/kg per week the response interval was prolonged by more than 2 weeks in the one-injection group, leading to a correction interval which was nearly twice as long as in the three-injection-group. Results were similar concerning the sustaining dose; this was about 100% higher with one rhEpo administration than with three. Although from these results the three-injection strategy should be favored, at least from the economic point of view, one must keep in mind that even a one-injection therapy is feasible in children with preterminal renal failure, which needs only a higher, i.e., double dose. Up to now we have not discovered any differences in side effects between these two regimens, but the study is still going on.

Taking the two regimens together, the response dose was comparable to that found on CAPD, but lower than that on HD. With similar response intervals, the sustaining dose was smaller in children with preterminal renal failure than those on dialysis, although the mean sustaining Hb level achieved was higher than that on CAPD or HD. These findings can easily be interpreted as a lesser involvement of the pathomechanisms leading to renal anemia, namely erythropoietin synthesis, hemolysis, and blood loss during this early stage of renal failure [18]. However, the only patient not achieving a response was on CT. This nonresponse originated not in the severity of renal failure but in the overshadowing of renal anemia by anemia of chronic inflammation, aggravated additionally by more or less constant bowel bleedings, so that in spite of a marked rhEpo-induced reticulocytosis blood transfusions remained inevitable. To what extend the previous cytotoxic and radiation treatment might have contributed to the nonresponse remains questionable since another patient in whom a comparable therapy – for the same underlying disease – did not lead to chronic inflammation and bleeding responded well, in the same way as did all the other patients.

Concerning the target level we tried to imitate physiologic conditions (Hb 12.0–14.0 g/dl) not only to diminish but actually to overcome the anemia. The choice of a rather high Hb target range was supported by the fact that in preterminal renal failure, as under peritoneal dialysis, there is neither a risk of fistula clotting nor of deterioration of dialyzer clearance due to an increase in blood viscosity when the Hb concentration normalizes.

The course of creatinine clearance in our children, followed over a period of 6 months, gave no evidence of an inappropriate decrease in renal function, and they are well in accordance with previous observations in adults (Table 5). The increase in glomerular filtration rate in two children with nephronophthisis and one child with renal dysplasia needs further elucidation

Table 5. Erythropoietin treatment in adult predialysis patients

Investi-gators	Year of study	*n*	Dos (U/kg per week)	Frequency of adminis-tration (times per week)	Duration (months)	Hematocrit% Be-fore	Hematocrit% After	Comments
Brown et al. [3]	1988	5	150 i.v.	3	12	29	39	SCr un-changed, BP constant
Stone et al. [35]	1988	8	50–150 i.v.	3	1.5	32	40	SCr un-changed, BP constant
Teehan et al. [36]	1988	11	50–150 i.v.	3	3	28	>35	SCr un-changed, BP constant
Eschbach et al. [6]	1989	17	50–100 i.v./s.c.	3	6	27	37	1/SCr un-changed, BP(↑)
Kleinmann et al. [13]	1989	14	100 s.c.	3	3	28	36	1/SCr un-changed, BP as in con-trols
Lim et al. [14]	1989	10	50–150 i.v.	3	12	28	36	1/SCr un-changed, BP constant
Hori et al. [9]	1989	7	4500 (total dose)	3	1		> 6.5	CiN/PAH unchanged, BP constant
Frenken et al. [8]	1990	8	–	–	2	24	39	CiN un-changed, BP constant
Schwartz et al. [31]	1990	9	–	–	12	29	39	RBC survival prolonged
Traindl et al. [37]	1990	19	150 s.c.	2	2	20	26	No SCr change

SCr, serum creatinine; BP, blood pressure; RBC, red blood cells; CiN, inulin clearance; CPAH, para-aminohippuric acid clearance before it can be attributed to improved renal tissue oxygenation due to rhEpo.

In this context it appears highly interesting that the SDS of growth velocity, which is the most reliable indicator of changes in body length in childhood, increased significantly in the more or less growth-retarded children when treated with rhEpo. These data are in contrast to those obtained previously in uremic rats [19] and in children with terminal renal failure [25, 33] in whom no catch-up growth has been detected so far, as was demonstrated in at least five of the children whom we analyzed. The reason for the positive effect of rhEpo on growth velocity in children on preterminal renal failure may be due to the early start of rhEpo treatment, combined with a physiologic Hb sustaining level. As the highest catch-up growth could be achieved in an infant i.e., during a period in which the uremic individual usually loses the highest amount of its growth capacity, rhEpo may in the future play a central role in adequately treating children with congenital renal failure.

With respect to the lower incidence of side effects in rhEpo-treated children with preterminal renal failure, its routine use in this group of patients seems rather promising and should become part of the supportive treatment. Finally, it remains unknown – in the lack of a control group – whether an increased need for antihypertensive drugs and intermittent hyperkalemia would have developed without rhEpo administration. It should be noted in this context, however, that hypertension as well as hyperkalemia is typical in uremic rats treated with rhEpo, in contrast to controls – even in those who were pair-fed [19, 22]. Whether an altered food intake was responsible for hyperkalemia in our patients cannot yet be definitely determined.

In children on CAPD there is the chance of totally avoiding any injection by adding rhEpo to the dialysis fluid, which is of great practical advantage especially in small children. However, the use of this method has not become established in centers treating adult patients by CAPD since the pain of a subcutaneous injection can be ignored in adults. The risk of peritonitis has been overestimated, and the pharmacokinetic calculations suggest a very low bioavailability [4, 5, 15, 17]. Indeed, comparing the dosage in our CAPD children with respect to the route of administration, the amount of rhEpo needed for intraperitoneal administration was about twice as high as that for the subcutaneous. This was true as well for the response as for the sustaining dose. This clinical observation is well in line with our pharmacokinetic data presented elsewhere (Brune et al., this volume) [25], indicating an rhEpo bioavailability of 10% after intraperitoneal and one of 25% after subcutaneous administration in CAPD children. These relatively positive results compared to experience in adult CAPD patients [2, 5, 7, 11] might be due to lower filling volumes and longer dwell times (of the overnight bag in CAPD and the last bag in CCPD) of the peritoneal dialysis fluid in pediatric patients. The dose needed for the intraperitoneal route was comparable to that used in an intravenous regimen once a week in a smaller pediatric CAPD study [26].

Although we initially had the impression of an increased risk of peritonitis in association with intraperitoneal rhEpo treatment, leading to a switch to the subcutaneous route, the retrospective analysis of our data revealed no difference in the incidence of peritonitis, which is in accordance with recent reports on long-term treatment results in children [30]. However, the incidence of peritonitis seems rather high, and the question arises whether this observation may be due to an immunomodulating effect of rhEpo on other cytokines, thus inhibiting the immunologic capacity of peritoneal macrophages.

In contrast to children on CT the use of rhEpo did not induce further amelioration of growth velocity in children with terminal renal failure on CAPD. One must assume that at this late stage of the disease no further influence on body growth by an increase in the Hb concentration is possible. From this point of view the start of Epo treatment with the beginning of dialysis seems too late for children with a delayed body growth.

Finally, another positive effect of rhEpo must be pointed out which is of special interest in childhood; the reduction in cytotoxic antibodies due to the fact that blood transfusions could be almost totally avoided. Since renal transplantation is the only final therapeutic goal in children with chronic renal failure, rhEpo could contribute substantially to a faster grafting in these patients, which is interesting insofar as no negative effect of rhEpo on graft survival has been detected up to now [27].

Unfortunately, we were unable in this study to test the amelioration of rhEpo on ultrafiltration, which has been reported in the literature [34]. The positive effects of rhEpo treatment include an improvement in overall quality of life of our young patients, which cannot easily be measured, especially over the whole period of childhood. For those responsible for the care of the patients it suffices that on rhEpo all children recovered laughing – not only before, but even during dialysis treatment. The difficult and hard way to transplantation has thus been made a great deal easier for our young candidates by a drug that contributes to more physiologic conditions in renal failure.

References

1. Bargman J, Breborowicz A, Rodela H et al. (1988) Intraperitoneal administration of recombinant human erythropoietin in uremic animals. Perit Dial Int 8:249–252
2. Boelaert J, Schurgers M, Matthys E et al. (1989) Comparative pharmacokinetics of recombinant human erythropoietin administered by the intravenous, subcutaneous and intraperitoneal routes in CAPD patients. Perit Dial Int 9:95–98
3. Brown CD, Delano BG, Lundin AP et al. (1988) Treatment of uremic and azotemic anemic patients with human recombinant erythropoietin raises whole blood viscosity (WBV) appropriately with respect to hematocrit. Clin Res 36:515 A
4. Cheng KP, Cy C, Chan MK, Yu L, Fang GX, Wei D (1991) Correction of anemia in patients on continuous ambulatory peritoneal dialysis with subcutaneous recombinant erythropoietin twice a week: a long-term study. Clin Nephrol 35:207–212

5. Coppens PJW, Frenken LAM, Struijk DG, Tiggeler RGWL, Krediet RT, Koene RAP (1990) Comparative study of intraperitoneal and subcutaneous administration of recombinant human erythropoietin (rhuEPO) in patients treated with continuous ambulatory peritoneal dialysis (CAPD). Kidney Int 37:326
6. Eschbach JW, Kelly MR, Haley NR et al. (1989) Treatment of the anemia of progressive renal failure with recombinant human erythropoietin. N Engl J Med 321:158–163
7. Frenken LAM, Coppens PJW, Tiggeler RGWL, Koene RAP (1988) Intraperitoneal erythropoietin. Lancet 2:1495
8. Frenken LAM, Wetzels JFM, Sluter HE et al. (1990) Renal hemodynamics and effects of captopril in pre-dialysis patients treated with recombinant human erythropoietin. Kidney Int 37:551
9. Hori K, Kugmagai H, Onoyama K et al. (1989) Effects of erythropoietin on anemia and hemodynamics in chronic renal failure without dialysis treatment. Kidney Int 35:227
10. Hughes RT, Cotes PM, Pippard MJ et al. (1990) Subcutaneous administration of recombinant human erythropoietin to subjects on continuous ambulatory peritoneal dialysis: an erythrokinetic assessment. Br J Haematol 75:268–273
11. Icardi A, Paoletti E, Molinelli G (1990) Efficacy of recombinant erythropoietin after subcutaneous or intraperitoneal administration patients on CAPD. Perit Dial Int 10:51–82
12. Kitagawa T, Ito K, Komatsu Y et al. (1988) The clinical studies of recombinant human erythropoietin in children with renal anemia. J Pediatr (Japan) 41:3251–3262
13. Kleinmann K, Schweitzer S (1989) The use of recombinant human erythropoietin in the correction of anemia in predialysis patients and its effect on renal function: a double-blind, placebo-controlled trial. Am J Kidney Dis 14:496–506
14. Lim VS, DeGodwin RL, Zavala D et al. (1989) Recombinant human erythropoietin treatment in predialysis patients: a double-blind placebo-controlled study. Ann Intern Med 110:108–114
15. Liu SF, Chung WWM, Leung CB, et al. (1990) Pharmacokinetics and pharmacodynamics of subcutaneous and intraperitoneal administration of recombinant human erythropoietin in patients on continuous ambulatory peritoneal dialysis. Clin Nephrol 33:47–51
16. Macdougall IC, Roberts DE, Neubert P et al. (1989) Pharmacokinetics of recombinant human erythropoietin in patients on continuous ambulatory peritoneal dialysis. Lancet 1:425–427
17. Miranda B, Selgas R, Rinon C, Fernandez-Zamorano A et al. (1990) Effects of h-r erythropoietin (EPO) through two different applications on CAPD patients
18. Müller-Wiefel DE (1982) Renale Anämie im Kindesalter – Untersuchungen zur Pathogenese und Kompensation. Thieme, Stuttgart
19. Müller-Wiefel DE, Mehls O, Schwehm P et al. (1987) Correction of renal anemia by recombinant human erythropoietin does not improve growth in experimental uremia. Nephrol Dial Transplant 2:397
20. Müller-Wiefel DE et al. (1988) Specific problems of renal anemia in childhood. Contrib Nephrol 66:71–84
21. Müller-Wiefel DE, Feist K, Twittenhoff W, Brune T (1988) Beseitigung der renalen Anämie mit rekombinantem humanem Erythropoietin (rhEPO). Monatsschr Kinderheilkd 137:514
22. Müller-Wiefel DE, Mehls O, Scigalla P, Schwehm P (1988) Effects of recombinant human erythropoietin (rhEPO) during correction of renal anemia (RA) in experimental uremia. Pediatr Res 24:274
23. Müller-Wiefel DE, Bosch A, Feist K, Bier V, Schärer K (1989) Erythropoietin treatment in pediatric patients with renal anemia: application forms, dosage and response. Pediatr Nephrol 3:C 69
24. Müller-Wiefel DE, Feist K, Bosch A (1990) Changes of hemoglobin-oxygen ($Hb\text{-}O_2$) affinity in pediatric patients under treatment with recombinant human erythropoietin (rhEPO). Blood Purif 8:109

25. Müller-Wiefel DE (1991) A review of a 3-year European experience with erythropoietin in pediatric peritoneal dialysis patients. 11th annual conference on peritoneal dialysis, 6–8 February, 1991, Nashville, Tennessee
26. Offner G, Hoyer PF, Latta K, Winkler L et al. (1990) One year's experience with recombinant erythropoietin in children undergoing continuous ambulatory or cycling peritoneal dialysis. Pediatr Nephrol 4:498–500
27. Rigden SPA, Montini G, Morris M, Clark KGA et al. (1990) Recombinant human erythropoietin therapy in children maintained by hemodialysis. Pediatr Nephrol:618–622
28. Schärer K, Müller-Wiefel DE (1982) Renal anemia in children, a review. Int J Pediatr Nephrol 3:193–198
29. Schärer K, Müller-Wiefel DE (1987) Hematological problems in renal insufficiency. In: Holliday MA, Barratt T, Vernier RL (eds) Pediatric nephrology. Williams and Wilkins, Baltimore, pp 805–813
30. Schröder CH (1991) Intraperitoneal administration of erythropoietin in children treated by continuous ambulatory peritoneal dialysis. 11th annual conference on peritoneal dialysis, 6–8 February, 1991, Nashville, Tennessee
31. Schwartz AB, Kelch B, Terzian L et al. (1990) One year of rHuEPO therapy prolongs RBC survival, increases marrow cellularity, iron utilization, reticulocytosis and Hct and may stabilize RBC membrane in chronic renal failure and uremia patients. Kidney Int 37:378
32. Scigalla P, Bonzel KE, Bulla M et al. (1989) Therapy of renal anemia with recombinant human erythropoietin in children with endstage renal disease. Contrib Nephrol 76:227–241
33. Sinai-Trieman L, Salusky IB, Fine RN (1989) Use of subcutaneous recombinant human erythropoietin in children undergoing continuous cycling peritoneal dialysis. J Pediatr 114:550–554
34. Steinhauer HB, Librich-Birkner I, Dreyling KW, Schollmeyer P (1989) Increased ultrafiltration after erythropoietin-induced correction of renal anemia in patients on continuous ambulatory peritoneal dialysis. Nephron 53:91–92
35. Stone WJ, Graber SE, Krantz SB et al. (1988) Treatment of the anemia of predialysis patients with recombinant human erythropoietin: a randomized, placebo-controlled trial. Am J Med Sci 296:171–179
36. Teehan B (1988) Hematologic and physiologic studies in predialysis patients receiving recombinant human erythropoietin. Dial Transpl 636–639
37. Traindl O, Franz M, Kovarik J (1990) Recombinant human erythropoietin in patients with chronic renal insufficiency in the predialysis period. Kidney Int 37:245

Bone Marrow Erythroid Precursor Cytoplasmic Ca^{2+} and 1,25-Dihydroxyvitamin D_3 Regulate the Response to Human Recombinant Erythropoietin in Uremia: In Vitro and In Vivo Studies

S. Carozzi, M.G. Nasini, A. Cantaluppi, M. Salit, and S. Lamperi

Introduction

Recent studies have shown that Ca^{2+} has a stimulatory effect as a second messenger or coupling factor in cell activation and secretion [1, 2]. In addition, 1,25-dihydroxyvitamin D_3 [1,25$(OH)_2D_3$] has been demonstrated to induce cell proliferation and maturation, most likely via a direct effect on the plasma membrane permeability to Ca^{2+} [3, 4]. In this regard, the Ca^{2+} concentration in the culture medium has recently been reported to condition the modulation of erythropoiesis by erythropoietin (EPO) [5].

To evaluate the influence of Ca^{2+} on erythropoiesis in uremia we measured the following parameters in some anemic patients with continuous ambulatory peritoneal dialysis (CAPD): in vitro, with the addition to the medium of increasing concentrations of Ca^{2+} alone or with 1,25$(OH)_2D_3$, bone marrow nonadherent mononuclear cell cytoplasmic Ca^{2+} concentration and the growth of bone marrow erythroid precursors burst- (BFU-E) and colony-forming units – erythroid (CFU-E); in vivo, bone marrow nonadherent mononuclear cell cytoplasmic Ca^{2+} concentration and erythroid colony growth before and after 24 weeks of recombinant human erythropoietin (rHuEPO) administration, and the duration of rHuEPO therapy necessary to correct anemia.

Patients and Methods

Patient Population and rHuEPO Schedule. Twelve anemic CAPD patients (six men, six women, mean age 45.6 ± 11.0 years) were studied before and after 24 weeks of rHuEPO therapy. The hormone dose was 50 U/kg b.w. for 2 weeks, 75 U/kg for 4 weeks, and subsequently 100 U/kg; rHuEPO was given three times a week as an intravenous bolus. Ten healthy subjects constituted the source of normal bone marrow cells.

In Vitro Studies. Before initiating rHuEPO therapy, bone marrow BFU-E and CFU-E development and bone marrow nonadherent cytoplasmic Ca^{2+} concentration were assayed in the presence of Ca^{2+} alone (0–5000 mM) and Ca^{2+} plus 1,25$(OH)_2D_3$ (Upjohn, Kalamazoo, USA)(0–10^{-6} *M*).

Pagel/Weiss/Jelkmann (Eds.)
Pathophysiology and Pharmacology of Erythropoietin

In Vivo Studies. BFU-E and CFU-E development and bone marrow nonadherent cytoplasmic Ca^{2+} concentration were assayed before and after 24 weeks of rHuEPO therapy, and the relationship was evaluated between this Ca^{2+} concentration and the time necessary to normalize hematological parameters (mean hematocrit values 36.1%±1.1%, mean hemoglobin values 12.2±1.2 g/dl, mean reticulocyte values 40.4±14.4 × 10^9/l).
Assay Methods. Nonadherent mononuclear bone marrow cells, were obtained from whole bone marrow cells (drawn by sternal aspiration) by centrifugation over Ficoll-Hypaque gradient followed by three adherence procedures to plastic petri dishes. BFU-E and CFU-E development was determined using the Iscove method by incubating 10^5 nonadherent bone marrow mononuclear cells in 35-mm culture plates with 0.8% methylcellulose, 30% fetal calf serum, 1% deionized bovine serum albumin with a total volume of 1 ml per plate [6]. Non adherent bone marrow mononuclear cell cytoplasmic Ca^{2+} concentration was determined by a Quin-2 fluorescent indicator as described by Tsien et al. [7].

Statistics. Values were expressed as mean ± standard error of the mean. Statistical significance was evaluated by Student's *t* test for paired observations and by the Wilcoxon matched-pair test. Relationships were analyzed by linear regression analysis. A *p* value less than 0.05 was considered statistically significant.

Results

The nonadherent bone marrow cell cytoplasmic Ca^{2+} concentration was lower in the anemic CAPD patients than in controls ($p<0.0005$). The addition of Ca^{2+} to the culture medium led to a dose-dependent increase in cytoplasmic Ca^{2+} which reached normal levels with 1000 μM ($p<0.0005$). This effect was potentiated by the addition of $1,25(OH)_2D_3$, such that with only 500 μM Ca^{2+} and 10^{-12} *M* vitamin D_3 a normal value was reached ($p<0.0005$). The BFU-E colony growth was also reduced in the anemic CAPD patients compared to controls. While the addition of Ca^{2+} led to a dose-dependent increase, this growth remained, however, below the normal range ($p<0.05$); vitamin D_3 potentiated this effect. CFU-E colony development behaved similarly, but it increased by more than six times to reach normal levels ($p<0.0005$).

The effects of 24 weeks of rHuEPO therapy were determined in the in vivo studies. Ca^{2+} concentration increased nearly three times, thus reaching normal values ($p<0.0005$). There was an increase of about 40% in the BFU-E colony growth increased by about 40% but it remained below the normal range ($p<0.05$), while CFU-E growth increased by about six times to reach normal levels ($p<0.0005$; Fig. 1). An inverse relationship was found to exist between the bone marrow mononuclear cell cytoplasmic Ca^{2+} concentration

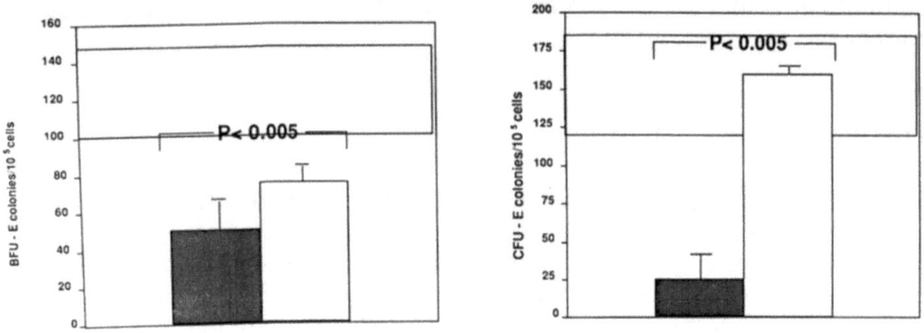

Fig. 1. Effects of rHuEPO therapy on BFU-E and CFU-E growth from nonadherent bone marrow mononuclear cells of 12 anemic CAPD patients. *Shaded bars*, before rHuEPO therapy. Hematocrit, 22.2% ±2.3%; hemoglobin, 6.5±0.89 g/dl; reticulocytes, 16.2±11.9 × 10^9/l. *Unshaded bars*, after 24 weeks of rHuEPO therapy. Hematocrit, 36.1% ±1.1%; hemoglobin, 12.2±1.2 g/dl; reticulocytes, 40.5±14.7 × 10^9/l

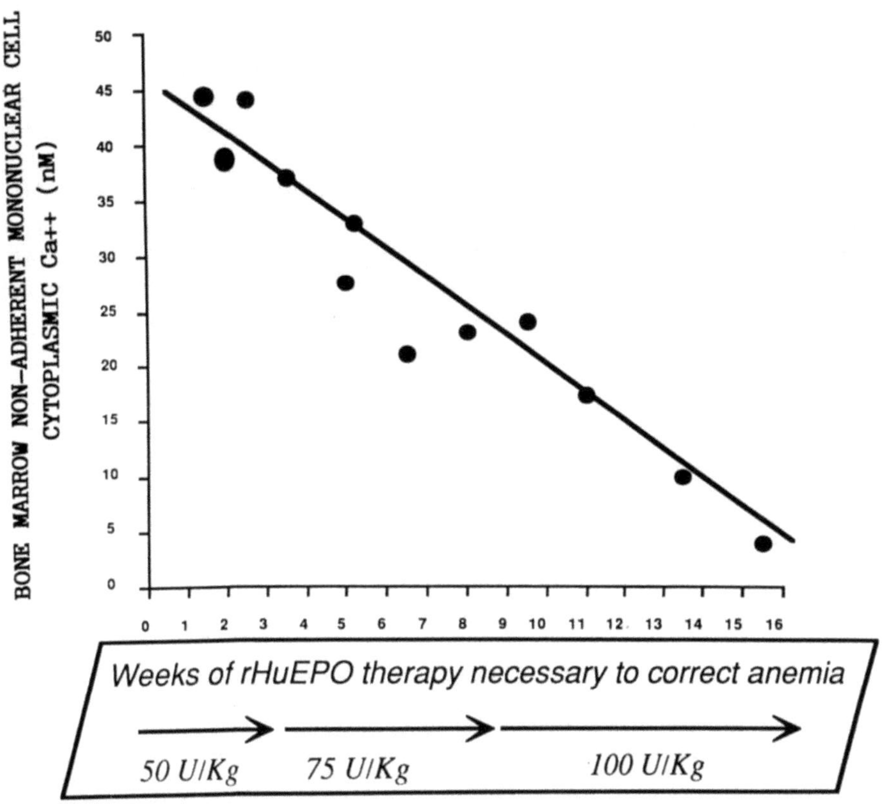

Fig. 2. Relationships between the weeks of rHuEPO therapy necessary to correct the anemia and the nonadherent bone marrow mononuclear cell cytoplasmic Ca^{2+} concentration in 12 anemic CAPD patients ($r=0.95$; $p<0.005$)

before treatment and the number of weeks of rHuEPO therapy necessary to correct the anemia (Fig. 2).

Discussion and Conclusions

Our studies show that in uremic patients before rHuEPO therapy the bone marrow nonadherent mononuclear cell cytoplasmic Ca^{2+} concentration and BFU-E and CFU-E growth were lower than in controls. A dose-dependent increase in these parameters followed the addition of Ca^{2+} to the medium, and vitamin D_3 potentiated this effect. rHuEPO therapy led to a normalization of bone marrow nonadherent mononuclear cell Ca^{2+} concentration and BFU-E and CFU-E growth. The Ca^{2+} concentration in these cells before treatment and the duration of therapy necessary to correct the anemia were inversely related.

For some time it has been known that in vitro increased RNA synthesis follows the addition of EPO to erythropoietic cells [8]. This occurs essentially with EPO-responsive erythropoietic cells which have specific membrane receptors. As soon as EPO binds to these receptors one of the first effects observed is an increased cell membrane uptake of Ca^{2+} [9]. This erythropoietic effect is similar to that seen in vitro when accessory hematologic cells are cultured in the presence of mitogens or ionophoric substances, in which case there is also an increased uptake of Ca^{2+} into the cytoplasmic pool. This event indicates Ca^{2+} as an early messenger of cellular activity, in that it leads to the stimulation of metabolic processes which involve certain cytoplasmic (such as calmodulin) and nuclear proteins [10]. It thus appears that cytoplasmic Ca^{2+} modifications induced by EPO represent an important step in the differentiation and maturation of erythroid precursors.

References

1. Tsien RY, Pozzan T, Rink TJ (1946) T-cell mitogens cause early changes in cytoplasmic free Ca^{2+} and membrane potential in lymphocytes. Nature 295:68–73
2. Metcalf JC, Pozzan T, Smith GA et al. (1980) A calcium hypothesis for the control of cell growth. Biochem Soc Symp 45:1–5
3. Bhalla AK, Amento EP, Clemens TL et al. (1983) Specific high affinity receptors for 1,25-dihydroxyvitamin D_3 in human peripheral blood mononuclear cells: presence in monocytes and induction in T-lymphocytes following activation. J Clin Endocrinol 57:1310–1314
4. Tsoukas CD, Provvedini D, Manolagas SC (1984) 1,25-dihydroxyvitamin: a novel immunoregulatory hormone. Science 224:1438–1442
5. Krantz SB, Sawyer ST, Kourny MJ et al. (1986) Use of purified erythropoietin responsive cells produced by the anemia strain of Friend virus to study the action of erythropoietin. In: Rich IN (ed) Molecular and cellular aspects of erythropoietin and erythropoiesis. Springer, Berlin Heidelberg New York, pp 89–93 (NATO ASI series H, vol 8)
6. Iscove NM, Sieber F, Winterhalter KH (1974) Erythroid colony formation in culture of mouse and human bone marrow: analysis of the requirement for erythropoietin by gel

filtration and affinity chromatography on agarose concanavalin A. J Cell Physiol 83:309–314

7. Tsien R, Pozzan T, Rink T (1982) Calcium homeostasis in intact lymphocytes: cytoplasmic free calcium monitored with a new intracellularly trapped fluorescent indicator. J Cell Biol 94:325–332
8. Goldwasser E (1981) Erythropoietin and red cell differentiation. In: Cunningham D, Goldwasser E, Watson H, Fox CF (eds) Control of cellular division and development, part A. Liss, New York, pp 487–492
9. Sawyer ST, Krantz SB (1984) Erythropoietin stimulated $^{45}Ca^{2+}$ uptake in Friend virus-infected erythroid cells. J Biol Chem 259:2769–2774
10. Chantler PD (1985) Calcium-dependent association of a protein complex with the lymphocyte plasma membrane: probable identity with calmodulin-calcineurin. J Cell Biol 101:207–213

Dyserythropoietic Abnormalities in Hemodialysis Patients on Long-Term Therapy with Human Recombinant Erythropoietin

S. Lamperi, S. Carozzi, M. G. Nasini, A. Cantaluppi, and M. Salit

Introduction

Erythropoietin (EPO) can be considered the primary regulator of erythropoiesis in physiological conditions [1, 2]. However, much is still unknown about the effects which situations of stress have on bone marrow cells, as EPO is part of a mechanism that operates under emergency situations. In these conditions the persistently high levels of the hormone may induce a further shortening of the medullary transit time, with a consequently inadequate stem cell maturation which leads to cellular defects [3]. This condition, which can be reproduced in animals by the administration of high doses of EPO, has become of great interest since pure recombinant human EPO (rHuEPO) has been used for the treatment of patients with severe anemia in chronic renal failure. In fact, a derangement of erythroid progenitor cell kinetics has been reported in chronic hemodialysis patients who undergo rHuEPO administration at an elevated dose rate for several months [4].

The aim of this study was to investigate the morphologic features of erythroid cells in the course of the stress of EPO stimulation – in uremic patients who had been treated for long periods with elevated hormone doses due to resistance to treatment.

Patients and Methods

Ten uremic patients (four women and six men, mean age 52.3 ± 4.2 years) on chronic, thrice weekly hemodialysis and treated with rHuEPO for severe anemia (mean hematocrit 20.1% ± 2.1%, hemoglobin 5.9 ± 0.2 g/dl) were studied. None of the patients were anephric, had other causes of anemia besides the uremia, had recently been transfused, or showed signs of iron overload. Ten healthy volunteers constituted the control group.

The rHuEPO was given thrice weekly as an intravenous bolus at the end of each dialysis session in progressively increasing doses (from 102.5 ± 24.6 to 554.5 ± 44.3 U/kg b.w. per week) according to response and tolerance, for 16 months.

Hematocrit and hemoglobin levels, reticulocyte count, and red blood cell mean corpuscular volume were determined weekly by standard methods.

Pagel/Weiss/Jelkmann (Eds.)
Pathophysiology and Pharmacology
of Erythropoietin

Red blood cell survival was evaluated before and after 16 months of rHuEPO therapy, by chromium-51 labeling [5]. We assayed the in vitro development of bone marrow erythroid burst-forming units (BFU-E) and erythroid colony-forming units (CFU-E) from bone marrow nonadherent mononuclear cells obtained by aspiration from the iliac crest and isolated from whole bone marrow cells by centrifugation on Ficoll-Hypaque gradient and subsequent adhesion to plastic petri dishes. This was performed before and after 2 weeks and 4, 8, 12, and 16 months of rHuEPO therapy using a serum-free methylcellulose culture technique [6]. Reticulocyte and bone marrow erythroblast morphology was evaluated, before and after 16 months of rHuEPO therapy by light and electron microscopy [7]. Bone marrow cell iron deposits were identified before and after 16 months of rHuEPO therapy by Perls' reaction.

Values (mean ± standard error) were analyzed by Student's *t* test for paired and unpaired data and the Wilcoxon matched-pairs test. A *p* value less than 0.05 was considered statistically significant.

Results

In seven hemodialysis patients treated for 4 months with standard doses of rHuEPO (102–272 U/kg per week) nonsignificant improvements were seen in both reticulocyte count and hemoglobin levels. The rHuEPO dose was therefore progressively increased. By the 5th month of therapy, when 310 U/kg rHuEPO per week was given, a transient rise (over the normal range) of reticulocyte count was seen. Thereafter, a decrease was seen starting from the 6th month. At the same time, hemoglobin levels did not improve significantly. The rHuEPO dose was therefore progressively increased to 554 U/kg per week by the 16th month. A dose-dependent rise in hemoglobin levels was therefore seen while the reticulocyte count remained stable within the normal range. We termed these patients EPO slow responders (Fig. 1a).

In three patients treated for an equal period of time with identical EPO doses, a different behavior of these hematological parameters was observed. By the 4th week of therapy, when 272 U/kg rHuEPO per week was given, a rise in both reticulocyte count and hemoglobin levels was seen. Thereafter, although the EPO dose was unchanged, both reticulocyte count and hemoglobin concentration decreased. The hormone dose was therefore progressively increased to 554 U/kg per week. The reticulocyte count, however, remained below the normal range, as did hemoglobin levels. We termed these patients EPO nonresponders (Fig. 1b).

In the seven EPO slow-responder hemodialysis patients no significant increases in bone marrow BFU-E or CFU-E development were seen in the first 3 months of EPO therapy when the hormone dose given ranged from 102 to 272 U/kg per week. By the 4th month, when EPO was increased to 340 U/kg per week a noticeable rise (over the normal range) in CFU-E

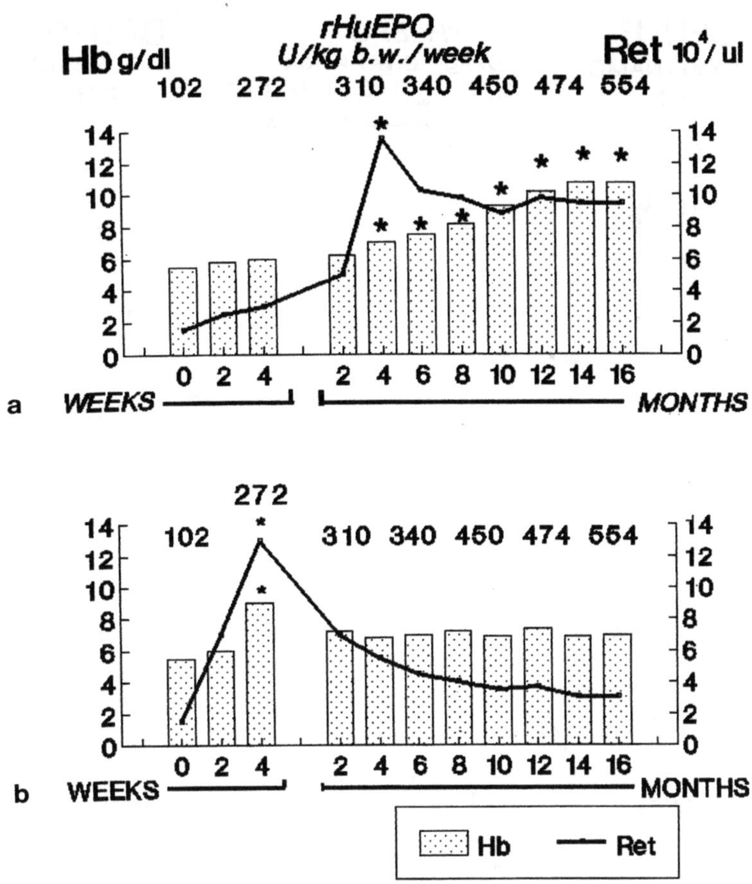

Fig. 1a, b. Hemoglobin (*Hb*) levels and reticulocyte (*Ret*) count in hemodialysis patients treated for 16 months with intravenous rHuEPO. **a** Seven slow responders to the hormone. **b** Three nonresponders to the hormone. * $p<0.05$, in comparison to basal values

development was observed. CFU-E colony growth was unchanged after a further increase in the dose of EPO to 554 U/kg per week. By the 4th month of EPO therapy BFU-E growth had also increased. The values observed, however, never exceeded the normal range. Thereafter, in spite of the dose of EPO being progressively increased, a drop in BFU-E growth (at the lower normal limit) was detected (Fig. 2a).

In the three EPO nonresponder hemodialysis patients a marked increase in CFU-E development (over the normal range) had already been was seen at the 2nd week of therapy; thereafter, despite the progressive rise in EPO dose, CFU-E growth decreased to values observed before treatment. In the same patients BFU-E development increased progressively starting from the 2nd month of EPO therapy and continuing through the period of observation, so that by the 16th month, values exceeded the normal range (Fig. 2b).

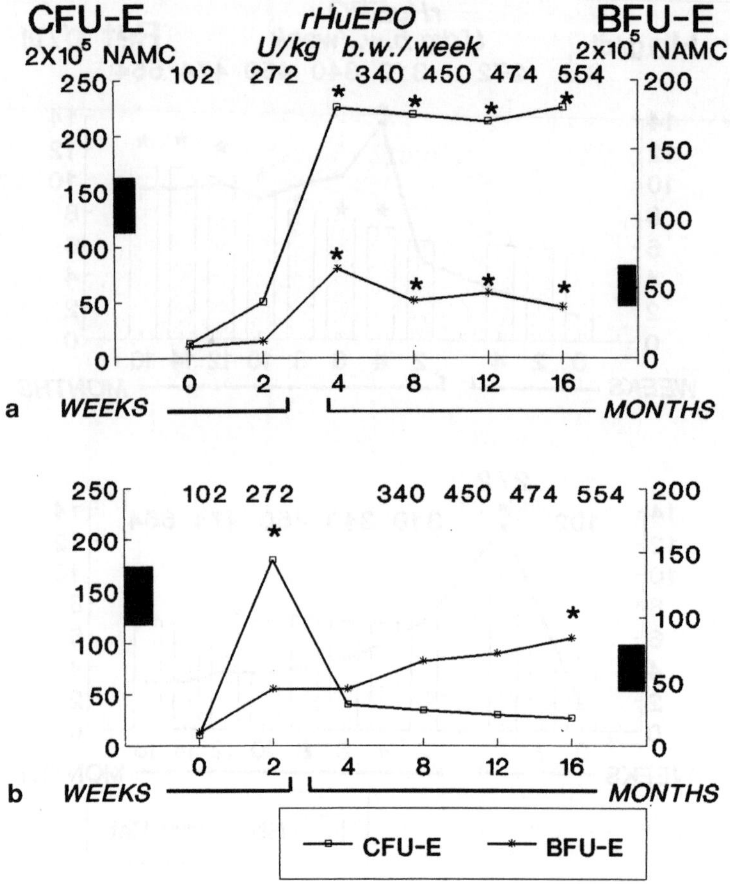

Fig. 2a, b. BFU-E and CFU-E development from bone marrow nonadherent mononuclear cells (*NAMC*; colonies per 2 × 10^5 NAMC plated) in hemodialysis patients treated for 16 months with intravenous rHuEPO. **a** Seven slow-responders to the hormone. **b** Three nonresponders to the hormone. *Dark areas*, normal range; * $p<0.05$, in comparison to basal values

In all the patients studied, before EPO therapy, the mean red cell corpuscular volume was within the normal range (88.2 ± 1.2 μm^3 in the uremic group, 87.9 ± 1.8 μm^3 in the control group). During the various months of therapy it remained stable in the seven EPO slow-responder patients to EPO (88.9 ± 1.1 μm^3), but it increased significantly in the three nonresponder patients (119.6 ± 3.7 μm^3).

Before EPO therapy the erythrocyte life span, expressed as $t_{1/2}$ of $_{51}Cr$ was slightly below the normal range (23.2 ± 1.2 days in the uremic group, 27.2 ± 3.4 in the control group). By the 16th month of EPO therapy it was unchanged in the seven EPO slow-responder patients, but it had significantly decreased in the three EPO nonresponder patients (17.1 ± 2.2 days).

In the three nonresponders to EPO the electron microscopy studies showed the presence in the bone marrow of giant cells, occasionally megaloblastic in appearance, with irregular or ragged cell borders and asyncronous nuclear and cytoplasmic maturation, bi- or multinucleated forms, cytoplasmic clefts, rents or vacuoles, an increase in type II and III sideroblasts and iron-laden macrophages. On the other hand, in the seven EPO slow-responder patients, no alterations in bone marrow erythroblast morphology were seen.

Discussion and Conclusions

Studies of anemic uremic patients treated with intravenous rHuEPO have shown that both bone marrow BFU-E and CFU-E are capable of responding, even if the CFU-E should be considered the real target of the hormone [8]. However, the consequences of the use of nonphysiological rHuEPO doses and of a persistent bone marrow cell hyperstimulation must yet be ascertained. During prolonged treatment with rHuEPO some authors found an unexpected depletion of the BFU-E compartment. This behavior has been correlated with an early expansion of the BFU-E compartment (stimulated by rHuEPO), after which a depletion followed as the cells matured and were insufficiently replaced from the pluripotential stem cell pool [9, 10]. The failure of the CFU-E number to decrease under the same conditions may be attributed to an increased entry of cells from the BFU-E compartment, thus balancing the rHuEPO-induced CFU-E maturation. This could reflect a higher proliferative rate in this compartement or a greater degree of self-renewal [11].

Although these data are still being discussed, an unbalanced cycling in the early and late erythroid cell line indicates that a number of disorders in erythroid precursor cell development are caused by prolonged treatment with rHuEPO [12]. Our data suggest that these disorders are linked with kinetic and morphologic alterations of the same cells. The abnormalities of the normoblasts which we found with electron microscopy studies are similar to those seen in a variety of diserythropoietic disorders and may indicate an aspecific mistake in the differentiation and maturation of red blood cell precursors [13].

It has been observed in animals that an abnormally rapid red blood cell maturation in response to intense stimulation (as the result of a shortened medullary transit time) leads to defective cell formation, which mainly becomes macrocytic [14]. It has also been observed that the size of reticulocytes is related to the intensity of erythropoietic stimulation, and that the larger reticulocytes mature into adult macrocytes, while the macrocyte response to the stress of erythropoietic hyperstimulation represents an altered pattern in red blood cell production [15]. This is analogous to our observation in bone marrow and peripheral red blood cells of uremic patients treated with high rHuEPO doses, suggesting that the more rapid maturation induced by skipped divisions of erythroid cells in response to rHuEPO-induced hyperstimulation leads to ineffective erythropoiesis and a shortened erythrocyte life span. The consequence is the need for ever-increasing doses of rHuEPO, with a progressive bone marrow erythroid cell derangement.

Therefore, caution is necessary when utilizing high doses of rHuEPO for long periods, especially in those few uremic patients who show a particular resistance to this therapy.

References

1. Erslev AJ, Caro J (1983) Pathophysiology of erythropoietin. In: Dunn C (ed) Current concepts of erythropoiesis. Wiley, New York, pp 1–19
2. Goldwasser E (1984) Erythropoietin and its mode of action. Blood Cells 10:147–162
3. Stohlman F, Brecher G, Moores RR (1962) Humoral regulation of erythropoiesis. VIII. The kinetics of red cell production and the effect of erythropoietin. In: Jacobson LO, Doyle M (eds) Erythropoietin. Grune and Stratton, New York, pp 162–172
4. Reid CDL, Fidler J, Oliver DO, Cotes MP, Piffard MJ, Wineals CG (1988) Erythroid progenitor cell kinetics in chronic hemodialysis patients responding to treatment with recombinant human erythropoietin. Br J Haematol 70:375–380
5. Stohlman F Jr (1961) Humoral regulation of erythropoiesis. VII. Shortened survival of erythrocytes produced by erythropoietin or severe anemia. Proc Soc Exp Biol Med 107:884–887
6. Casadevall N, Vainchencher W, Lacombe C (1982) Erythroid progenitors in polycythemia vera: demonstration of their hypersensitivity to erythropoietin using serum free cultures. Blood 59:447–451
7. Fresco R (1981) Electron microscopy in the diagnosis of bone marrow disorders of the erythroid series. Semin Haematol 18:279–291
8. Gregory CJ, Eaves AC (1978) Three stages of erythropoietic progenitor cell differentiation distinguished by the number of physical and biological property. Blood 51:527–537
9. Iscove N, Sueber F, Wintherhalter KH (1973) Erythroid colony formation in culture of mouse and human bone marrow: analysis of the requirement for erythropoietin by gel filtration and affinity chromatography on Sepharose concanavalin A1. J Cell Physiol 83:309–320
10. Dessypris EN, Graber SE, Stone WJ (1987) Effects of recombinant erythropoietin on human marrow hematopoietic progenitors in vivo. Blood 70 (Suppl 1):132
11. Peschle C, Cillo C, Migliaccio G, Lettieri F (1980) Fluctuations of BFU-E and CFU-E cycling after erythroid perturbation: correlation with variations of pool size. Exp Hematol 8:96–102
12. Hara H, Ogawa H (1988) Erythropoietin precursors in mice under erythropoietic stimulation and suppression. Exp Hematol 5:141–145
13. Lewis SM, Nelson DA, Pitcher CS (1972) Clinical and ultrastructural aspects of congenital dyserythropoietic anemia type I. J Haematol 23:113–119
14. Brecher G, Stohlman F (1962) The macrocytic response to erythropoietin. In: Jacobson LO, Doyle M (eds) Erythropoiesis. Grune and Stratton, New York, pp 216–221
15. Heimpel H, Fortiza-Vila J, Queisser W (1971) Electron and light microscopy study of the erythroblast of patients with congenital dyserythropoietic anemia. Blood 37:299–310

Cardiological Findings in Patients with End-Stage Renal Failure Under Treatment with Erythropoietin

G. Schütterle, W. Kramer, R. Schäfer, J. Kaufmann, and V. Wizemann

The extent of left ventricular myocardial hypertrophy is generally considered to be a predictor of cardiac morbidity and mortality (Devereux et al. 1984; Levy et al. 1988, 1990). The evolution of left ventricular hypertrophy has been causally related to coronary artery disease (CAD), arterial hypertension, and increased cardiac output (Devereux 1990; Messerli and Schmieder 1990; Koren et al. 1991). This is apparently also the case with patients with end-stage renal disease undergoing chronic intermittent hemodialysis treatment (Bullock et al. 1984; Kramer et al. 1986a,b; Harnett et al. 1988; Silberberg et al. 1989). Whether secondary hyperparathyroidism significantly contributes to myocardial fibrosis in patients on hemo- or peritoneal dialysis does, however, appear doubtful at the present time (London et al. 1987a,b; Mall et al. 1990; Hüting et al. 1991a,b).

In the absence of CAD and hypertension the increase oft left ventricular muscle mass (LVMM) appears to be due to increased cardiac output. Increased work load is imposed upon the left ventricle by an uncontrolled volume and weight gain in the interdialytic periods and by renal anemia (independent of blood transfusions) (Löw et al. 1989; Silberberg et al. 1989; Klaus et al. 1991).

The study presented here was designed to clarify whether partial correction of renal anemia by administration of recombinant human erythropoietin is able to induce regression of left ventricular muscle mass and improve tolerance to exercise-induced ischemia in patients with (group B, $n=7$) and without (group A, $n=28$) CAD. The presence or absence of CAD was established by coronary angiography: The incidence of stenoses of major coronary vessels was 78 ± 16% in group B.

Patients were selected in the following way: echocardiography was performed between dialyses in 131 patients; 74 of these fulfilled the criteria of the American Society of Echocardiography (Sahn et al. 1978) and 46 patients were excluded because of CAD, hypertension, valvular heart disease, or diabetes mellitus. The 28 patients (medium age 45 ± 12 years) allocated to group A fulfilled the following entering criteria: LVMM, 125 g/m^2; normal left ventricular function; ejection fraction > 55%; anemia (hematocrit < 25%, no transfusions).

There was no correlation between LVMM and the serum levels of parathyroid hormone and alkaline phospatase. In patients with renal insufficiency LVMM was significantly higher as compared to a group of

Pagel/Weiss/Jelkmann (Eds.)
Pathophysiology and Pharmacology
of Erythropoietin

apparently healthy individuals (93±7 g/m² versus 198±34.7 g/m², $p<0.001$). An inverse relationship could be established between hematocrit and LVMM on the one hand, and hematocrit and left ventricular end-diastolic diameter (LVEDD) on the other.

The parameters measured and calculated before and during erythropoietin treatment in group A patients are presented in tables 1 and 2. The target hematocrit value of 35% was achieved after 4 months of treatment: At this point of time there was a tendency for LVEDD, left ventricular end-systolic diameter (LVESD), LVMM index, heart rate, and cardiac output to be decreased, whereas the left ventricular ejection time increased.

Table 1. Patients' characteristics before and during rhEPO treatment[a]

Parameter		Controls[b]	rhEPO (4 mths)[b]	rhEPO (16 mths)[b]
Hematocrit	(%)	21.4 ± 2.4	36.5 ± 1.5*	35.6 ± 0.7
Hemoglobin	(g/ml)	7.1 ± 0.8	11.4 ± 0.5*	11.1 ± 0.3
Body weight	(kg)	64.0 ± 8.2	63.9 ± 8.3	63.8 ± 8.4
Body surface area	(m²)	1.72± 1.5	1.72± 1.4	1.72± 1.4
Mean arterial pressure	(mmHg)	93.2 ±12.9	92.2 ±11.9	93.0 ±11.3
Heart rate	(beats/min)	80.5 ±12.2	72.9 ±11.1*	73.1 ± 8.3

rhEPO, recombinant human erythropoietin; * $p<0.01$
[a] Group A, $n=28$.
[b] Mean values ± SD.

Table 2. Echocardiographic results during treatment with rhEPO[a]

Parameter		Controls[b]	rhEPO (4 mths)[b]	rhEPO (16 mths)[b]
LVEDO	(mm)	52.6 ± 3.4	49.6 ± 3.4	47.9 ± 2.9**
LVESD	(mm)	32.6 ± 3.5	31.4 ± 3.5	30.4 ± 3.1*
LVPW	(mm)	12.8 ± 0.8	12.3 ± 0.8	12.1 ± 0.9
IVS	(mm)	13.2 ± 1.4	12.8 ± 1.4	12.8 ± 1.1
LVMM index	(g/m²)	198.9 ±34.7	172.6 ±32.6	160.2 ±27.6**
Ejection fraction	(%)	67.5 ± 5.4	66.5 ± 5.0	66.2 ± 4.9
Heart rate	(beats/min)	80.5 ±12.2	72.9 ±11.1	73.1 ± 8.3**
Stroke volume	(ml)	90.3 ±12.9	77.3 ± 9.9	70.9 ± 8.6**
Cardiac output	(l/min)	7.22± 1.25	5.60± 0.87	5.17± 0.73**
LVET	(msec)	280 ±18	290 ±20	290 ±17*

rhEPO, recombinant human erythropoietin; LVEDD, left ventricular end-diastolic diameter; LVESD, left ventricular end-systolic diameter; LVPW, left ventricular posterior wall thickness; IVS, interventricular septal thickness; LVMM, left ventricular muscle mass; LVET, left ventricular ejection time; * $p<0.05$; ** $p<0.01$
[a] Group A, $n=28$.
[b] Mean values ± SD.

The changes in these parameters after 16 months were significant as compared to pretreatment values. Thickness of the posterior wall decreased but the thickness of the ventricular septum remained unchanged.

In seven patients (group B) clinically relevant CAD was established by exercise tolerance test, Holter monitoring with ST segment trend analysis, radionuclide scintigraphy, and coronary angiography. In these patients tolerance to exercise – induced ischemia was significantly improved by elevating the hematocrit to about 35%. Diminished ST segment depression despite prolongation of exercise, and an increase of exercise capacity (watt per kilogram body weight) was observed. The product of systolic pressure × heart rate as an index of myocardial oxygen consumption did not change significantly. The observation that redistribution in the thallium scan was no longer present in four patients is compatible with the above-mentioned findings.

The following conclusions can be drawn from our study. Partial correction of renal anemia induces regression of left ventricular hypertrophy despite a concomitant increase in blood viscosity. Left ventricular muscle mass is reduced to normal after 16 months. This applies to patients in whom left ventricular hypertrophy is not due to CAD, hypertension, valvular heart disease, and/or diabetes mellitus. Exercise tolerance is improved in patients with significant CAD.

These results suggest that anemia in patients with end-stage renal disease treated with intermittent hemodialysis only partly accounts for the development of left vertricular hypertrophy. It appears that the additional volume load imposed on the left ventricle by introduction of an arteriovenous fistula and an uncontrolled weight and volume gain during the interdialytic period also plays an important pathophysiologic role.

The important questions regarding the point of time during the evolution of chronic renal failure at which left ventricular hypertrophy begins, and the parameters which can be correlated to its development remain to be elucidated in future clinical trials.

References

Bullock RE, Amer HA, Simpson I, Ward MK, Hall RJC (1984) Cardiac abnormalities and exercise tolerance in patients receiving renal replacement therapy, Br Med J 289:1479–1484

Devereux RB (1990) Does increased blood pressure cause left ventricular hypertrophy or vice versa? Ann Intern Med 112:157–158

Devereux RB, Drayer JIM, Chien S, Pickering TG, Letcher RL, DeYoung JL, Sealey JE, Laragh JH (1984) Whole blood viscosity as a determinant of cardiac hypertrophy in systemic hypertension. Am J Cardiol 54:592–595

Harnett JD, Parfrey PS, Griffiths SM, Gault MH, Barre P, Guttmann RD (1988) Left ventricular hypertrophy in end-stage renal disease. Nephron 48:107–115

Hüting J, Kramer W, Wizemann V, Schütterle G (1991a) Effects of secondary hyperparathyroidism on left ventricular hypertrophy and function in patients on different renal replacement therapies. In: Timio M, Wizemann V (eds) Cardionephrology. Wichtig, Milan, pp 119–122

Hüting J, Kramer W, Kühn K, Schütterle G, Wizemann V (1991b) Risk factors for diastolic left ventricular dysfunction in patients on continuous ambulatory peritoneal dialysis (CAPD) In: Timio M, Wizemann V (eds) Cardionephrology. Wichtig, Milan, pp 321–324

Klaus D, Schwarze D, Lederle RM, Saul F (1991) Behandlung mit rekombinantem humanem Erythropoietin. Nieren Hochdruckkrankheiten 20:28–35

Koren MJ, Devereux RB, Casale PN, Savage DD, Laragh JH (1991) Relation of left ventricular mass and geometry to morbidity and mortality in uncomplicated essential hypertension. Ann Intern Med 114:345–352

Kramer W, Wizemann V, Thormann J, Kindler M, Mueller K, Schlepper M (1986a) Cardiac dysfunction in patients on maintenance hemodialysis. I. The importance of associated heart disease in determining alteration of cardiac performance. Contrib Nephrol 52:97–109

Kramer W, Wizemann V, Lämmlein G, Thormann J, Kindler M, Schlepper M, Schütterle G (1986b) Cardiac dysfunction in patients on maintenance hemodialysis. II. Systolic and diastolic properties of the left ventricle assessed by invasive methods. Contrib Nephr 52:110–124

Levy D, Anderson KM, Savage DD, Kannel WB, Christiansen JC, Castelli WP (1988) Echocardiographically detected left ventricular hypertrophy: prevalence and risk factors. Ann Intern Med 108:7–13

Levy D, Garrison RJ, Savage DD, Kannel WB, Castelli WP (1990) Prognostic implications of echocardiographically determined left ventricular mass in the framingham heart study. N Engl J Med 322:1561–1566

London GM, DeVernejoul MC, Fabiani F, Marchais SJ, Guerin AP, Metivier F, London AM, Llach F (1987a) Secondary hyperparathyroidism and cardiac hypertrophy in hemodialysis patients. Kidney Int 32:900–907

London GM, Fabiani F, Marchais SJ, DeVernejoul MC, Guerin AP, Safar ME, Metivier F, Llach F (1987b) Uremic cardiomyopathy: an inadequate left ventricular hypertrophy. Kidney Int 31:973–980

Löw I, Grützmacher P, Bergmann M, Schoeppe W (1989) Echocardiographic findings in patients on maintenance hemodialysis substituted with recombinant human erythropoietin. Clin Nephrol 31:26–30

Messerli FH, Schmieder RF (1990) Arterielle Hypertonie, linksventrikuläre Hypertrophie und plötzlicher Herztod. Dtsch Med Wochenschr 115:1682–1687

Mall G, Huther W, Schneider J, Lundin P, Ritz E (1990) Diffuse intermyocardocytic fibrosis in uraemic patients. Nephrol Dial Transplant 5:39–44

Sahn DJ, DeMaria A, Kisslo J, Weyman A (1978) Recommendations regarding quantitation in M-mode echocardiography: results of a survey of echocardiographic measurements. Circulation 58:1072–1083

Silberberg JS, Barre PE, Prichard SS, Sniderman AD (1989a) Impact of left ventricular hypertrophy on survival in end-stage renal disease. Kidney Int 36:286–290

Silberberg JS, Rahal DP, Patton DR, Sniderman AD (1989b) Role of anemia in the pathogenesis of left ventricular hypertrophy in end-stage renal disease. Am J Cardiol 64:222–224

Long-Term Treatment with Recombinant Human Erythropoietin in Haemodialysis Patients: Effects on Left Ventricular Performance

S. Casati, G.C. Ambroso, C. Ponticelli, A. Pierini, R. Mangiarotti, and C. Pini

Introduction

Left ventricle enlargement and dilation is a frequent result of the hyperdynamic circulatory condition induced by anaemia in the uraemic patient [1–3]. Thus, an adequate correction of anaemia may improve the haemodynamic status in these patients.

Studies on cardiovascular changes during therapy with recombinant human erythropoietin (rHuEPO) are now becoming available [3–7]. However, only short-term follow-ups have been reported.

In the present paper we report the long-term effects of the correction of anaemia with rHuEPO on cardiovascular function, evaluated by M-mode echocardiography, in a group of chronic haemodialysis patients.

Patients and Methods

Twenty-two chronic haemodialysis patients were selected for treatment with rHuEPO because of severe anaemia (mean haemoglobin 6.5±0.8 g/dl). No causes other than uraemia accounted for the anaemia. The drug was given as an intravenous bolus at the end of dialysis, three times a week. The starting dose was 24 U/kg body weight. During the follow-up, rHuEPO doses were adjusted as described elsewhere [8] to maintain haemoglobin (Hb) between 10 and 12 g/dl.

Patients with clinical, electrocardiographic or echocardiographic signs of valvular diseases, past or current evidence of pericarditis or congestive heart failure, as well as obese, diabetic patients and all those with "inadequate echocardiographic window" were excluded from the study. Thus, six patients mean age 42±12 years (range 21–60) on chronic haemodialysis for a mean 9±6 years were considered eligible for the study. Their pre-rHuEPO mean Hb was 6.5±0.8 g/dl. Four of the six were on maintenance antihypertensive therapy (beta-blockers, calcium antagonists, converting enzyme inhibitors).

Echocardiographic examinations were performed both before and immediately after mid-week dialysis. To minimize the influence of water load, the test was performed only if the patients' interdialytic weight gains differed by

Pagel/Weiss/Jelkmann (Eds.)
Pathophysiology and Pharmacology
of Erythropoietin

less than 10% from weight gains at baseline. The time schedule was: (a) before rHuEPO treatment (baseline); (b) after short-term correction of anaemia (3 and 6 months); and (c) after long-term correction of anaemia (1, 2 and 3 years). M-mode echocardiographic examinations were performed according to the recommendations of the American Society of Echocardiography [9] employing a two-dimensional sectorial image for correct positioning of the transducer and careful measuring. A Sigma 1 SC (Kontron) with a 3.5 MHz transducer and an MSV 110 (Piker) with a 3.0 MHz transducer were used. Echocardiograms were tape recorded and afterwards data were obtained with an image analysing computer (Cardio 200, Kontron) from an average of at least three consecutive cardiac cycles.

Left ventricular end-diastolic diameter (LVED) and left ventricular fractional shortening (SHRT) (normal values: 40±1.5%) were measured and calculated. The mean velocity of circumferential fibre shortening (MVCF) was estimated using the method of Cooper et al. [10]. Stroke volume was calculated from a regression equation as described by Teichholz et al. [11]. Left ventricular systolic work (LVSW) (normal values: 110±15 g-m/beat) was calculated and assumed to equal external cardiac work (i.e., the product of stroke volume and mean aortic pressure) [12]. Diastolic arterial pressure (DAP) was recorded and considered an index of after load to evaluate peripheral perfusion [13].

For the statistical analysis, we used a statistical software for data description and comparison of groups with paired *t*-tests and multiple linear regression, looking for correlations between haemoglobin and haemodynamic parameters. The functions for cubic spline interpolation were also computed using a statistical software program. All data are expressed as mean ± standard deviation.

Results

Hb changes between baseline (6.5±0.8 g/dl) and subsequent times were always statistically significant ($p<0.001$) (Table 1). The highest Hb level (12±0.9 g/dl) was reached after 6 months of treatment (mean dose of rHuEPO: 400±75 U/kg per week), while at third year the mean Hb was 10.5±0.4 g/ml (mean rHuEPO dose: 135±45).

Pre-dialysis heart rates tended to be lower at 3 and 6 months of treatment and significantly higher at 2 years (Table 1).

Pre-dialysis DAP rose significantly between 3 months and 1 year. Post-dialysis DAP was higher at 3 months than at 6 months and 1 year (Table 1). As a consequence, antihypertensive drug prescription had to be increased for all hypertensive patients when the Hb rose from 6.5±0.8 to 11.9±0.6 g/dl. In contrast, at 3 years, when the Hb level was 10.5 g/dl, antihypertensive drugs had to be reduced by 50% from that before rHuEPO treatment, to keep blood pressure values unchanged.

Table 1. Echocardiographic Parameters at Every Step Before and After Dialysis[a]

	LVED (mm)	SHRT (%)	DAP (mmHg)	HR (beats/min)	CO (L/min)	HB (g/ml)
Pre-dialysis						
baseline	52.5±4.9	47±6	85± 8	73±12	9.8±2.3	6.5±0.8
3 months	51.1±4.6	44±9	85± 5	66±10	7.9±0.9	10.2±0.7
6 months	50.5±6.8	42±8	85±10	67±10	7.6±1.7	12.0±0.9
1 year	52.2±5.3	44±5	89± 4	69± 8	9.7±2.1	11.9±0.6
2 years	52.1±7.9	44±3	87± 9	74± 6	9.3±2.4	11.5±0.3
3 years	52.7±6.2	39±5	85±10	67±12	7.8±1.7	10.5±0.4
Post-dialysis						
baseline	48.5±8.7	46±5	80± 6	78±10	9.1±3.7	
3 months	45.3±6.9	41±2	71±13	75±13	6.7±1.4	
6 months	45.3±6.3	47±7	77± 8	77±18	7.7±3.1	
1 year	48.7±4.3	45±6	84±4	75± 9	8.3±1.9	
2 years	47.2±6.5	46±9	75±11	82±10	8.9±3.0	
3 years	49.8±7.4	45±4	78±15	78±11	9.2±3.0	

Statistical analysis (paired *t*-test with two-tailed probability)

LVED

3 months	pre vs post	$p=0.029$
1 year	pre vs post	$p=0.006$
2 years	pre vs post	$p=0.036$

HR

3 months	pre vs post	$p=0.030$
1 year	pre vs post	$p=0.035$
2 years	pre vs post	$p=0.010$
3 years	pre vs post	$p=0.024$
pre-dialysis	3 months vs 2 years	$p=0.022$

SHRT

pre-dialysis		
	baseline vs 3 years	$p=0.005$
	1 yr vs 3 years	$p=0.049$
post-dialysis		
	3 months vs 3 years	$p=0.024$

CO

pre-dialysis		
	3 months vs 1 year	$p=0.04$
	6 months vs 1 year	$p=0.03$
	1 year vs 3 years	$p=0.001$
post-dialysis		
	3 months vs 1 year	$p=0.035$
	3 months vs 3 years	$p=0.048$

DAP

pre-dialysis		
	3 months vs 1 year	$p=0.04$
post-dialysis		
	3 months vs 1 year	$p=0.025$
	6 months vs 1 year	$p=0.035$

[a] Mean and standard deviation of LVED (left ventricular diastolic diameter); SHRT (left ventricular fractional shortening); DAP (diastolic arterial pressure); HR (heart rate); CO (cardiac output).

Neither pre- nor post-dialysis mean LVED values changed significantly. However, statistically significant differences between pre- and post-dialysis mean LVED were observed at 3 months, 1 year and 2 years (Table 1).

Before rHuEPO treatment was started, pre-dialysis SHRT was 47±6%. After 6 months of rHuEPO, SHRT lowered to 42±8% and tended to increase again to 44±5% at 1 year at a Hb level of 11.9±0.6 g/dl. At year 3 of rHuEPO treatment, when Hb was stabilized at 10.5 g/dl, the mean SHRT became normalized (39±5%). In post-dialysis evaluation, the mean SHRT was significantly increased between 3 months and 3 years (Table 1).

Pre-dialysis and post-dialysis cardiac outputs (CO) were also evaluated. At baseline, pre-dialysis CO was 9.8±2.3 l/min; and was 9.1±3.7 after dialysis. After 6 months of rHuEPO treatment, pre-dialysis CO fell to 7.6±1.7 l/min and post-dialysis CO fell to 7.7±3.1 to 6 months. At 3 years pre-dialysis CO was 7.8±1.7 l/min and post-dialysis CO was 9.2±3.0. Data and statistics are fully reported in Table 1.

Shortening data were plotted against Hb and diastolic pressure. Our data indicated that, in pre-dialysis evaluations, DAP worsened at Hb levels higher than 11.5 g/dl, while both SHRTand DAP were markedly improved at a mean Hb value of 10.5 g/dl (Fig. 1). In post-dialysis evaluations, SHRT and DAP were markedly reduced at 3 months, while at the highest Hb level both parameters returned to baseline values. At 3 years, SHRTand DAP remained unchanged despite a marked reduction of Hb (Fig. 2).

LVSW was also evaluated and plotted against LVED. At baseline, pre-dialysis LVSW was 203±44 g-m/beat; this fell to 158±75 after dialysis.

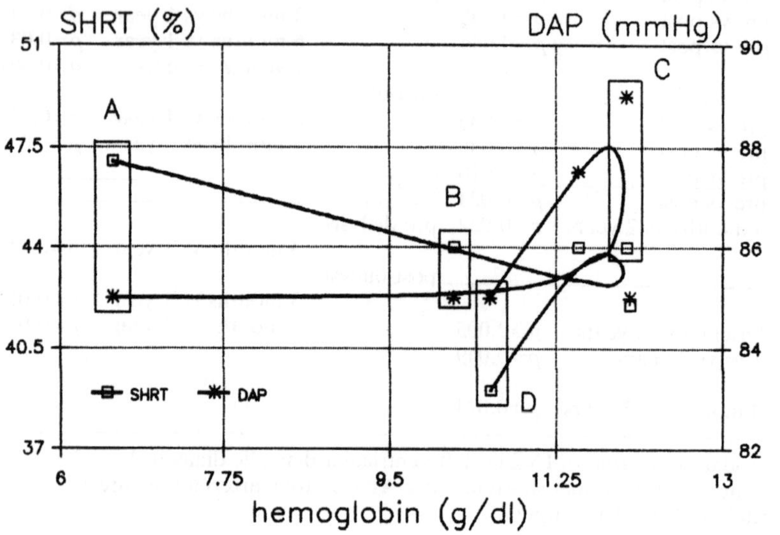

Fig. 1. Spline curves between pre-dialysis SHRT, DAP and Hb. *A* box indicates baseline values; *B* box, 3rd month values; *C* box, 1st year values; *D* box, 3rd year values. Symbols (□; *) are the measured values. *Curves* indicate the functions computed for the interpolated values

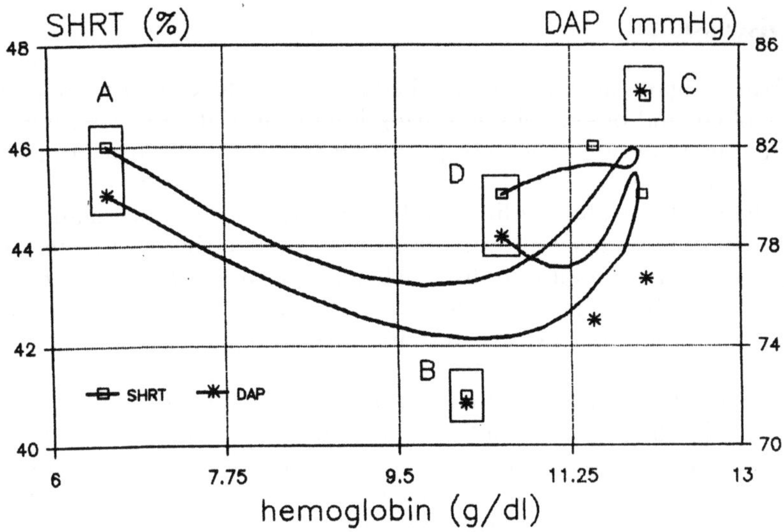

Fig. 2. Spline curves between post-dialysis SHRT, DAP and Hb. *A* box indicates baseline values; *B* box, 3rd month values; *C* box, 1st year values; *D* box, 3rd year values. Symbols (□; *) are the measured values. *Curves* indicate the functions computed for the interpolated values

During rHuEPO treatment, LVSW did not change significantly. At 3 years, pre-dialysis LVSW was 167±60 and post-dialysis LVSW was 173±37 g-m/beat (Fig. 3).

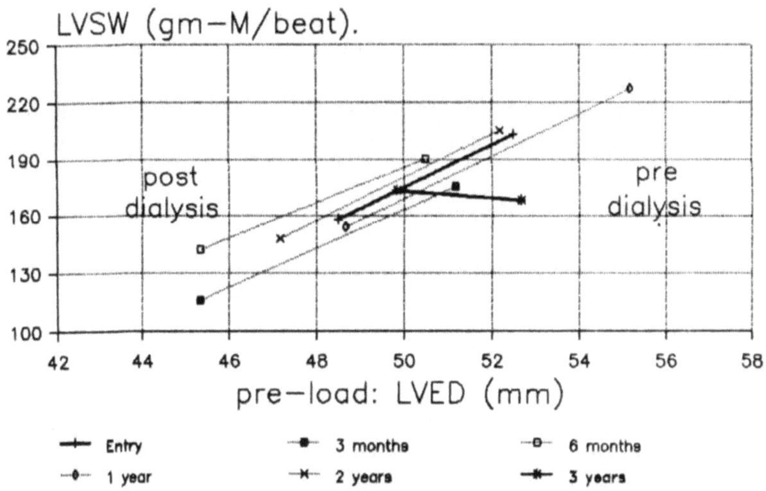

Fig. 3. Pre- and post-dialysis LVSW plotted against LVED in six chronic haemodialysis patients under rHuEPO treatment

Discussion

To investigate the impact on cardiac function induced by the changes in blood rheology and blood volume during long-term rHuEPO treatment echocardiographic evaluations were performed sequentially in a group of chronic haemodialysis patients. It is well-known that echocardiographic data can be influenced by water loading conditions [14]. For this reason and to minimize the bias between serial evaluations we recorded the data from our patients when loading conditions for each patient were similar to the respective baseline. It should be remembered that echocardiography is, at present, the most reliable non-invasive test for serial evaluations of myocardial function.

Pre-dialysis LVED, which is considered an index of left ventricle preload [15, 16] showed a substantial homogeneity at any time point considered, indicating that during the correction of anaemia with rHuEPO no evident changes in preload occurred in our patients. The occurrence of a constant preload level in the highest range of efficiency under rHuEPO might be an effect of venoconstriction, resulting in adequate venous return and cardiac filling, as suggested by London et al. [3].

A significant increase in diastolic blood pressure was observed in our previously hypertensive patients after 1 year of rHuEPO treatment. This rise was attributed to the correction of the hypoxia-related vasodilatation and to the increase in peripheral resistance [17] which follows the sharp rise in red cell number during the early phases of rHuEPO treatment. The reported worsening of hypertension indirectly confirms the observation of functional abnormalities of resistance vessels in hypertensive patients [18]. After 3 years of rHuEPO treatment DAP was lower, despite reduction of antihypertensive drug prescriptions. This phenomenon may indicate that with time some homeostatic mechanisms set peripheral resistances at a lower level despite Hb concentrations much higher than at baseline. We cannot tell whether our findings might have been influenced by the changes in antihypertensive drug prescriptions we were compelled to make. It seems, though, that these modifications should not have significantly influenced patients' response [7].

As expected, pre-dialysis SHRT was raised before rHuEPO treatment, in accordance with the high output level due to anaemia. After an initial decrease, the SHRT tended to increase again, corresponding to the higher Hb level. This increase was attributed to the worsened afterload. When Hb was finally set at lower levels, SHRT stabilized at a more economic level. These data are in agreement with studies on chronic anaemia in non-uraemic conditions [19, 20].

A good haemodynamic efficiency was reached at 3 months, at a Hb level of 10.5 g/dl. This result was confirmed by the analysis of the SHRT plotted against the oxygen transport capacity (Hb) and the peripheral perfusion level (DAP). Further improvement in myocardial performance was observed at 3 years for identical values of DAP and Hb. A possible explanation for this phenomenon might be that the increased oxygen content per unit of blood had allowed a reduction of the systolic volume without altering the total amount of oxygen delivered to peripheral tissues.

A similar relationship was searched for in post-dialysis conditions. Data showed poor haemodynamic adaptation during the acute phase of correction of anaemia. On the contrary, haemodynamic conditions seemed to improve at 2 years, with the increase of DAP and SHRT.

The degree of correction of anaemia obtained at 3 years (10.5±0.4 g/dl) kept preload unchanged, resulting in a reduction in LVSW. This resulted in an improved heart performance, as indicated by the shift of preload and ventricular work function to a higher curve during dialysis. Thus, our data indicate that an adequate correction of anaemia with rHuEPO allows the cardiovascular system to work at its best and at the lowest risk for development of hypertension and related side effects.

References

1. Neff MS, Kim KE, Persoff M et al. (1971) Hemodynamics of uremic anemia. Circulation 43:876–883
2. Capelli JP, Kasparian H (1977) Cardiac work demands and left ventricular function in end-stage renal disease. Ann Intern Med 86:261–267
3. London GM, Zins B, Pannier B et al. (1989) Vascular change in haemodialysis patients in response to recombinant human erythropoietin. Kidney Int 36:878–882
4. Verbeelen D, Bossuyt A, Smitz J et al. (1989) Hemodynamics in patients with renal failure treated with recombinant human erythropoietin. Clin Nephrol 1:6–11
5. Low I, Grützmacher P, Bergmann M, Schoeppe W (1989) Echocardiographic findings in patients on maintenance hemodialysis substituted with recombinant human erythropoietin. Clin Nephrol 1:26–30
6. Cannella G, La Canna G, Sandrini M et al. (1990) Renormalization of high cardiac output and of left ventricular size following long-term recombinant human erythropoietin treatment of anemic dialyzed uremic patients. Clin Nephrol 34:272–278
7. Tagawa H, Nagano M, Saito H et al. (1991) Echocardiographic findings in hemodialysis patients treated with recombinant human erythropoietin: proposal for a hematocrit most beneficial to hemodynamics. Clin Nephrol 35:35–38
8. Casati S, Passerini P, Campise M et al. (1987) Benefits and risks of protracted treatment with human recombinant erythropoietin in patients having haemodialysis. Br Med J 295:1017–1020
9. Feigenbaum H (1981) Echocardiography, 3rd edn. Lea and Febiger, Philadelphia, pp 549–563
10. Cooper RH, O'Rourke RA, Karliner JS et al. (1972) Comparison of ultrasound and cineangiographic measurements of the mean rate of circumferential fiber shortening. Circulation 46:914
11. Teichholz LE, Kreulen T, Herman MV, Gorlin R (1976) Problems in echocardiographic volume determinations: echocardiographic-angiographic correlations in the presence or absence of asynergy. Am J Cardiol 37:7–11
12. Braunwald E, Ross J, Sonnenblick EH (1976) Determinants of cardiac performance. In: Braunwald E, Ross J, Sonnenblick EH (eds) Mechanism of contraction of the normal and the failing heart. Little, Brown, Boston, 92–123
13. Tarazi RC, Gifford RW jr (1985) Systemic arterial pressure. In: Sodeman WA, Sodeman TM (eds) Pathological physiology. Saunders, Philadelphia pp 228–260

14. Tomson CRV (1990) Echocardiographic assessment of systolic function in dialysis patients. Nephrol Dial Transplant 5:325–331
15. Sonnenblick EH, Strobeck JE (1977) Current concepts in cardiology. Derived indexes of left ventricular and myocardial function. N Engl J Med 296:978–982
16. Redwood DR, Henry WL, Epstein SE (1974) Evaluation of the ability of the echocardiography to measure acute alterations in left ventricular volume. Circulation 50:901–904
17. Eschbach JW, Adamson JW (1988) Recombinant human erythropoietin: implications for nephrology. Am J Kidney Dis 3:203–209
18. Panza JA, Quyyumi AA, Brush JE, Epstein SE (1990) Abnormal endothelium dependent vascular relaxation in patients with essential hypertension. N Engl J Med 323:22–27
19. Lewis RP, Sandler H (1971) Relationship between changes in left ventricular dimensions and the ejection fraction in man. Circulation 44:548–557
20. Florenzano F, Diaz G, Regonesi C, Escobar E (1984) Left ventricular function in chronic anemia: evidence of non-catecholamine positive inotropic factor in the serum. Am J Cardiol 54:638–645

Retinal Hypoxia and Anemia in Chronic Renal Failure: Effect of Erythropoietin

C. COSTAGLIOLA, M. RINALDI, P. SORICE, and A. DI BENEDETTO

Introduction

Patients suffering from chronic renal failure (CRF) show a normocytic and normochromic anemia [1–3]. Erythropoietin deficiency seems to be one of the main causes of anemia in these patients [4].

The reduced amounts of oxygen availability, secondary to the very low hemoglobin concentration, are responsible for pathological manifestations at the level of nervous, muscle, and heart tissue, which are expressions of a state of chronic tissue hypoxia [5].

Eye involvement in CRF is essentially represented by the very high incidence of cataract [6], alteration of the adaptometric curves (Costagliola and Di Benedetto 1991, unpublished data) and by the presence, in the more advanced stages of the disease, of areas of choroidal atrophy [7].

While cataract does not depend on the anemic status [8], defective dark adaptation and choroidal lacunae could be an expression of ischemic damage at the level of the choroidal vascular bed (Costagliola and Di Benedetto 1991, unpublished data) [9]; similarly, retinal circulation could also be affected by a deficiency of oxygen.

Oscillatory potentials (OP) currently represent one of the better approaches for verifying the integrity of retinal flow [10] and, thus, of retina oxygen supply [11].

The aim of this study has been to ascertain in CRF patients: (a) the presence of retinal flow alterations induced by the anemic status, and (b) whether the correction of anemia by erythropoietin (rHuEPO) administration was also able to improve the retina oxygen supply.

Patients and Methods

Ten patients (five men and five women) affected by CRF, aged between 35 and 64 years (mean age 47 years) were studied. The causes of renal failure were glomerulonephritis, interstitial nephritis, nephrosclerosis, and polycystic renal disease. CRF patients suffering from diabetic nephropathy were excluded from the study, because diabetes per se affects OP results [12]. All patients were outpatients undergoing hemodialysis three times a week. The

Pagel/Weiss/Jelkmann (Eds.)
Pathophysiology and Pharmacology
of Erythropoietin

mean treatment duration was 80 months (32–251 months). Informed consent for entry into the study had been obtained from all subjects after a detailed description of the aim of the work and the procedure used had been given.

Average hemoglobin (Hb) concentration before commencement of study was 6.8 + 0,5 g/100 ml and hematocrit (Ht) value was 21% + 2%. Patients did not receive blood transfusion therapy within one month prior to study initiation. Moreover, the clinical condition of patients was stable and they did not have any serious complication apart from renal anemia.

All patients underwent a detailed eye examination (visual acuity, biomicroscopy, tonometry, and funduscopy). OP were recorded in the dark, after an adequate period of adaptation (20 mins) and the flash was administered every 15 s. To determine amplitude size we used the Fourier mathematical analysis system. A EREV 85 (Lace Elettronica, Pisa, Italy) electroretinographic computerized system was adopted.

rHuEPO (EPREX, Cilag SpA Biotech, Italy) was administered to each patient i.v. three times a week at the end of hemodialytic treatment at a dosage of 50 U/kg for 60 days.

Results

Table 1 demonstrates that rHuEPO therapy has induced a significant increase in Hb concentration and in Ht value in all patients, as early as after 30 days. After 2 months of administration the results were even more significant.

All patients, after the treatment period, showed a significant increase in the OP amplitude, for all their four components (Table 2). In our recording, we adopted the Fourier mathematical analysis method: the signal was fragmentated into sinusoidal components with various frequencies (harmonics) of which we utilized only those necessary for its recomposition. In this way, it is possible to eliminate most of the "background noise" (constituted by the electroretinogram waves) and to obtain a more accurate measurement of the OP amplitude.

Table 1. Hemoglobin and hematocrit variations before and after rHuEPO administration

	No. of patients	Before rHuEPO	30 days[a]	60 days[a]
Hb (g/100 ml)	10	6.8±0.5	8.5±0.7*	10.3±0.6**
Ht (%)	10	21 ±2	29 ±3**	34 ±2**

* $p<0.05$; ** $p<0.01$ (Student's *t*-test)
[a] 30 and 60 days after rHuEPO administration.

Table 2. OP amplitude variations before and after rHuEPO administration

	Amplitude (μV)	
OP components	Before therapy	After therapy
01	9.41±0.11	12.54±1.31*
02	32.94±2.32	55.33±3.25**
03	21.95±1.12	43.92±3.87**
04	12.54±1.23	30.11±2.25**

* $p<0.05$; ** $p<0.01$ (Student's *t*-test)

Discussion

Our results show that the correction of the anemic status in CRF patients improves retina oxygen supply, as demonstrated by the enhanced amplitude in the recorded OP.

The term OP was first used by Yonemura [13] in 1962. Since then, it has been employed to identify in humans a series of three to four rhythmic wavelets having almost equal amplitude and an interpeak interval of about 6.5 ms [13, 14], which are superimposed on the ascending slope of the conventional electroretinogram. OP have clinical value in the assessment of retinal function. For example, in the early stage of diabetic retinopathy, the a- and b-waves of the electroretinogram may be of normal amplitude, whereas OP may be selectively abnormal [12, 14, 15]. This selective abnormality of OP is considered to be indirect evidence that OP are generated independently from a- and b-waves of the electroretinogram. The site for generation of OP is not yet known, but experimental evidence indicates that OP reflect activity of inhibitory feedback synaptic circuits within the retina [16]. Thus, OP reflect neuronal activity in the inner nuclear layer of the retina and are probably mediated by the amacrine cells or the interplexiform cells [17, 18]. The presence, in this layer, of areas void of capillaries around the retinal arterioles, because of the high local oxygen tension which affects vascular development during maturation, would render the local retinal structure extremely sensitive to ischemia. Therefore, the reduction in OP amplitude is due to photoreceptor dysfunction, secondary to choroidal ischemia or to an increased oxygen demand by the inner retina [11, 19]. However, hypoxia of the inner retina may contribute per se to alter OP amplitude [11].

In conclusion, rHuEPO administration, by improving the CRF-induced anemia, also enhances the oxygen availability for the retina. Moreover, OP provide a way to monitor retinal involvement in CRF, even in a preclinical phase, when no evident structural alterations have occurred.

References

1. Kaye M (1958) The anemia associated with renal disease. J Lab Clin Med 52:83–100
2. Hirshman GH, Wolfson M, Mosimann JF, Clark CB, Dante LM, Wineman RJ (1981) Complications of dialysis. Clin Nephrol 15:66–74
3. Costagliola C, Romano L, Sorice P, Di Benedetto A (1989) Anemia and chronic renal failure: the possible role of oxidative state of glutathione. Nephron 52:11–14
4. Fisher JW, Ohono Y, Berone J, Martinez M, Rege AB (1978) Role of erythropoietin and inhibitor of erythropoietin in anemia and renal insufficiency. Dial Transplant 7:472–481
5. Duke M, Abelmann WH (1969) The hemodynamic response to chronic anemia. Circulation 39:503–507
6. VanHeyhingen R, Harding JJ (1986) Do aspirin-like analgesic protect against cataract ? Lancet 1:1111–1113
7. Kitaoka M, Havasaka S, Mizuno K (1986) Focal chorioretinal atrophic lesions in patients treated with hemodialysis. Ophthalmologica 192:143–147
8. Costagliola C, Iuliano G, Menzione M, Simonelli F, Tortori A, Masturzi B, Di Benedetto A, Rinaldi E (1990) Systemic human diseases as oxidative risk factors in cataractogenesis. II. Chronic renal failure. Exp Eye Res 51:631–635
9. Hilton AF, Harrison JD, Lamb AM, Petrie JJ, Hardie I (1982) Ocular complications in hemodialysis and renal transplant patients. Aust J Ophthalmol 10:247–253
10. Miyake Y, Shirogaura N, Ota I, Horiguchi M (1988) Oscillatory potentials in electroretinograms of the human macular region. Invest Ophthalmol Vis Sci 29:1631–1635
11. Yonemura D, Aoki T, Tsuzuki K (1962) Electroretinogram in diabetic retinopathy. Arch Ophthalmol 68:19–23
12. Peachev NS, Charles HC, Lee CM, Fishmann GA, Cuhna-Vaz JG, Smith RT (1987) Electroretinogram findings in sickle cell retinopathy. Arch Ophthalmol 105:934–938
13. Yonemura D (1962) The oscillatory potentials of the electroretinogram. Acta Soc Ophthalmol 66:1566–1570
14. Speros P, Price AJ (1981) Oscillatory potentials: history, techniques and potential use in evaluation of disturbances of retinal circulation. Surv Ophthalmol 25:237–42
15. Algvere P, Gjottemberg M (1974) The diagnostic value of the oscillatory potentials of the ERG and fluorescein angiography in diabetic proliferative retinopathy. Acta Ophthalmol (Copenh) 168:97–101
16. Haymen N, Wachtmeister L, VanNorren D (1985) Origin of the oscillatory potentials in the primate retina. Vision Res 10:1365–1371
17. Yonemura D, Kavasaki K (1978) Electrophysiological study on activities of neuronal and non-neuronal retinal elements in man with reference to its clinical application. Jpn J Ophthalmol 22:195–202
18. Dowling JE, Ehinger B, Hedden WL (1976) The interplexiform cells: a new type of retinal neuron. Invest Ophthalmol Vis Sci 15:916–922
19. Peachev NS, Gagliano DA, Jacobson MS, Derlacki DJ, Fishman GA, Cohen SB (1990) Correlation of electroretinographic findings and peripheral retinal non-perfusion in patients with sickle cell retinopathy. Arch Ophthalmol 108:1106–1109

Experience with Subcutaneous Versus Intravenous Administration of Erythropoietin in Haemodialysis Patients

K. W. Dreyling, K. Ehlen, P. Schollmeyer, and W. H. Hörl

Introduction

The efficacy of recombinant human erythropoietin (rhEPO) in correcting the anaemia in patients with end-stage renal failure has been demonstrated in several clinical studies [1, 6, 9, 18, 19]. In these studies rhEPO was given intravenously. The optimal route of rhEPO administration has not been clearly established. In the study of MacDougall and coworkers in patients on continuous ambulatory peritoneal dialysis identical rhEPO doses were administered intravenously (i.v.) and subcutaneously (s.c.). Peak levels and bioavailability of s.c. rhEPO are lower than those of i.v. rhEPO [13]. Bommer et al. [3] reported s.c. administration of rhEPO to be effective at a smaller weekly dose than i.v. rhEPO. Similar results have been published by Zehnder and Blumberg [20]. However, these findings were challenged [14].

To date there is no unanimous judgement on the potential saving of rhEPO by s.c. versus i.v. application. To address this question a prospective multicentric study was performed. Intravenous rhEPO application three times per week was compared with s.c. administration once, three and seven times per week. We report the results of this study obtained from patients treated in the Freiburg area.

Patients and Methods

A total of 27 chronic haemodialysis patients, 13 females and 14 males, aged 44.3±2.6 years (mean±SEM) participated in the study. They were on regular haemodialysis for 7.5±1.2 years. All patients were on i.v. rhEPO therapy for at least 3 months (mean duration of previous rhEPO treatment 13.9±1.2 months, range 5–29 months).

Patients were divided into four groups (A, B, C, D). In group A patients, rhEPO was administrated i.v. three times per week. Group B, C and D patients were switched to s.c. rhEPO application. Group B patients received s.c. rhEPO once per week and group C patients received s.c. rhEPO three times per week. In group D, s.c. rhEPO was given daily; these patients received self-injection training.

Pagel/Weiss/Jelkmann (Eds.)
Pathophysiology and Pharmacology
of Erythropoietin

Study duration was 16 weeks. The initial rhEPO dose in all groups was identical to the dose given i.v. before start of study. Target haemoglobin (Hb) range was between 9.0 and 11.0 g/dl. If Hb exceeded 11.0 g/dl rhEPO was discontinued until serum Hb had fallen below 11.0 g/dl. Then, rhEPO therapy was restarted with a dose reduced by 25 U/kg body wt. per week. If Hb was lower than 9.0 g/dl, the rhEPO dose was increased by 25 U/kg body wt. per week if this was considered safe from the clinical point of view. RhEPO was provided by Cilag, Schaffhausen, Switzerland.

All patients gave their informed consent to the study protocol. Approval of the local ethical committee had been obtained.

Laboratory Tests

Baseline study of the patients included history and physical examination and determination of electrolytes, creatinine and parameters of liver function. Haematological parameters were checked weekly (Technicon H6000, Technicon Corporation, Tarrytown, USA). Extensive laboratory controls including iron parameters were performed every 4 weeks (Autoanalyzer Hitachi 737 and ES600, Boehringer Mannheim, FRG).

For statistical evaluation the Student's paired *t*-test was applied. Values are expressed as mean ± SEM.

Results

A total of 27 patients entered the study protocol. The group distribution and iron parameters of all patients are given in Table 1. In seven patients iron supplementation was performed. Two patients had been on oral iron therapy at the start of the study for 2 months and 2 years respectively. In these patients iron supplementation was continued unaltered throughout the study protocol. In five patients i.v. iron supplementation was begun during the study. Serum iron, transferrin and ferritin concentrations in groups A, B, C and D did not significantly change during the study interval (data not shown).

Twenty-two patients out of 27 completed the study protocol. Three patients went to successful renal transplantation (see Table 2). One patient refused to continue s.c. injections daily after 2 weeks. One patient died as a consequence of cerebral apoplexia.

Hb concentrations and rhEPO doses remained stable in the patients treated with i.v. rhEPO three times per week and in the patients treated with s.c. rhEPO once per week. In the patients treated with s.c. rhEPO three times per week we measured a rise in Hb values, and there was a slight but insignificant decrease in the adjusted rhEPO doses. The same phenomenon was observed in the patients treated with s.c. rhEPO seven times per week. Hb values and rhEPO doses of the patients with complete study protocols are shown in Figs. 1 and 2. The course of haematological values and rhEPO dose

Table 1. Iron parameters in study patients

Group	patient initials	Iron (mg/ml)	Transferrin (mg/ml)	TS (%)	Ferritin (ng/ml)	Iron supplementation
A	H.R.	69	281	25	187	
A	O.S.	35	286	12	186	Iron i.v. 120 mg/week from week 2 – week 15
A	E.T.	33	293	11	96	Patient refused iron therapy
A	H.F.	94	349	27	140	
A	T.K.	57	286	20	165	
A	M.K.	124	183	68	383	
A	E.R.	72	239	30	361	
A	G.S.	89	175	51	1543	
B	S.B.	58	241	24	212	
B	E.G.	54	246	22	866	
B	J.G.	53	246	22	388	
B	W.W.	74	304	24	184	
B	V.S.	48	293	16	154	
B	H.G.	43	308	14	41	
B	I.K.	44	223	20	365	
B	H.W.	49	243	20	91	
C	J.C.	53	233	23	753	
C	J.F.	73	328	22	117	Iron i.v. 120 mg/week from week 1 – week 8
C	A.H.	38	335	11	17	Iron i.v. 120 mg/week from week 1 – week 8
C	I.L.	75	265	28	112	
C	F.H.	85	294	29	310	Iron per os 160 mg/day for 2 years before and throughout study
D	L.D.	96	210	46	667	
D	U.K.	89	323	28	29	Iron i.v. 120 mg/week from week 2 – week 12
D	J.S.	124	231	54	1835	
D	H.W.	36	225	16	198	Iron i.v. 120 mg/week, starting week 1
D	E.W.	40	319	13	24	Iron per os 100 mg/day for 2 months before and throughout study
D	F.S.	194	211	92	7350	
Mean		70.3	265.6	28.4	621	
SEM		6.8	9.1	3.6	272	
No. of patients		27	27	27	27	

TS, transferrin saturation

Table 2. Changes in rhEPO dose and antihypertensive therapy during the study

Group	Patient initials	Starting Hb (g/dl)	rhEPO dose	Antihyper-tensive therapy	Remarks
A	H.R.	10.6	±	+	
A	O.S.	10.0	±	+	
A	E.T.	9.7	±		Transplantation after 8 weeks
A	H.F.	9.1	±		
A	T.K.	11.7	↓		
A	M.K.	8.5	↑		rhEPO dose increased to reach target Hb, apoplexia week 9.5
A	E.R.	8.5	↑(↓)	++	rhEPO dose increased to reach target Hb range, later on reduced because of hyper-tension
A	G.S.	9.5	±	+	
B	S.B.	10.0	±		
B	E.G.	9.7	↓(↑)		rhEPO dose at first reduced, later on increased
B	J.G.	9.7	±		
B	W.W.	9.4	±		
B	V.S.	10.3	±		Transplantation after 4 weeks
B	H.G.	8.6	±	+	
B	I.K.	9.0	↑	++	
B	H.W.	8.6	↓	+	
C	J.C.	9.9	↓	+	
C	J.F.	8.2	±	+	
C	A.H.	9.7	↓		
C	I.L.	8.9	↓		
C	F.H.	8.4	±	+	
D	L.D.	9.0	↓		
D	U.K.	9.2	↓		Shunt thrombosis and pulmonary embolism week 13
D	J.S.	9.8	±		
D	H.W.	9.1	±		Study terminated after 2 weeks
D	E.W.	9.9	↓	++	Transplantation after 10 weeks
D	F.S.	8.5	↑		rhEPO dose increased to reach target Hb range

↑, rhEPO dose increased; ↓, rhEPO dose decreased; +, antihypertensive therapy before start of study; ++, antihypertensive therapy intensified during the study

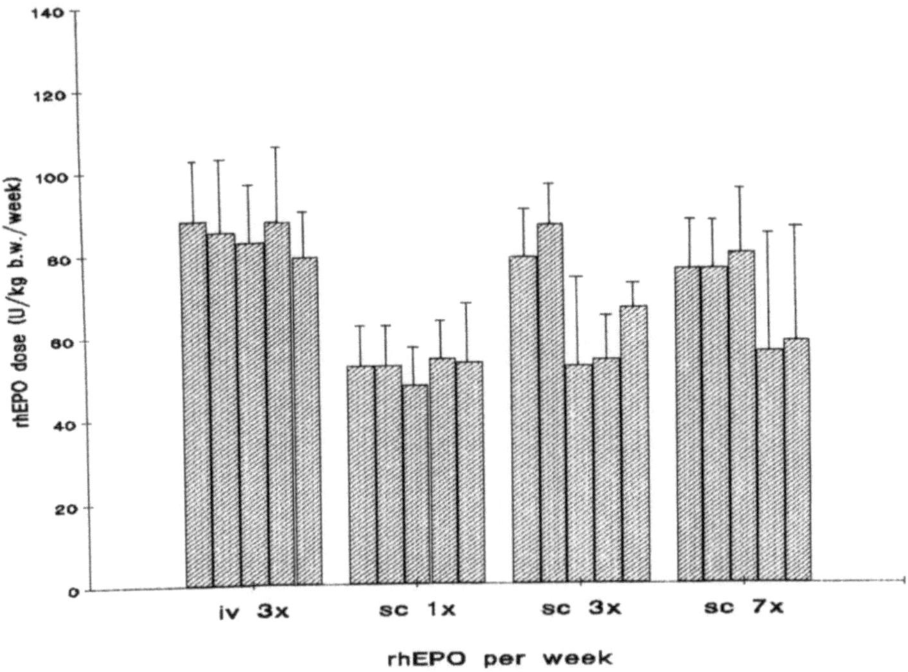

Fig. 1. Courses of haemoglobin concentrations (mean±SEM) at the beginning of study and after 4, 8, 12 and 16 weeks

adjustment is described for each treatment group separately. Initials of the patients are given to make identification possible in Tables 1 and 2.

Group A. Eight patients with starting Hb values range 8.5–11.7 were treated with i.v. rhEPO three times a week. Two patients failed to complete the study protocol. One patient (E.T.) was transplanted 8 weeks after the start of study. Until then, rhEPO dose had been unchanged. In two patients starting Hb values were lower than the target Hb value of 9.0 g/dl. Patient E.R. started with Hb 8.5 g/dl; the rhEPO dose was increased. Because of aggravation of hypertension rhEPO dose was then readjusted to the initial dose; intensivation of the preexisting antihypertensive therapy was necessary.

In patient M.K., the rhEPO dose was increased and Hb rose from initial 8.5 g/dl to 9.7 g/dl at week 9. After 9.5 weeks apoplexia occurred, the patient died subsequently (see below). Six patients completed the study protocol; for five of these rhEPO doses at the end of study were the same as at the beginning. The sixth patient began with an Hb value of 11.7 g/dl; the rhEPO dose was reduced by 25 U/kg body wt. per week. At the end of the study her Hb concentration was 10.4 g/dl.

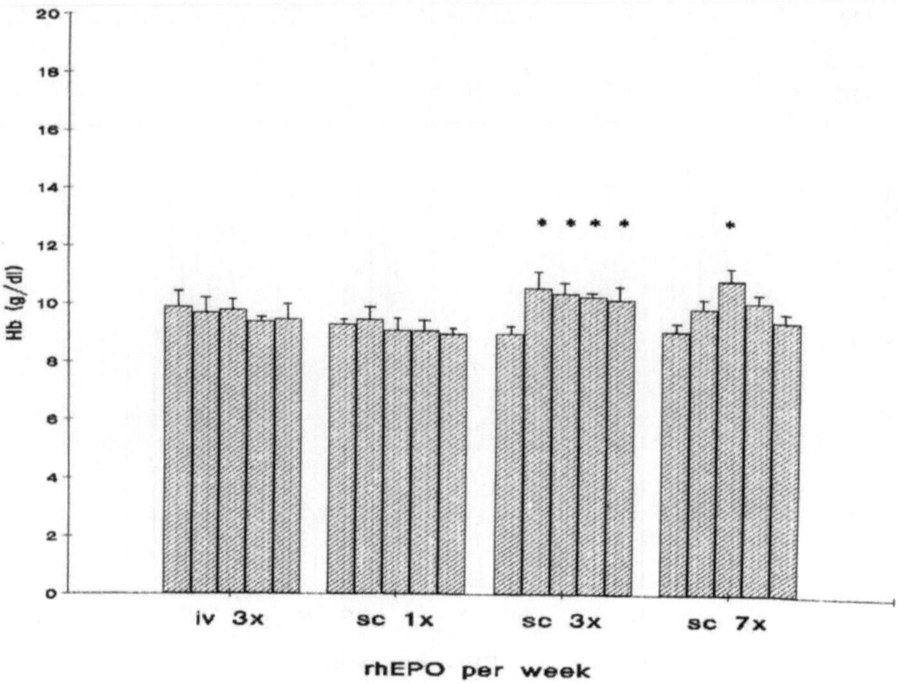

Fig. 2. Courses of mean rhEPO doses (mean±SEM) at the beginning of study and after 4, 8, 12 and 16 weeks. *$p<0.05$ compared to starting values

Group B. Eight patients were admitted to group B (s.c. rhEPO once a week). One patient (V.S.) was transplanted after 4 weeks; until then rhEPO dose and Hb values had been unchanged. In one patient (E.G.) Hb exceeded 11.0 g/dl; the rhEPO dose was reduced. Later tooth extraction with blood loss resulted in Hb decrease; the rhEPO dose was enhanced. In another patient (I.K.) rhEPO dose had to be increased to reach target Hb concentration. However, hypertension was aggravated and pre-existing antihypertensive medication had to be modified. In another patient (H.W.) Hb exceeded 11.0 g/dl in week 9; the rhEPO dose was discontinued and reduced subsequently. For the remaining four patients the rhEPO dose at the end of study was identical to the dose at the start of study.

Group C. Five patients entered group C (s.c. rhEPO three times a week), and all patients completed the study protocol. In all patients Hb values rose compared with starting values obtained with the identical i.v. rhEPO dose given previously. The rhEPO dose was reduced in three patients in whom Hb increased above 11.0 g/dl.

Group D. Six patients were treated in group D (s.c. rhEPO seven times a week). One patient (H.W.) refused to continue the study protocol after 2

weeks because of psychological stress induced by self injection. One patient (E.W.) was transplanted after 10 weeks. In this patient because of a high serum Hb after 2 weeks the rhEPO dose had already been reduced. Four patients completed the study protocol. In two of them rhEPO dose were reduced because of Hb increase above 12.0 g/dl. In patient F.S. with starting Hb concentration of 8.5 g/dl, rhEPO dose enhancement was necessary to reach the target Hb range. In one patient (J.S.) Hb concentrations as well as rhEPO dose remained unchanged.

Adverse Effects

Eleven patients out of 27 were treated with antihypertensive drugs at the start of the study. In three of them antihypertensive therapy had to be intensified: in two patients rhEPO dose had been increased to reach target values; in another patient hypertension was aggravated after switching i.v. rhEPO treatment from i.v. three times per week to s.c. seven times per week.

Nineteen patients were treated with s.c. rhEPO; 11 patients reported slight local burning pain after injection which was considered tolerable by all patients. In three patients haematomata in the area of injection were observed.

Two serious events occurred:

Case 1. In group A a 70-year-old male (M.K.) started with a Hb concentration of 8.5 g/dl. The patient was anephric. In 1950 the left kidney had been removed because of renal tuberculosis. In 1987 the diagnosis of hypernephroma was made and the right kidney was also removed. Since then the patient had been on regular haemodialysis treatment; he became transfusion dependent. In 1988 i.v. rhEPO treatment was started. Afterwards transfusions were no longer necessary. The patient was not hypertensive. To reach target Hb value of 9.0 g/dl, the rhEPO dose was increased in two steps from 86 U/kg body wt. per week to 137 U/kg body wt. per week. Hb rose from 8.5 to 9.7 g/dl. The patient remained normotensive. At week 9.5 apoplexia with left-sided hemiplegia occurred. When the patient arrived at the hospital blood pressure was not elevated. Within the next days clinical deterioration was observed. Because of the poor mental and physical condition of the patient the last haemodialysis treatment was performed 3 days after the event; the patient died 6 days later.

Case 2. The 42-year-old female (U.K.) started in group D with an initial Hb value of 9.2 g/dl under an i.v. rhEPO dose of 74 U/kg body wt. per week. I.v. iron supplementation was given from week 2 – week 12. When the patient had been switched to s.c. rhEPO seven times a week Hb rose up to 12.1 g/dl in week 8 and rhEPO therapy was stopped. In week 13 shunt thrombosis occurred. The patient was still on rhEPO pause and her Hb concentration was 10.7 g/dl. Since she presented with dyspnea and hypotension a lung perfusion and ventilation scan was done which revealed pulmonary embolism on the right. Doppler ultrasound investigation of the legs gave no support for deep venous thrombosis. The pulmonary embolism was probably caused by thrombotic material from the arteriovenous fistula. Anticoagulation therapy with heparin was given for 9 days. Later on a new vascular access was created. Complete recovery was obtained and rhEPO therapy was restarted.

Discussion

In patients with pre-terminal renal insufficiency or patients undergoing regular haemodialysis treatment, the peak concentration, the area under the curve, and the bioavailability have been described to be clearly lower after s.c. rhEPO administration than after i.v. administration [12]. On the other hand, the protracted resorption and elimination after s.c. rhEPO administration leads to a concentration plateau which exceeds i.v. rhEPO concentrations after 24 h. Similar results have been obtained in patients on continuous peritoneal haemodialysis [13]. In spite of the lower bioavailability of s.c. administered rhEPO it has been reported that to obtain the same therapeutic effect even smaller rhEPO doses are required with s.c. than with i.v. administration [3]. These results and our data indicate that for the therapeutic efficacy of rhEPO the extremely high peak values after i.v. administration are not necessary.

The concentration course after s.c. application is nearer the physiological continuous stimulation than the pulse stimulation by i.v. rhEPO administration. One might argue that part of the i.v. rhEPO dose is wasted without biological effect because the receptor sites in the bone marrow are saturated.

In haemodialysis patients, s.c. rhEPO administrated seven times per week has been reported to be more effective by requiring only 30% of the three times per week i.v. dose [10]. For i.v. rhEPO therapy it has been demonstrated that less rhEPO was necessary to maintain the target haematocrit when rhEPO was given in three injections per week compared to 1–2 injections per week [2]. We could not detect a difference between s.c. rhEPO three times per week and s.c. rhEPO seven times per week. However, because of the small number of patients only limited conclusions can be drawn.

Our data support an increased efficacy of s.c. rhEPO versus i.v. rhEPO administration. The trends observed have been confirmed by the results obtained from the complete multicentre study (Becker, CILAG, personal communication).

Slight local burning pain in the area of s.c. injection was reported by 11 out of 19 patients. However, this was not considered to be a severe problem by all patients. The local pain has been reported to be induced rather by additives than by rhEPO itself [4].

Two serious clinical events were observed: one patient suffered from shunt thrombosis with pulmonary embolism and in another patient apoplexy occurred.

Shunt thromboses have been reported as a side effect in rhEPO-treated haemodialysis patients [1, 6, 18, 19]. An increased incidence of thrombosis of vascular access might be induced by increased blood viscosity, elevated platelet counts and improved thrombocyte function [1, 5, 7]. A pronounced shortening of the prolonged bleeding time parallel to the progressive rise of haematocrit values has been described in rhEPO-treated uraemic patients [15]. In the double-blind, randomised, placebo-controlled study of the Canadian Erythropoietin Study Group, there was a positive association between treatment with rhEPO and the incidence of clotting of the vascular access [5]. Patients with previous vascular access-related problems are especially prone to shunt clotting [17]. While shunt thrombosis is a well-known complication of rhEPO therapy, to our knowledge pulmonary embolism as a consequence of a clotted arteriovenous fistula has not been reported in rhEPO-treated patients so far.

Neurological symptoms, especially seizures have been described as an adverse effect of rhEPO therapy in a various incidence [6, 8, 9, 16, 18]. The development of seizures seems not to be related to the dose of rhEPO nor to the rate of haematocrit increase. Usually the seizures are preceded by headache and hypertension inspite of sufficient previous antihypertensive treatment [16]. However, in a European multicentre study of 150 patients, in only one of four patients with seizures hypertension was present. In the same study one patient died from intracerebral bleeding, and in another patient treatment was temporarily discontinued due to hypertension and cerebral thrombosis [18]. Casati et al. [6] reported a patient treated with antihypertensive drugs who during dialysis developed aphasia and confusion. Cerebral computed tomography showed a small area with an ischaemic lesion. In a German multicentre study, during rhEPO treatment there was a patient with headache and spastic hemiplegia due to subdural haematoma. Since this complication occurred 11 days after the beginning of treatment in the presence of a normal blood pressure and a minor increase in haematocrit, a causative relationship to rhEPO treatment was considered unlikely by the authors [11]. In the patient reported by us neither cerebral computed tomography nor autopsy was performed. Therefore definite discrimination between haemorrhages and an ischaemic lesion is not possible. Since there was a clear increase of Hb concentration after raising rhEPO, an ischaemic apoplectic stroke seems more likely.

Generally s.c. rhEPO application is well accepted by the patients. Our data indicate that, in comparison with the i.v. route, s.c. rhEPO administration offers an economical benefit by its dosage-saving effect. Another advantage might be the more sustained rhEPO levels obtained by s.c. administration caused by slow release from subcutaneous depots.

Acknowledgement. We appreciate the cooperation of F. Hagemann, G. Schaeffer, Th. Rippich, R. Jontofsohn, H. Boeckle and R. Kuhn from dialysis centres in the area of Freiburg.

References

1. Bommer J, Alexiou C, Müller-Bühl U, Eifert J, Ritz E (1987) Recombinant human erythropoietin therapy in haemodialysis patients-dose determination and clinical experience. Nephrol Dial Transplant 2:238–242
2. Bommer J, Kugel M, Schoeppe W, Brunkhorst R, Samtleben W, Bramsiepe P, Scigalla P (1988) Dose-related effects of recombinant human erythropoietin on erythropoiesis. Contrib Nephrol 66:85–93
3. Bommer J, Ritz E, Weinreich T, Bommer G, Ziegler T (1988) Subcutaneous erythropoietin. Lancet 1:406
4. Bommer J, Weinreich T, Ritz E, Zeier M, Bommer G (1989) Efficacy of subcutaneous or intravenous recombinant human erythropoietin therapy in dialysis patients. Nephrol Dial Transplant 4:471A
5. Canadian Erythropoietin Study Group (1990) Association between recombinant human erythropoietin and quality of life and exercise capacity of patients receiving haemodialysis. Br Med J 300:573–578
6. Casati S, Passerini P, Campise MR, Graziani G, Cesana B, Perisic M, Ponticelli C (1987) Benefits and risks of protracted treatment with human recombinant erythropoietin in patients having haemodialysis. Br Med J 295:1017–1020
7. Dreyling KW, Steinhauer HB, Geiger H, Hörl WH, Schollmeyer P (1989) Platelet function under recombinant human erythropoietin therapy in haemodialysis patients. Nephrol Dial Transplant 4:472A
8. Edmunds ME, Walls J, Tucker B, Baker LRI, Tomson CRV, Ward M, Cunningham J, Moore R, Winearls CG (1989) Seizures in haemodialysis patients treated with recombinant human erythropoietin. Nephrol Dial Transplant 4:1065–1069
9. Eschbach JW, Egrie JC, Downing MR, Browne JK, Adamson JW (1987) Correction of the anemia of end-stage renal disease with recombinant human erythropoietin. N Engl J Med 316:73–78
10. Granolleras C, Branger B, Beau MC, Deschodt G, Alsabadani B, Shaldon S (1989) Experience with daily self-administrated subcutaneous erythropoietin. Contrib Nephrol 76:143–148
11. Grützmacher P, Bergmann M, Weinreich T, Nattermann U, Reimers E, Pollok M (1988) Beneficial and adverse effects of correction of anaemia by recombinant human erythropoietin in patients on maintenance haemodialysis. Contrib Nephrol 66:104–113
12. Kampf D, Kahl A, Passlik J, Pustelnik A, Eckardt K-U, Ehmer B, Jacobs C, Baumelou A, Grabensee B, Gahl GM (1989) Single-dose kinetics of recombinant human erythropoietin after intravenous, subcutaneous and intraperitoneal administration. Contrib Nephrol 76:106–111
13. MacDougall IC, Neubert P, Coles GA, Roberts DE, Dharmasena AD, Williams JD (1989) Pharmacokinetics of recombinant human erythropoietin in patients on continuous ambulatory peritoneal dialysis. Lancet 1:425–427
14. Meisl F, Manker W, Loibl U, Buxbaum M, Stockenhuber F, Steinbach K (1990) No therapeutic or economic advantage of erythropoietin application by the subcutaneous route. 11th International Congress of Nephrology, Tokyo, 202A
15. Moia M, Vizzotto L, Cattaneo M, Mannucci PM, Casati S, Ponticelli C (1987) Improvement in the haemostatic defect of uraemia after treatment with recombinant human erythropoietin. Lancet 2:1227–1229

16. Raine AEG (1990) Seizures and hypertension events. Semin Nephrol 10(2) [Suppl 1]:40–50
17. Samtleben W, Baldamus CA, Bommer J, Grützmacher P, Nonnast-Daniel B, Scigalla P, Gurland HJ (1989) Indications and contraindications for recombinant human erythropoietin treatment. Contrib Nephrol 76:193–200
18. Sundal E, Kaeser U (1989) Correction of anaemia of chronic renal failure with recombinant human erythropoietin: safety and efficacy of one year's treatment in a European multicentre study of 150 haemodialysis-dependent patients. Nephrol Dial Transplant 4:979–987
19. Winearls CG, Pippard MJ, Downing MR, Oliver DO, Reid C, Cotes PM (1986) Effect of human erythropoietin derived from recombinant DNA on the anaemia of patients maintained by chronic haemodialysis. Lancet 2:1176–1179
20. Zehnder C, Blumberg A (1991) Recombinant human erythropoietin in anemic patients on maintenance hemodialysis: comparison between intravenous and subcutaneous administration. Nephron 57:485–486

Pharmacodynamic Aspects of Recombinant Human Erythropoietin – Long-Term Results in Patients with Hemodialysis Treatment

H.W.F. Wilms, J. Berger, U. Gattig, G. Warnecke, J. Kloppstech, T. Döll, O. Stenger, C. Hansen, P. Kusche, and M. Bonato

Introduction

According to Eschbach [1], some important questions about the clinical use of recombinant human erythropoietin (EPO) have not been answered although the first patients were treated 5 years ago, and an estimated 175 000 patients worldwide are currently on EPO therapy. What is the optimal initial and constant dose? Does the dose depend on age or quality of dialysis (kind of treatment)? What is the reason for an obviously quite different individual response to exogenous EPO?

EPO was administered to about 150 of approximately 350 patients in total who had various diseases, were of various ages, and had been on dialysis care for different tirespars; all underwent intermittent hemodialysis (HD). EPO was substituted because of complaints due to anemia (decrease in physical and mental powers, dyspnea at rest and on exertion, anginosis, fatigue, intermittent claudication, dizziness, inappetence, signs of depression) if all other known principles of therapy had already been carried out.

EPO was adjusted individually to the patients hematocrit response and was administered three times a week after dialysis treatment to achieve an increased packed cell volume keeping the hematocrit between 30% and 33%.

Therapy was commenced carefully by individual adjusted dose according to presumed personal risk (e.g., hypertension, coronary heart disease, seizures in history).

A computer-based system was used to monitor all clinical data and laboratory findings.

Results of this paper are part of a continuous clinical report.

Methods

In view of the individual doses needed to reach the target range of hematocrit three groups were defined:

IV: Commencement of EPO therapy by intravenous (i.v.) administration

Pagel/Weiss/Jelkmann (Eds.)
Pathophysiology and Pharmacology of Erythropoietin

SC: Commencement of EPO therapy by subcutaneous (s.c.) administration

IVSC: Conversion from i.v. to s.c. administration under otherwise constant conditions

Patients

All observed patients were outpatients and stable on dialysis treatment during the observation period. Patients with episodes of blood loss, acute disease causing hospitalization, or changes in dialysis treatment protocol, were excluded from the study. The three groups comprised:

IV: 89 patients, age 18–88 years

SC: 35 patients, age 22–76 years

IVSC: 19 patients out of the i.v. response group, meeting the following criteria: constant hematrocrit; no clinical signs of infection or blood loss; constant dialysis conditions; no change in medication; patients had been observed for another period of more than 20 weeks. EPO was converted from i.v. to s.c. application under otherwise constant therapeutic conditions, and another observation period of 20 weeks followed. EPO dosage was studied in the different routes of administration.

At the onset, the mean hematocrit was 25.79% (2.74 SD) in the IV group, and 27.48% (4.25 SD) in the SC group.

Mean hematocrit was constant in the i.v. period of the IVSC group with 30.93±3.24 (mean, SD) and did not change significantly with time.

Monitoring and Laboratory Tests

All patients underwent regular physical examination every 3 months minimum and were seen by the physician at the onset and end of every dialysis. They were monitored and served only by nurses who had received special training on dialysis treatment.

All data for therapy monitoring were determined at meaningful intervals:

- Weight after HD (kg), blood pressure before HD (mmHg), parameters of dialysis treatment (kind of dialyzer, HD time), clinical behavior, number of blood transfusions or phlebotomy (at every dialysis)
- Hematocrit (vol.%) (1 h before end of dialysis treatment once weekly minimum)
- Serum Creatinine, urea, sodium, and potassium (before and after HD on clinical request, once monthly)

- Ferritin, parathormone (PTH, mid terminal), iron, blood count (every 3 months minimum)
- Folic acid, vitamin B 12 (on request where clinical suspicion)

The individual data on sex, age, primary disease, associated diseases, time on dialysis treatment when starting EPO therapy, kind of osteopathy (Delling classification, aluminum positive or negative), were also considered for retrospective statistical analysis.

Statistical Analysis

Calculations were performed by paired *t*-tests.

Results

No significant difference in EPO dosage was found in the IV group ($n=89$) for the parameters of sex (male: 53.9%; *female:* 46.1%); age (< 40 years: 30.3%; 41–50 years: 16.9%; > 50 years: 52.8%); primary disease (end-stage renal failure by chronic glomerulonephritis, interstitial nephritis including analgetic nephropathy, vascular forms including diabetic kidneys, others);

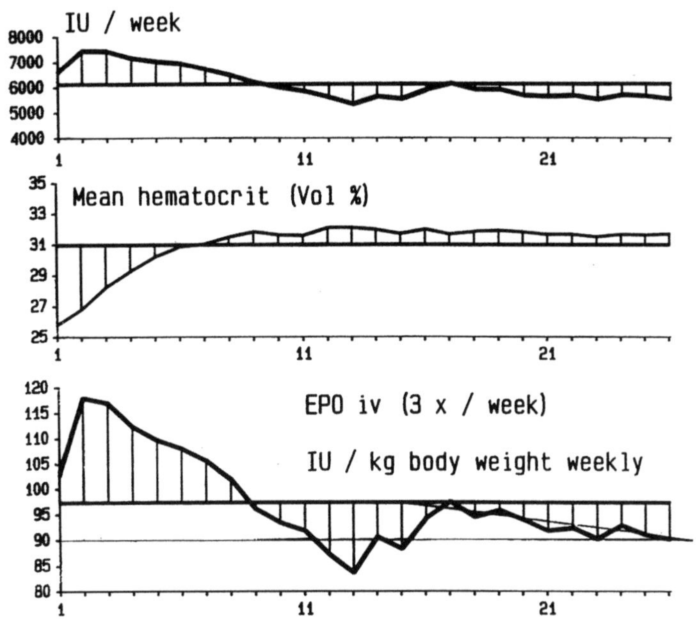

Fig. 1. Mean dosage of i.v. EPO to achieve target hematocrit of 31% vol.%. Observation period of 26 weeks after commencement of therapy.

time on dialysis treatment before EPO therapy (< 12 months; 13–24 months; 25–36 months; > 36 months), detection of aluminum in bone marrow biopsy specimens (true; false); parathyroid hormone level (< 500 ng/l; 500–1000 ng/l; 1000–2000 ng/l; > 2000 ng/l).

Figure 5 suggests a higher dosage of EPO in the group: ferritin was higher than 300 ng/ml.

EPO Therapie by i.v. Administration (IV Group). The target hematocrit level of 31% was achieved within 9 weeks after onset of EPO therapy. EPO dose of 5878±2951 (mean, SD) IU/week (93.3±46 IU/kg body weight) was required to keep the hematocrit constant after 9 weeks.

Mean dosage was higher in the beginning of therapy as shown in Figs. 1 and 3.

EPO therapy by s.c. Administration (SC Group). The target hematocrit level of 31% was achieved within 5 weeks after onset of EPO therapy. EPO dose of 4249±1849 (mean, SD) IU/week (67.4±29.3 IU/kg body weight) was required to keep the hematocrit constant after 5 weeks.

Mean dosage was higher in the beginning of therapy as shown in Figs. 2 and 3.

Conversion from i.v. to s.c. Administration (IVSC Group). In the period of i.v. administration of EPO the hematocrit level was stable 30.93% ±3.24 (mean, SD), and 31.25% ±3.09 (mean, SD) in the following s.c. application period (see Fig. 4).

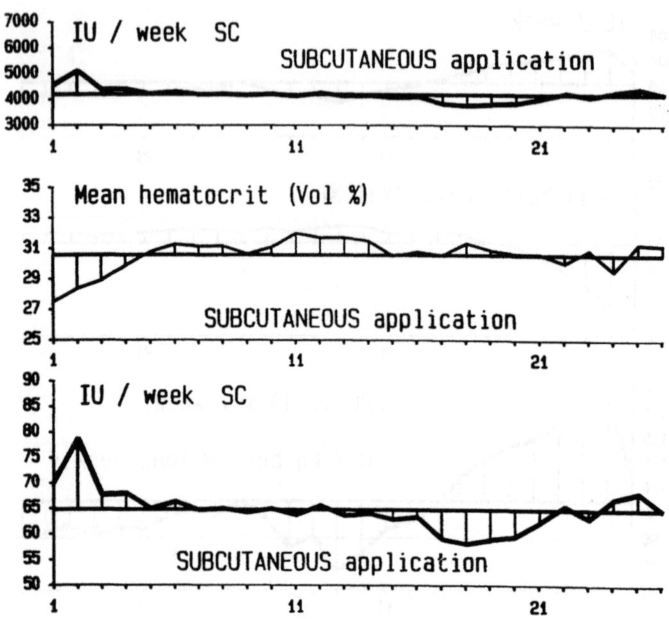

Fig. 2. Mean dosage of s.c. EPO to achieve target hematocrit of 31% vol. %. Observation period of 26 weeks after commencement of therapy.

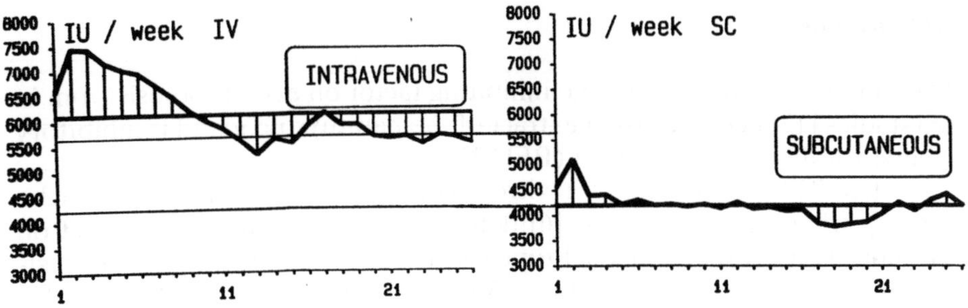

Fig. 3. Comparison of the mean dosage of i.v. EPO and s.c. EPO to achieve target hematocrit of 31%vol.%. Observation period of 26 weeks after commencement in every plot.

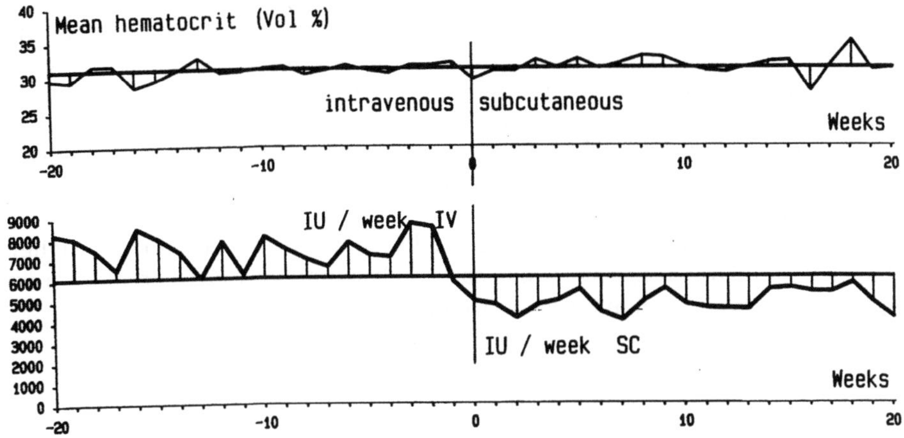

Fig. 4. Conversion of mode of administering EPO from intravenous (–20 – 0 weeks) to subcutaneous (0 – 20 weeks) to achieve constant hematocrit level

Mean EPO dose was 7429±2521 (mean, SD) IU/week intravenously and, after changing the administration route, 4851±2980 IU/week subcutaneously.

Discussion

There was no dependence of EPO doses on primary disease or on parameters of age, sex, or time on dialysis before EPO therapy.

Detection of aluminum in bone marrow biopsy specimens did not influence the maintenance dose significantly in statistical analysis, although a clear dependence of EPO response on renal anemia was observed in some patients.

Parathormone

PTH is incriminated as a possible inhibiting factor on erythropoiesis from in vitro studies [12] because crude extract of the parathyroid gland is inhibitory on burst-forming unit erythroid (BFU-E).

Purified PTH fails to inhibit in vitro erythropoiesis according to Delwiche et al. [2], whereas Hampl et al. [6] demonstrated an influence of PTH on exogenous EPO-stimulated erythropoiesis in hemodialysis patients suggesting reduced reticulocyte production. A clinical study by Hollomby et al. [8] found no correlation between time to response on EPO therapy and pre-study PTH, and no inhibitory effects of PTH could be observed by Grützmacher et al. [6] in an 8-week period.

We did not find any dose dependence on parathyroid hormone levels.

Ferritin. Hollomby et al. [8] found no correlation between time to response on EPO therapy and pre-study ferritin in 20 patients, but from an empiric view, there is a clear dependence of EPO response and maintenance dose on serum ferritin levels. Schaefer et al. [13] have suggested replacement therapy when serum ferritin levels have fallen to 100–150 μg/l.

Regarding Fig. 5, ferritin levels seem to be best adjusted at about 200 ng/ml, and EPO requirement seems to be higher at lower and higher levels as well.

The need for an increased dosage of EPO in the group with ferritin higher than 300 ng/ml needs further single case analysis and may be of great interest in regard to iron supplementation therapy.

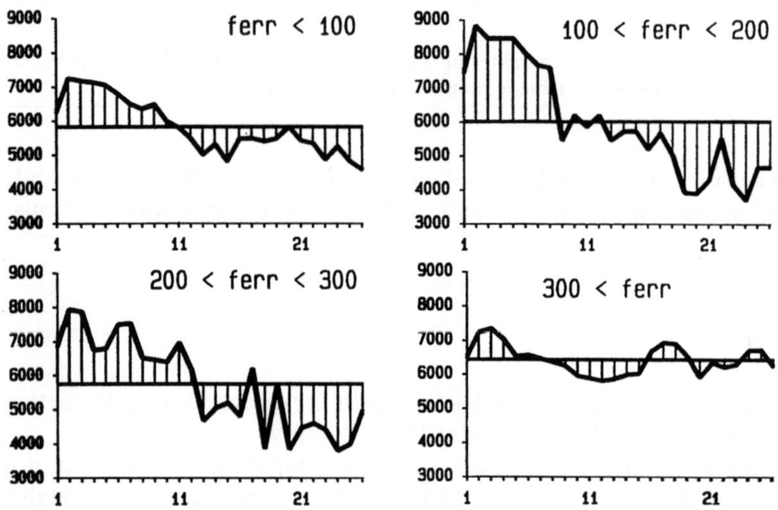

Fig. 5. EPO mean dose intravenously to achieve constant hematocrit by levels of ferritin. Observation period of 26 weeks

S.c. versus i.v. Administration

In view of pharmacokinetic data, EPO given subcutaneously may have advantages over i.v. injection, because after s.c. administration maximal plasma levels are achieved after about 8 [3] – 12 h [10] and maintained for at least 16 h [3] and 3–4 days [10]. As mentioned by some authors, this would lead to reduced dosage [1, 5] and costs [1].

Although Meisl et al. [11] do not give any therapeutic or economic advantages of s.c. EPO administration, in our hands, in a long-term study of a well-monitored group of patients under otherwise constant conditions, there was a clear difference in maintenance dose of EPO for the two routes of application, leading to reduced mean doses of 72% when commencing EPO therapy, and 65% after conversion from i.v. to s.c. administration.

Target hematocrit was achieved earlier (i.v.: 9 days; s.c.: 5 days), and variation in mean dose was lower in the s.c. response group. The effectiveness of EPO is greater when given s.c. than i.v. unlike Kelly et al. [9] state, although a relatively low bioavailability (11%–31%) is determined [10].

We have come to the conclusion, that s.c. administration of EPO obviously has a number of advantages over i.v. administration.

References

1. Bommer J, Ritz E, Weinreich T, Bommer G, Ziegler T (1988) Subcutaneous erythropoietin. Lancet 2(8607):406
2. Delwiche F, Garrity MJ, Powels SJ, Robertson RP, Adamson JW (1983) High levels of the circulating form of parathyroid hormone do not inhibit in vitro erythropoiesis. J Lab Clin Med 102:613–620
3. Egrie JC, Eschbach JW, McGuire T, Adamson JW (1988) Pharmacokinetics of recombinant human erythropoietin. Kidney Int 33(1):262
4. Eschbach JW (1991) Erythropoietin. Biology, clinical use and impact on patients. International symposium Feb 8–9, Nashville
5. Flaharty KK, Besarab A, Caro J, Medina F, Vlasses PH (1990) Intravenous and subcutaneous pharmacokinetics and pharmacodynamics of human recombinant erythropoietin r-Hu-EPO for anemia in end-stage renal disease. Clin Pharmacol Ther 47(2):144
6. Grützmacher P, Ehmer B, Messinger D, Scigalla P (1990) Response to recombinant erythropoietin in aluminum overload and hyperparathyroidism. Kidney Int 37(1):299
7. Hampl H, Riedel E, Wendel G, Stabell U, Kessel M (1988) Influence of parathyroid hormone (PTH) on exogenous erythropoiesis in hemodialysis patients (abstr). Kidney Int 33(1):224
8. Hollomby DJ, Muirhead N, Hodsman AB, Cordy PE, Slaughter D (1990) The role of aluminum (Al) and PTH in erythropoietin (EPO) resistance in hemodialysis-patients (abstr). Kidney Int 37(1):301
9. Kelly MR, Haley NR, Adamson JW, Blagg CR, Abels RI, Eschbach JW (1990) Home subcutaneous (SC) recombinant human erythropoietin (r-Hu-EPO) is as effective and safe as given intravenously (IV) (abstr). Kidney Int 37(1):240
10. MacDougall IC, Neubert P, Coles GA, Roberts DE, Dharmasena AD, Williams JD (1989) Pharmacokinetics of recombinant human erythropoietin in patients on continuous ambulatory peritoneal dialysis. Lancet 1:425–427

11. Meisl F, Manker W, Loibl U, Buxbaum M, Stockenhuber F, Steinbach K (1990) No therapeutic or economic advantage of erythropoietin application by the subcutaneous route (abstr). Kidney Int 37(1):311
12. Meytes D, Bogin E, Ma C, Dukes PP, Massry SG (1981) Effects of parathyroid hormone on erythropoiesis. J Clin Invest 67:1263–1269
13. Schaefer RM, Kuerner B, Zech M, Denninger G, Borneff C et al. (1988) Treatment of the anemia of hemodialysis patients with recombinant human erythropoietin. Int J Artif Organs 11:249–254

Comparison of Three Routes of Administration of Recombinant Human Erythropoietin Expressed in C-127 Mouse Cells

T. Brune, H. Heinrichs, G. Reiner, R. Zerban, J.-U. Pauly, L. Fontaine, D.E. Müller-Wiefel, W. Oster, and F. Schindel

Introduction

Erythropoietin (EPO) is a highly glycosylated hormone produced by peritubular cells of the kidney which triggers the differentiation and proliferation of progenitor cells of the erythropoiesis in mammals and humans. By means of recombinant DNA technology sufficient amounts of purified recombinant human erythropoietin (rHuEPO) have recently become available. Clinical studies have shown that rHuEPO is able to correct the anemia of end-stage renal disease [1, 2]. The pharmacokinetic properties of rHuEPO (produced in Chinese hamster ovary cells, CHO cells) have been investigated for three routes of administration: intravenous (IV), subcutaneous (SC), and intraperitoneal (IP). The results of these pharmacokinetic studies with rHuEPO showed considerable differences in these approaches. The bioavailability was 20%–80% after SC administration and 10%–35% after IP administration. The serum peak level (C_{max}) of rHuEPO was higher after IV injection than after SC and IP administration. The SC and IP routes were characterized by a prolonged absorption phase [3–8].

Here we performed two clinical studies to analyze the pharmacokinetic properties and bioavailability of rHuEPO synthesized in another expression system, C-127 mouse fibroblast cells transfected with the human erythropoietin gene [9]. Absorption kinetics, bioavailability, and other pharmacokinetc parameters of rHuEPO after IV, SC, and IP administration were analyzed. The aim of study 1 was to assess in human volunteers the pharmacokinetic properties of rHuEPO expressed in C-127 mouse cells after single IV and SC administrations of 100 U/kg body weight. Absorption kinetics and the bioavailability of the SC administration were of specific interest. Study 2 examined the pharmacokinetics and bioavailability of rHuEPO after IP administration in eight patients undergoing continuous ambulatory peritoneal dialysis (CAPD). In addition, absorption kinetics and the possible role of the peritoneal serosa were of special interest. Experiments in the rat were performed to determine the role of the peritoneal serosa after IP administration or rHuEPO.

Pagel/Weiss/Jelkmann (Eds.)
Pathophysiology and Pharmacology
of Erythropoietin

Materials and Methods

RHuEPO was provided in vials containing 1 ml solution, corresponding to 8000 U (Behringwerke, Marburg, FRG).

Study 1. Twelve healthy volunteers received single IV bolus and SC doses of 100 U rHuEPO/kg in a randomized two-period cross-over study. A washout period of 2 weeks between IV and SC administrations was observed. Blood samples were taken at the following time points: IV, preinjection, 5, 10, 20, 30, 45 min, 1, 2, 3, 4, 6, 8, 12, 24, and 48 h after injection; SC, preinjection, 30 min, 1, 2, 4, 6, 8, 10, 12, 16, 20, 32, 48, 72 h, and 1 week after injection.

Study 2. Eight pediatric CAPD patients in two medical centers received IP doses of 75 – 225 U rHuEPO/kg in an open, non-randomized clinical study. RHuEPO doses were added to the dialysate fluid of the overnight bag (CAPD II, Fresenius). The dwell time was 18 h. Blood samples and dialysate samples were taken at the following time points: 10, 20, 40 min, 5, 7, 12, 18, and 30 h after administration. We also examined the immunofluorescence pattern of rat peritoneal serosa directly, 3 and 6 h after IP administration of rHuEPO.

For IV administration, systemic clearance, steady-state volume, half-life of the drug, and its mean residence time were derived from mono- and biexponential curves, which were adjusted to individual courses of the serum level by means of nonlinear regression with weighted least squares. For SC administration, the Bateman function together with a lag time was used. Bioavailability was calculated as the ratio of individual AUC values after IV and SC administration. For IP administration, only descriptive methods were used, as persisting serum levels after 18 h did not allow a reliable calculation of the AUC. Only peak time and peak levels were determined.

Results

The median initial half-life of rHuEPO after SC administration clearly had the highest value, with 3.8 h compared to IV administration with 0.5 h (Fig. 1; Table 1). The terminal half-life or rHuEPO after SC administration was higher than that after IV administration (median 17.7 h SC versus 5.0 h IV). The distribution of rHuEPO was essentially confined to serum, as indicated by the median steady-state volume of 3.8 l after IV administration. The median peak time after SC administration was 10 h compared to 5 h after IP administration. The median peak time after IV administration was 0.17 h (Fig. 1, 2; Table 1). The median C_{max} of rHuEPO was 3136 mU/ml after IV administration which was clearly higher than that after SC administration

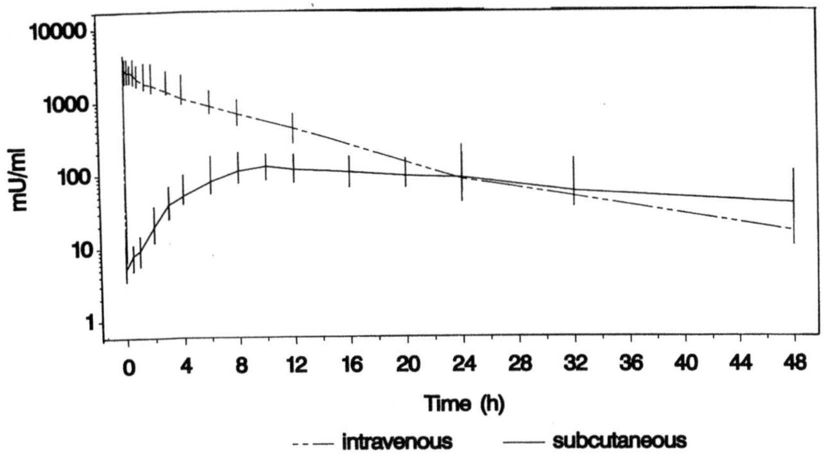

Fig. 1. Erythropoietin serum levels in healthy volunteers: median course and range after IV and SC administration of rHuEPO

(124.4 mU/ml) and IP administration (64.0 mU/ml). The bioavailability of rHuEPO after SC administration was only 25%. The bioavailability of rHuEPO after IP administration could not be assessed, as the serum concentration levels of EPO remained nearly stable for 18 h, and no AUC value could be determined (Fig. 2; Table 1).

Immunofluorescence experiments with rat peritoneal serosa after IP administration of rHuEPO showed that directly after IP administration of

Table 1. Median of pharmacokinetic variables of rHuEPO

	Healthy volunteers (100 U/kg) IV (n=12)	SC (n=11)	CAPD patients (75–225 U/kg) IP (n=8)
Half-life (h)			
Initial	0.5	3.8	n.a.
Terminal	5.0	17.7	n.a.
Mean time (h)	6.8	33.2	n.a.
V-ss or V-sys/f (l)	3.8	71.5	n.a.
Clearance/f (ml/min)	9.2	37.6	n.a.
C_{max} (mU/ml)	3136	124.4	64.0[a]
t_{max} (h)	0.17	10.0	5.0
Lag time (h)	–	1.0	–
AUC (mU × h/ml)	15530	3822	n.a.
Bioavailability (%)	100	25.1	n.a.

The mean value for the bioavailability of the SC administration was 26.9% (95% CI: 20.2% – 33.6%).
n.a.: Not assessed
[a] Standardized for 100 U/kg.

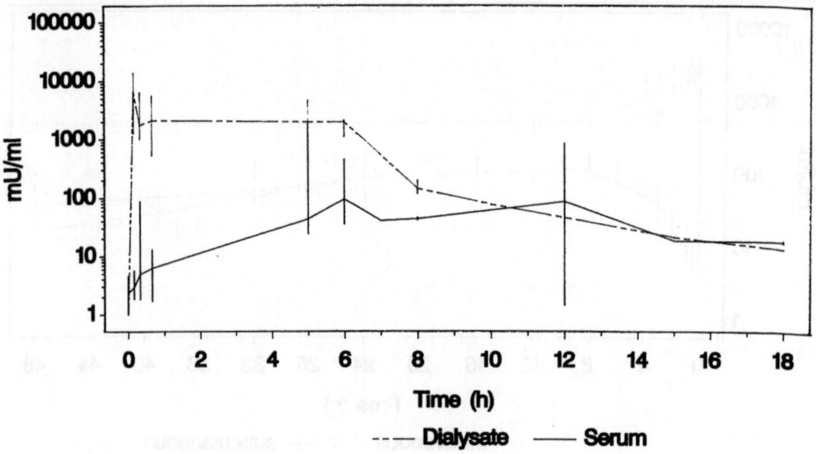

Fig. 2. Erythropoietin levels in CAPD patients: median course and range in serum and dialysate after IP administration of rHuEPO

rHuEPO and 3 h later the peritonal serosa did not show any binding of a monoclonal anti-rHuEPO antibody. In contrast, the peritoneal serosa was strongly positive with the same antibody 6 h after IP administration of rHuEPO (Fig. 3, 4).

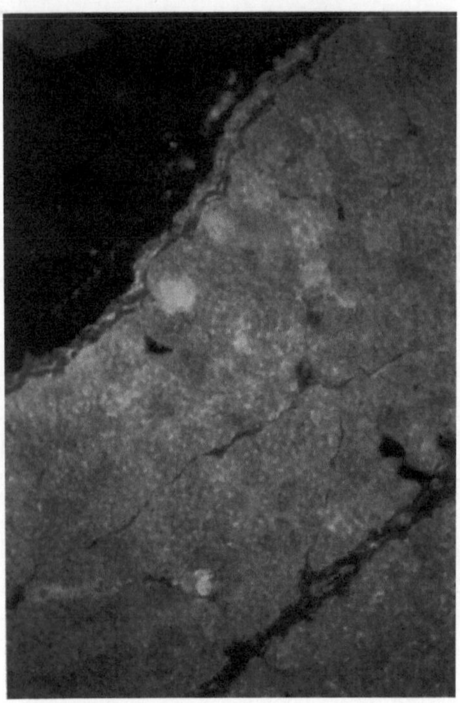

Fig. 3. Immunofluorescence in rats 3 h after IP administration of rHuEPO. Negative Result

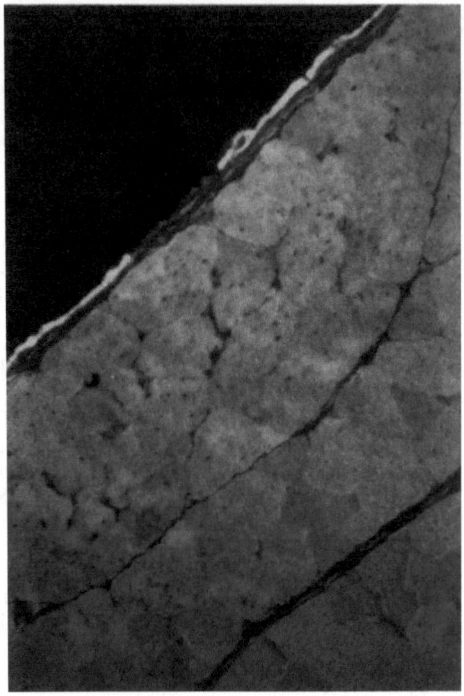

Fig. 4. Immunofluorescence in rats 6 h after IP administration of rHuEPO. Whereas the muscle tissue is negative, the peritonal serosa is clearly positive with a monoclonal antibody against rHuEPO

Conclusions

We conclude that the terminal half-life of rHuEPO after SC administration reflects the kinetics of absorption from the subcutaneous tissue. It seems that SC administration of rHuEPO leads to more physiologic serum levels as compared to IV administration of rHuEPO. In spite of a diminished bioavailability of rHuEPO after SC application of 25% the SC administration of rHuEPO seems to be even more effective clinically in reaching and maintaining appropriate hematocrit values as compared to the IV administration (unpublished results and [10, 11]). Although the bioavailability of rHuEPO after IP administration is probably much lower than that of SC administration, the IP administration could be an alternative route for patients unsuitable for IV and SC administration of rHuEPO. The peritoneal serosa seems capable to adsorb rHuEPO after IP administration and might play a major role for absorption kinetics of rHuEPO after IP administration.

There seems to be no substantial pharmacokinetic difference among the different rHuEPOs derived from CHO cells and C-127 mouse cells. The glycosylation pattern of rHuEPO derived from C-127 mouse cells leads to no major pharmacokinetic differences as compared to rHuEPO from CHO cells. Whereas the IV and SC routes are proven and reliable administration modes of rHuEPO, IP administration is still under clinical evaluation.

References

1. Winearls CG, Oliver DO, Pippard MJ, Reid C, Downing MR, Cotes PM (1986) Effect of human erythropoietin derived from recombinant DNA on the anaemia of patients maintained by chronic haemodialysis. Lancet 2:1175–1178
2. Eschbach JW, Egrie JC, Downing MR, Browne JK, Adamson JW (1987) Correction of the anemia of end-stage renal disease with recombinant human erythropoietin. Results of a combined phase I and II clinical trial. N Engl J Med 316: 73–78
3. Boelart J, Schurgers M, Matthys E, Daneels R, De Cre M, Bogaert M (1988) Recombinant human erythropoietin pharmacokinetics in CAPD patients: comparison of the intravenous, subcutaneous and intraperitoneal routes. Nephrol Dial Transplant 3: 493
4. Egrie JC, Eschbach JW, McGurie T, Adamson JW (1988) Pharmacokinetics of recombinant human erythropoietin (rHuEPO) administered to hemodialysis patients. Kidney Int 33:262 (abstract)
5. Kampf D, Kahl A, Passlick J, Pustelnik A, Eckardt K-U, Ehmer B, Jacobs C, Baumelou A, Grabensee B, Gahl GM (1989) Single-dose kinetics of recombinant human erythropoietin after intravenous, subcutaneous and intraperitoneal administration. Preliminary results. In: Baldamus CA, Scigalla P, Wieczorek L. Koch KM (eds) Erythropoietin: from molecular structure to clinical application. Contrib Nephrol 76:106–111
6. Lui SF, Chung WWM, Leung CB, Chan K, Lai KN (1990) Pharmacokinetics and pharmacodynamics of subcutaneous and intraperitoneal administration of recombinant human erythropoietin in patients on continuous ambulatory peritoneal dialysis. Clin Nephrol 33: 47–51
7. Flaharty KK, Caro J, Erslev A, Whalen JJ, Morris EM, Bjornsson TD, Vlasses PH (1990) Pharmacokinetics and erythropoietic response to human recombinant erythropoietin in healthy men. Clin Pharmacol Ther 47:557–564
8. McMahon FG, Vargas R, Ryan M, Jain AK, Abels RI, Perry B, Smith IL (1990) Pharmacokinetics and effects of human erythropoietin after intravenous and subcutaneous injections in healthy volunteers. Blood 76:1718–1722
9. Wizemann V, Brune T, Kramer W, Schäfer R, Schütterle G (1991) Recombinant human erythropoietin expressed in C-127 mouse cells: efficacy, side-effects and cardiovascular actions. Nephrol Dial Transplant 2 (Suppl):122–125
10. Bommer J, Ritz E, Weinreich T, Bommer G, Ziegler T (1988) Subcutaneous erythropoietin. Lancet ii:406
11. Granolleras C, Branger B, Deschodt G, Alsabadani B, Shaldon S (1989) The optimal form of erythropoietin for end stage renal disease patients. Nephrol Dial Transplant 4:473 (abstract)

Peritoneal Permeability in Continuous Ambulatory Peritoneal Dialysis After the Correction of Anemia by Erythropoietin

C. Castelnovo, A. De Vecchi, A. Scalamogna, and S. Casati

Introduction

Anemia due to insufficient production of erythropoietin is a common feature of patients with chronic renal failure undergoing dialysis [1–6]. Although anemia in patients on continuous ambulatory peritoneal dialysis (CAPD) is less frequent and severe than in hemodialysis (HD) [5] it can be a relevant clinical problem [6].

The administration of recombinant human erythropoietin (rHuEPO) has been proved to be effective in correcting anemia both in CAPD and in HD patients [1–6]. Some authors suggest that the drug per se or the subsequent correction of anemia could increase the peritoneal ultrafiltration in patients on peritoneal dialysis. In the present study, in order to assess the effect of rHuEPO on peritoneal dialysis efficiency, we performed peritoneal equilibration test (PET) before and after the correction of anemia.

Patients and Methods

Seven patients undergoing CAPD were studied. Four were women and three men; the mean age was 49.4±5.9 years (range 40–56). They had been on CAPD for 43.1±20.5 months (range 7–59). Residual renal function was 2.30±2.69 ml/min and urine volume was 430±244 ml/24 h. Characteristics of the patients at the beginning of the study are reported in Table 1.

Table 1. Characteristics of patients at the beginning of the study

	Sex	Age (years)	CAPD (months)	Renal disease
1	F	57	51	Non-diagnosed
2	F	46	47	Chronic pyelonephritis
3	M	40	7	Non-diagnosed
4	F	52	38	Non-diagnosed
5	M	55	73	Non-diagnosed
6	F	57	19	Non-diagnosed
7	M	47	39	Cryoglobulinemic glomerulonephritis

Pagel/Weiss/Jelkmann (Eds.)
Pathophysiology and Pharmacology of Erythropoietin

RHuEPO was given subcutaneously twice weekly with an initial dose of 100 U/kg body weight per week. Every 6 weeks the rHuEPO dose was increased by 20% till the target hemoglobin of 9.5–10.5 g/ml was achieved. The dose of rHuEPO was decreased by 20% in the case of further increases in hemoglobin above the target level.

PET was performed 1–30 days before starting rHuEPO and 1–60 days after the correction of anemia. PET was performed according to the method of Verger [7] with a standard 4.25%-glucose, 4-h, 2-liter exchange. Plasma and dialysate urea, creatinine and glucose were determined after 0.2, and 4 h. Residual volume was determined according to Chandran and Flynn [8]. Twenty-four hour peritoneal and urinary volumes and clearances of creatinine and urea were measured by routine methods. No patient showed episodes of peritonitis during the 3 months before entering the study. The number of hypertonic exchanges was not modified during the study period. Data are reported as means ± SD. Statistical calculations were performed by the Student's *t*-test for paired data.

Results

The correction of anemia was achieved in all patients within 86±37 days. During this period no patients had peritonitis episodes. At the beginning 4-h

Table 2. Pet before and after rHuEPO treatment

	Before rHuEPO	After rHuEPO
4-h UF (ml)	771 ±210	983 ±186
4-h D/P creatinine	0.61±0.08	0.68±0.10
4-h D/P urea	0.84±0.03	0.79±0.19
4-h D/Do glucose	0.28±0.05	0.21±0.04

UF, ultrafiltration; D/P; plasma/dialysate ratio; D/DO D = dialysate glucose at 1 dwell time, DO = dialysate glucose at 0 dwell time

Table 3: 24-h peritoneal function before and after rHuEPO treatment

	Before rHuEPO	After rHuEPO
UF (ml)	9457 ±609.1	9566 ±642
Creatinine, clearance (ml/min)	4.60±0.80	4.87±0.84
Urea clearance (ml/min)	5.9 ±0.53	5.5 ±1.28
Protein loss (g/day)	5.68±1.97	7.52±3.11
Calcium clearance (ml/min)	3.24±1.49	3.86±0.66
Phosphate clearance (ml/min)	3.08±1.53	3.95±0.65

UF, ultrafiltration

ultrafiltration was below 500 ml in two patients. Normal values of equilibration were defined as the mean ± SD of our CAPD population [9]. Hyperabsorption was observed in only one patient and no one showed a reduced peritoneal permeability. Mean levels of 4-h ultrafiltration and urea and creatinine plasma/dialysate ratios did not show significant changes (Table 2) after the correction of anemia. Residual volume did not show variations of more than 15% in any patient. Ultrafiltration increased more than 20% in four patients. There was no change in 24-h peritoneal ultrafiltration, creatinine, and urea clearances after the correction of anemia (Table 3). Residual renal function measured as 24-h urine volume and creatinine clearance did not change after rHuEPO treatment (Table 4).

Discussion

RHuEPO is a safe and effective drug in the treatment of anemia in patients either on HD or CAPD [1–6].

In HD patients the correction of anemia was not followed by any changes in dialysis efficiency [3], whereas in CAPD some authors reported an increase in 4-h peritoneal ultrafiltration after the correction of anemia [10]. It has been hypothesized [10] that the correction of anemia improves cardiac function and may ameliorate dialysis efficiency by increasing mesenteric perfusion. These authors found a significant rise in 4-h ultrafiltration followed by an augmented clearance of urea, while clearance of creatinine did not change. On the contrary, others [4, 6] did not find significant changes in ultrafiltration volumes, plasma/dialysate creatinine ratios, or 0–4-h dialysate glucose concentration ratios.

In our series we observed a nonsignificant increase in 4-h ultrafiltration without changes in plasma/dialysate urea, creatinine ratios, and dialysate glucose concentration ratio while 24-h ultrafiltration did not change significantly. The failure to show a statistically significant increase in ultrafiltration is probably due to the small number of patients, rather than to the lack of efficacy of rHuEPO in increasing ultrafiltration in CAPD patients.

Table 4. 24-h residual renal function before and after rHuEPO treatment

	Before rHuEPO	After rHuEPO
Urine volume (ml/24 h)	430 ±244	275 ±43
Creatinine clearance (ml/min)	2.30±2.69	0.72±0.15
Urea clearance (ml/min)	0.38±0.34	0.37±0.15
Protein loss (g/day)	0.36±0.30	0.17±0.22
Calcium clearance (ml/min)	0.15±0.03	0.09±0.04
Phosphate clearance (ml/min)	0.48±0.33	0.30±0.16

The discrepancy between our and others results could be explained by differences in the characteristics of the patients (age, sex, time of dialysis, severity of anemia, etc.), or in methods used for performing PET (glucose concentration, volume etc.), or in rHuEPO administration protocol. In our series only patients with serum hemoglobin below 7.5 g/ml were treated by rHuEPO. In addition we performed PET by a standard 2-liter exchange with 4.25% glucose concentration [9].

In conclusion our data show that the correction of anemia by rHuEPO treatment does not seem to be associated with a significant increase in peritoneal function, although other studies on larger patient populations are needed to further assess this hypothesis.

References

1. Casati S, Passerini P, Campise MR, Graziani G, Cesana B, Perisic M, Ponticelli C (1987) Benefits and risks of protracted treatment with human erythropoietin in patients having hemodialysis. Br Med J 295:1017–1020
2. Eschbach JW, Adamson JW (1988) Correction of the anemia of hemodialysis patients with recombinant human erythropoietin. Results of a multicenter study. Kidney Int 33:189
3. Ponticelli C, Casati S (1989) Correction of anemia with recombinant human erythropoietin. Nephron 52:201–208
4. Macdougall IC, Davies ME, Hutton RD, Cavill I, Lewis NP, Coles GA, Williams JD (1990) The treatment of renal anaemia in CAPD patients with recombinant human erythropoietin. Nephrol Dial Transplant 5:950–955
5. De Vecchi A, Castelnovo C, Scalamogna A, Guerra L, Casati S (1991) Erythropoietin in CAPD patients: an Italian experience. In: Peritoneal Dialysis. G La Grece, C Rones, M Ferciani, S Chiaramonte, P Conz (Eds). Wichtig Editore, Milano pp 215–222
6. Piraino B, Johnston JR (1990) The use of subcutaneous erythropoietin on CAPD patients. Clin Nephrol 33:200–202
7. Verger C (1985) Relationship between peritoneal membrane structure and its permeability: clinical implications. Advances in Continuous Ambulatory Peritoneal Dialysis, Toronto University pp 87–95
8. Chandran PKG, Flynn CT (1985) Residual intraperitoneal volumes in CAPD patients. Advances in CAPD, 18–25
9. De Vecchi A, Scalamogna A, Guerra L, Soriani N, Zubani R, Ponticelli C (1987) Analisi delle curve di ultrafiltrazione in 60 pazienti in CAPD. In: Atti del IV Convegno Nazionale di Dialisi Peritoneale. Wichtig, Milan, pp 55–57
10. Steinhauer HB, Lubrick-Birkner I, Dreyling KW, Horl WH, Schollmeyer P (1989) Increased ultrafiltration after erythropoietin induced correction of renal anemia in patients on continuous ambulatory peritoneal dialysis. Nephron 53:91–92

V. Primary Hemopoietic Disorders and Chemotherapy

Serum Erythropoietin Levels in Aplastic Anemia Determined by ELISA

H. Schrezenmeier, G. Noé, I.N. Rich, and A. Raghavachar

Introduction

Aplastic anemia (AA) is a disorder characterized by bone marrow failure leading to pancytopenia. The underlying cause remains unknown in most cases. Several mechanisms have been implicated in the pathogenesis of the disease, including an intrinsic stem cell defect, lymphocyte-mediated suppression of hemotopoiesis, dysfunction of the bone marrow microenvironment, and abnormalities in regulation of hematopoiesis by humoral factors (Nissen-Druey 1989). Elevated levels of interferon-γ in the serum (Zoumbos et al. 1985) and increased production of tumor necrosis factor-α by mononuclear cells (Hinterberger et al. 1988), both potent inhibitors of hematopoiesis, have been described in AA. Monocyte interleukin-1 (IL-1) production is decreased in AA patients (Nakao et al. 1988). However, elevated levels of serum burst-promoting activity (BPA) and colony-stimulating activity (CSA) have been reported (Adams and Barrett 1985; Nissen et al. 1985; Yen et al. 1985). In previous studies, we demonstrated that elevated CSA is attributable mainly to granulocyte-macrophage colony-stimulating factor (GM-CSF) in the serum (Schrezenmeier et al. 1989). Elevated levels of BPA, CSA and GM-CSF are correlated with severity and prognosis of the disease (Nissen et al. 1985; Schrezenmeier et al. 1989).

There are several reports demonstrating high levels of erythropoietin (EPO) in AA sera compared to other types of anemia in patients with normal renal function (Hammond et al. 1968; Pavlovic-Kentera et al. 1979; Takeichi et al. 1988; Urabe et al. 1987; Wognum et al. 1989; Mason-Garcia et al. 1990). However, the main interest of previous studies was to compare EPO concentration in AA with EPO levels in other hematological disorders. In these former studies AA was considered a homogeneous disease. With few exceptions, sera investigated were drawn at diagnosis, and there was poor follow-up of patients during disease progression or recovery. Finally, EPO levels were correlated with hematocrit rather than with further clinical features of the patients.

The objectives of this study were: (a) to evaluate whether there is at least a subgroup of AA patients who show inadequately elevated EPO serum levels in response to anemia; (b) to determine whether, as with CSA, EPO

Pagel/Weiss/Jelkmann (Eds.)
Pathophysiology and Pharmacology
of Erythropoietin

can serve as a marker of disease severity and as a prognostic parameter; and (c) to monitor EPO serum levels during the course of the disease.

Material and Methods

We analyzed sera from 66 AA patients (median age 36 years), 42 of whom had severe aplastic anemia according to the criteria of Camitta et al. (1982) and 24 nonsevere AA. Of these, 52 were treated with immunosuppressive regimens using antilymphocyte globulin (ALG) and cyclosporin A either alone or in combination (Frickhofen et al. 1991). ALG was given for 8 days; 26 patients achieved a partial or complete remission and 26 remained pancytopenic. Blood samples were drawn at diagnosis, during the 14 days before initiation of ALG treatment, within 3 days after the last day of ALG treatment, and 3 months or later after initiation of immunosuppression.

The control groups consisted of 24 healthy individuals and 16 patients diagnosed as having iron deficiency or hemolytic anemia.

For the assay of EPO, serum was prepared by centrifugation at 2000 *g* for 10 min and stored at –20°C. EPO serum levels were measured by a "sandwich" EPO enzyme-linked immunosorbent assay (ELISA) using purified monospecific polyclonal anti-EPO antibodies. Details of this specific and highly sensitive ELISA are described by Noé et al. (1991). The mean EPO serum level in healthy controls was 9.4 mU/ml (range 2–17.9 mU/ml).

Differences in EPO serum levels between the groups were compared by the Mann-Whitney test. Where appropriate, results are given as the mean ± standard error of the mean. The relationship between EPO serum levels and hematocrit values was assessed with exponential regression analysis employing the least-squares method. Regression slopes were compared using an analysis of covariance.

Results

Figure 1 presents a semilogarithmic plot of the exponential reverse relationship between the EPO serum levels and hematocrit in the group of patients with iron deficiency or hemolytic anemia ($n=16$) and in the AA patients ($n=66$). There was a wide scattering of values, but EPO serum levels correlated negatively with hematocrit in the AA patients ($r=-0.6$; $p=0.00001$). Hematocrit values were not significantly different in the two groups: 25.3% ±0.7% in AA patients versus 28.8% ±1.8% in iron deficiency or hemolytic anemia patients ($p=0.063$). However, mean EPO serum levels were significantly higher in AA patients (2331±510 mU/ml) compared to anemic controls (341±178 mU/ml; $p<0.001$). In addition, comparison of the ($\log_{10}$EPO)/hematocrit relationship in the two groups revealed a significantly different slope of the two regression lines ($p<0.05$).

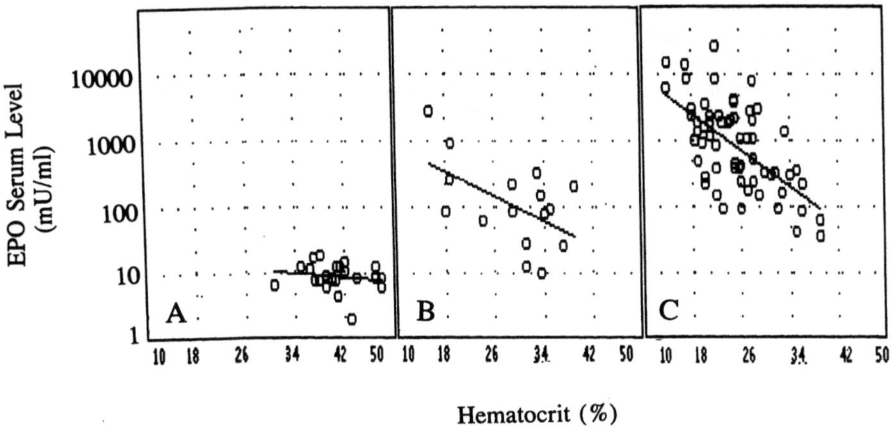

Fig. 1A–C. Relationship between hematocrit and EPO serum levels. **A** Healthy controls (n=24). **B** Patients with iron deficiency or hemolytic anemia (n=16). **C** AA patients (n=66). Equations for regression lines: $\log_{10}EPO = -0{,}0066x + 1.2$ (**A**), $\log_{10}EPO = -0.0474x + 3.4$ (**B**), and $\log_{10}EPO = -0.0681x + 4.6$ (**C**)

Comparison of EPO serum levels and hematocrit values of patients with severe (n=42) or nonsevere (n=24) AA is given in Table 1. EPO serum levels in the two groups did not differ significantly, nor did the regression lines.

Table 1. Hematocrit and EPO serum levels at diagnosis of patients with severe or nonsevere aplastic anemia

Severitiy of disease	n	EPO serum levels (mU/ml)	Hematocrit (%)
All patients	66	2331± 510	25.3±0.7
Severe AA	42	1887± 418	25.2±1.0
Nonsevere AA	24	3107±1197	25.6±1.1

Differences not significant (Mann-Whitney test).

Table 2. Hematocrit and EPO serum levels at diagnosis of patients responding or not responding to immunosuppressive treatment

Response to immunosuppression	n	EPO serum levels (mU/ml)	Hematocrit (%)
Responders	26	3190±1123	24.4±1.2
Nonresponders	26	1837± 384	25.2±0.9

Differences not significant (Mann-Whitney test).

Table 3. Hematocrit and EPO serum levels before and after immunosuppressive treatment with antilymphocyte globulin

Time of measurement	EPO serum levels (mU/ml)	Hematocrit (%)
Before treatment	2331±510	25.3±0.7
After treatment	664±188	26.1±1.0

Blood was collected during the 14 days before initiation of ALG treatment. Blood for measurement after treatment was drawn, in most cases, 1–3 days after the last day of an 8-day course of ALG. Differences in serum EPO significant, those in hematocrit not significant (Mann-Whitney test).

Comparing pretreatment values of the 26 responders and the 26 nonresponders to immunosuppressive treatment, we observed no significant difference in EPO levels or in the regression lines (Table 2).

As shown in Table 3, immunosuppressive treatment by ALG induced a significant decline in EPO serum levels, from a mean of 2330 mU/ml before treatment to 664 mU/ml immediately after treatment ($p=0.03$). Since there was no significant change in hematocrit, the decrease in EPO serum levels was not attributable to an increase in red blood cell mass. The ALG-induced decline in EPO serum levels occurred regardless of whether patients received concomitant medication with cyclosporin A. EPO serum levels after ALG treatment were not significantly inversely correlated with their respective hematocrit values ($r=-0.32$; $p=0.16$).

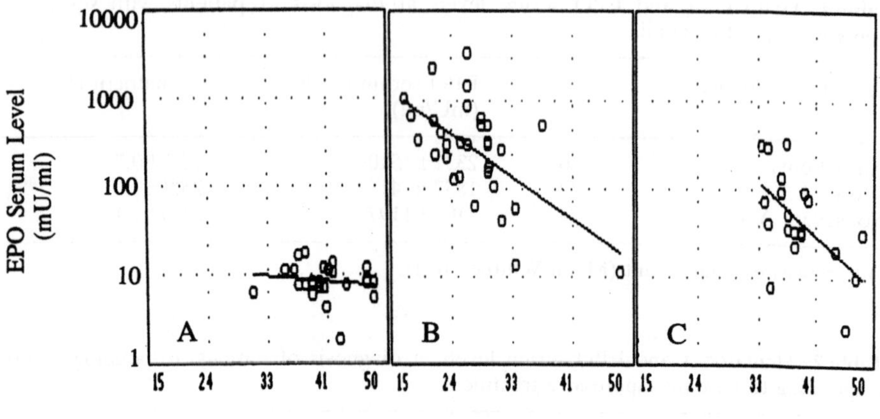

Fig. 2A–C. Relationship between hematocrit and EPO serum levels. **A** Healthy controls ($n=24$). **B** Nonresponders ≥ 3 months after the start of immunosuppression ($n=26$). **C** AA patients in remission ≥ 3 months after initiation of immunosuppression ($n=20$). Equations for regression lines: $\log_{10}EPO = -0,0066x + 1.2$ (**A**) and $\log_{10}EPO = -0.0654x + 4.3$ (**C**). The difference between the slopes in **A** and **C** is significant ($p<0.05$)

During hematopoietic recovery, EPO serum levels decreased but remained elevated. Twenty-one patients achieving a remission after immunosuppression with normal or only moderately decreased hematocrit still showed elevated EPO serum levels compared to healthy controls (n=24; Fig. 2). The difference between EPO serum levels (85.6±22.7 mU/ml in AA patients versus 9.4±0.8 mU/ml in controls) was significant (p<0.001).

Discussion

EPO is elevated in response to anemia in all nonrenal anemias (Erslev et al. 1987). The relationship between hematocrit and EPO serum levels in AA is a matter of controversy. Some authors reported relatively high serum EPO bioactivity for the degree of anemia in patients with AA compared to patients with iron deficiency or hemolytic or megaloblastic anemia (Hammond et al. 1968; Pavlovic-Kentera et al. 1979; Takeichi et al. 1988). In contrast, others (Alexanian 1973; Erslev et al. 1987) observed EPO levels which were strictly dependent on hematocrit, irrespective of the specific type of nonrenal anemia. Using a specific and sensitive EPO ELISA and using the serum from a large number of patients, we confirm that EPO serum levels are higher in AA than in nonhypoplastic types of anemia with comparable degree of anemia.

There is evidence that EPO production is modulated by cytokines; IL-1 and tumor necrosis factor-α inhibit the pO_2-dependent formation of EPO in hepatoma cell cultures (Jelkmann et al. 1990). Since IL-1 production by mononuclear cells is decreased in AA (Nakao et al. 1988), the lack of negative regulation of EPO production by IL-1 might contribute to the relatively high EPO serum levels in AA. It is generally accepted that renal production of EPO is controlled by the red blood cell mass via hypoxia (Jelkmann 1986). Stohlman and Brecher (1959) postulated that, in addition to this feedback control, the proliferative activity of the erythroid bone marrow may influence EPO production. Our observations on AA patients support this hypothesis.

ALG treatment leads to a decrease in EPO levels. This is not caused by a corresponding change in hematocrit. Since this phenomenon is present both in patients with and in those without concomitant medication with cyclosporin A, it is not attributable to a side effect of this known nephrotoxic drug. ALG has a variety of effects, including complement-mediated cytotoxicity on lymphocytes, induction of lymphocyte proliferation, direct action on hematopoietic progenitor cells, and stimulation of growth factor release (Kawano et al. 1988; Barbano et al. 1988; Taniguchi et al. 1990). It is not known which of these is responsible for a decline in EPO serum levels. One can speculate that the ALG effect on EPO is mediated by cytokines that are released by different target cells by ALG stimulation (Kawano et al. 1988; Taniguchi et al. 1990). In the early phase after ALG treatment, EPO levels and respective hematocrit are no longer correlated. Therefore, factors other than red blood cell mass seem to be involved in EPO regulation at that time. The phenomenon of ALG-induced decrease in EPO serum levels deserves careful follow-up because we cannot rule out that decrease in EPO serum levels in the early phase after ALG treatment interferes with hematopoietic recovery.

Surprisingly, EPO serum levels in AA patients in remission continued to be higher than in healthy controls. It is known that in AA patients even in remission the number of progenitor cells in the bone marrow, as assessed by colony assay in semisolid media or in long-term bone marrow cultures, is reduced (Kern et al. 1977; Marsh et al. 1990). The diminished population of progenitors cells might lead to increased EPO levels even in remission patients.

EPO does not serve as an indicator of disease severity, nor does it predict outcome after immunosuppression.

Studies on EPO serum levels in different types of anemia can lead to better understanding of regulation of EPO production. We conclude that, in addition to the oxygen-regulated feedback control, other mechanisms such as the effect of cytokines and the erythropoietic potential of the erythron are involved in the regulation of EPO serum levels under pathophysiological conditions.

References

Adams JA, Barett AJ (1982) Haematopoietic stimulators in the serum of patients with severe aplastic anaemia. Br J Haematology 52:327

Alexanian R (1973) Erythropoietin excretion in bone marrow failure and hemolytic anemia. J Lab Clin Med 82:438

Barbano GC, Schenone A, Roncella S, Ghio R, Corcione A, Mori PG, Ferrarini M, Pistoia V (1988) Anti-lymphocyte globulin stimulates normal human T cells to proliferate and to release lymphokines in vitro. A study at the clonal level. Blood 72:956

Camitta BM, Storb R, Thomas ED (1982) Aplastic anemia: pathogenesis, diagnosis, treatment and prognosis. N Engl J Med 306:645

Erslev AJ, Wilson J, Caro J (1987) Erythropoietin titers in anemic, non-uremic patients. J Lab Clin Med 109:429

Frickhofen N, Kaltwasser JP, Schrezenmeier H, Raghavachar A, Vogt HG, Herrmann F, Freund M, Meusers P, Salama A, Heimpel H (1991). A prospective trial of antilymphocyte globulin and methylprednisolone with or without cyclosporine for treatment of aplastic anemia. N Engl J Med 324:12972

Hammond D, Shore N, Movassaghi N (1968) Production, utilization and excretion of erythropoietin. I. Chronic anemias. II. Aplastic crisis. III. Erythropoietic effects of normal plasma. Ann NY Acad Sci 149:516

Hinterberger W, Adolf G, Aichinger G, Dudczak R, Geißler K. Höcker P, Huber C, Kalhs P, Knapp W, Köller U, Lechner K, Volc-Platzer B (1988) Further evidence for lymphokine overproduction in severe aplastic anemia. Blood 72:266

Jelkmann W (1986) Renal erythropoietin: properties and production. Rev Physiol Biochem Pharmacol 104:139

Jelkmann W, Wolff M, Fandrey J (1990) Modulation of the production of erythropoietin by cytokines: in-vitro studies and their clinical implications. In: Schäfer RM, Heidland A, Hörl WH (eds) Erythropoietin in the 90s. S. Karger, Basel (Contributions to Nephrology, vol 87), p 68

Kawano Y, Nissen C, Gratwohl A, Speck B (1988) Immunostimulatory effects of different antilymphocyte globulin preparations: a possible clue to their clinical effect. Br J Haematology 68:115

Kern P, Heimpel H, Heit W, Kubanek B (1977) Granulocytic progenitor cells in aplastic anemia. Br J Haematol 35:613

Marsh JCW, Chang J, Testa NG, Hows JM, Dexter TM (1990) The hemopoietic defect in aplastic anemia assessed by long-term bone marrow culture. Blood 76:1748

Mason-Garcia M, Beckman BS, Brokins JW, Powell JS, Lanham W, Blaisdell S, Keay L, Su-Chen L, Fisher JW (1990) Development of a new radioimmunoassay for erythropoietin using recombinant erythropoietin. Kidney Internat 38:969

Nakao S, Matsushima K, Young N (1988) Decreased interleukin 1 production in aplastic anaemia. Br J Haematology 71:431

Nissen C, Moser Y, Speck B, Gratwohl A, Weis J (1985) Stimulatory serum factors in aplastic anaemia. Br J Haematology 61:499

Nissen-Druey C (1989) Pathophysiology of aplastic anemia. Baillière's clinical Haematology 2:37

Noé G, Riedel W, Kubanek B, Rich IN (1992) A sensitive sandwich ELISA for measuring erythropoietin in human serum. Br J Haematology 80:285

Pavlovic-Kentera V, Milenkovic P, Rudivic R, Jovanvic V, Biljanovic-Paunovic L (1979) Erythropoietin in aplastic anemia. Blut 39:345

Schrezenmeier H, Raghavachar A, Heimpel H (1989) GM-CSF levels in the sera of patients with aplastic anemia. Blut 59:339

Stohlman F Jr, Brecher G (1959) Humoral regulation of erythropoiesis. V. Relationship of plasma erythropoietin level to bone marrow activity. Proc Soc Exp Biol Med 100:43

Takeichi N, Umemura T, Nishimura J, Motomura S, Kozuru M, Ibayashi H (1988) Regulation of erythropoietin and burst-promoting activity production in patients with aplastic anemia and iron deficiency anemia. Acta Haematol 80:145

Taniguchi Y, Frickhofen N, Raghavachar A, Digel W, Heimpel H (1990) Antilymphocyte immunologlobulins stimulate peripheral blood lymphocytes to proliferate and release lymphokines. Eur J Haematol 44:244

Urabe A, Saito T, Fukamachi H, Kubota M, Takaku F (1987) Serum erythropoietin titers in the anemia of chronic renal failure and other hematological states. Int J Cell Clon 5:202

Wognum AW, Lansdorp PM, Eaves AC, Krystal G (1989) An enzyme-linked immunosorbent assay for erythropoietin using monoclonal antibodies, tetrameric complexes, and substrate amplification. Blood 74:622

Yen YP, Zabala P, Doney K. Clemons G, Gillis S, Powell JS, Adamson JW (1985) Hematopoietic growth factors in human serum. Erythroid burst-promoting activity in normal subjects and in patients with severe aplastic anemia. J Lab Clin Med 106:384

Zoumbos NC, Gascon P, Djen JY, Young NS (1985) Interferon is a mediator of hematopoietic suppression in aplastic anemia in vitro and possibly in vivo. Proc Natl Acad Sci USA 82:188

Serum Erythropoietin in Myelodysplastic Syndromes

A. Grossi, D. Rafanelli, A.M. Vannucchi, G. Longo, P. Bacci, A. Del Genovese, and P.R. Ferrini

The myelodysplastic syndromes (MDS) are a group of clonal hematological disorders characterized by a highly ineffective hemopoiesis affecting all bone marrow lineages. They are commonly defined according to the French-American-British (FAB) classification [1]. The most common feature of these syndromes is an anemia of varying degree, which is unresponsive to common hematinics, thus defined as refractory anemia. Laboratory studies have shown poor erythroid progenitor growth in culture [2], low reticulocyte counts in relation to hematocrit, and highly ineffective erythropoiesis when ferrokinetic studies were performed [3]. The recent availability of immunoassays for erythropoietin (Epo) has made possible the measurement of its serum levels in many clinical conditions characterized by anemia. This has led to the interesting observation that, besides the expected low levels unrelated to the degree of anemia that are observed in chronic renal failure, Epo levels were also lower than expected in diseases such as sickle cell disease [4], rheumathoid arthritis [5, 6], cancer [7], multiple myeloma [8], and AIDS [9].

This data has prompted a number of clinical trials. Serum Epo has also been measured in MDS, and nonuniform results have been found in the few studies reported so far, including the present paper. Vadhan-Raj et al. [10] found that Epo levels are markedly elevated and appropriate to the degree of anemia in 14 patients with MDS, although distributed over a large range (63–3700 mU/ml), and the slope of correlation between serum Epo and hematocrit was superimposable upon that of iron deficiency anemia. Also, Bessho et al. [11] reported very high levels in eight patients. The largest study was published by Jacobs et al. [12], who observed an overall negative correlation between Epo and hemoglobin in 46 MDS patients, with no difference in terms of FAB classification, but warned that the response was variable over a broad range of concentrations (32–4900 mU/ml) even in patients with similar hemoglobin concentrations, suggesting that the result might be due to the number of observations. A weak negative correlation was also found with the percentage of erythroblasts in the bone marrow.

We studied ten patients with MDS and one with agnogenic myeloid metaplasia, another clonal disease characterized by ineffective erythropoiesis. None of them presented clinical or laboratory signs of renal failure. All these patients have later been enrolled in a therapeutic trial with recombinant human Epo. Serum Epo was in the range of 33–4000, and very scattered

Pagel/Weiss/Jelkmann (Eds.)
Pathophysiology and Pharmacology of Erythropoietin

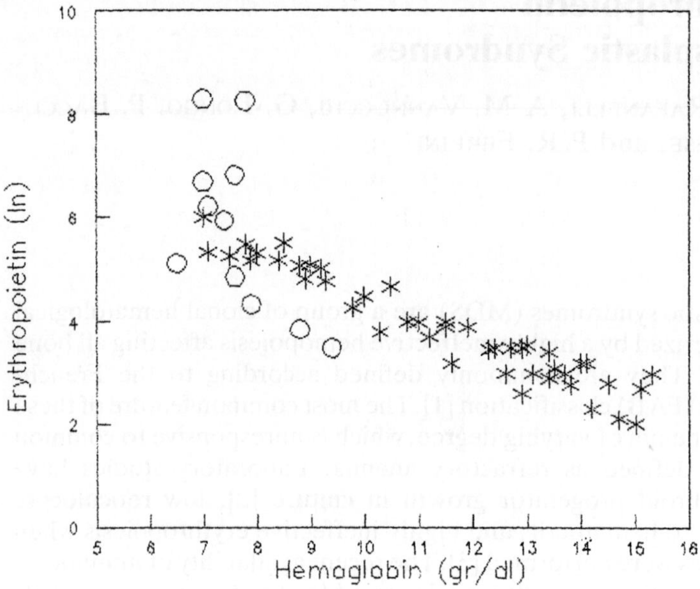

Fig. 1. Serum Epo concentrations of patients (O) and controls (*) plotted against hemoglobin levels. The parameters are not significantly correlated in MDS patients (r=–.579;p=.0619)

Table 1. Laboratory parameters in 11 MDS patients

Pat. no.	Diag	Hb (g/dl)	Epo (mU/ml)	Ret (109/l)	LFR (%)	MFR (%)	HFR (%)	Resp
1	RAS	7.8	3858	37.2	89.1	6.4	4.5	OT
2	RAS	7.6	129	42.5	78.4	13.9	7.7	PR
3	RA	7.0	4100	27.6	78.4	13.9	7.7	OT
4	RA	8.4	500	35.8	75	17.3	7.7	CR
5	RAEB	7.0	803	287	84	9	6.7	NR
6	RAEB	8.8	47	59.5	76.3	18.3	5.4	CR
7	RA	7.6	915	22.2	84.4	13.7	1.9	OT
8	RA	6.5	167	35.9	77.5	14.8	7.7	OT
9	RA	7.4	380	8.7	88.0	6.4	5.6	NR
10	AMM	7.9	77.2	75.2	78.7	15.3	6.0	CR
11	RAEB	9.4	33	122.5	75.7	21.2	3.1	CR

Diag, Diagnosis; RA, refractory anemia; RAS, refractory anemia with ringed sideroblast; RAEB, refractory anemia with excess of blasts; AMM: agnogenic myeloid metaplasia; Hb, hemoglobin; Reticulocyte absolute number; LFR, low fluorescence ratio reticulocytes; MFR, medium fluorescence ratio reticulocytes; HFR, high fluorescence ratio reticulocytes. Resp, responsiveness to Epo therapy; OT, on therapy; CR, complete remission; PR, partial remission; NR, nonremission.

values were found with similar hemoglobin concentrations. No correlation existed between Epo and hemoglobin, as is evident in Fig. 1 (r=–.579; p=.0619). We considered hemoglobin to be a more appropriate parameter for comparison than hematocrit because the latter is influenced by the macrocytosis either due to reticulocyte number in the red blood cell population, mainly those newly released from bone marrow, or to DNA-impaired synthesis [13]. We analyzed Epo levels in relation to reticulocyte absolute count, and again we found no correlation (Table 1).

We also investigated whether Epo levels are correlated with a particular maturation pattern of reticulocytes. For this we used a Sysmex R-1000 (TOA Medical Electr., Hamburg, FRG) reticulocyte-dedicated flow-cytometer, analyzing the RNA-derived fluorescence [14]. Cells were divided into three classes: high, medium, and low fluorescence ratio. This classification is age dependent in that the most fluorescent cells are the most immature while the mature cells give almost no fluorescence due to the negligible RNA content. We found that Epo levels were not correlated with any particular distribution of the fluorescence classes (Table 1). The growth of bone marrow burst colony-forming units–erythroid (BFU-E) and colony-forming units–erythroid (CFU-E) was in general rather poor (Table 2) and unrelated to Epo concentrations, confirming the data of others [11, 12]. Finally, it was also noted that FAB classification did not distinguish particular subsets of patients in relation to serum Epo. Seven of these patients have recently completed a 3-month pilot study in which 150 U/kg per day (Eprex, Cilag, Milan, Italy) were administered subcutaneously. Six of them, including the one with agnogenic myleloid metaplasia, had a partial or complete response, defined, respectively, as a 50% reduction in transfusion requirement and a 1 g/dl unsupported increase in hemoglobin concentration. Responsiveness was unrelated to pretreatment serum Epo levels, reticulocyte number or representation of the classes of maturation, and the growth of BFU-E and

Table 2. Serum Epo levels and erythroid progenitor growth before and after therapy

Pat. no.	Diagnosis	Epo (mU/ml)	BFU-E Before	After	CFU-E Before	After
1	RAS	3858	–	–	–	–
2	RAS	129	3	20	–	–
3	RA	4000	0	–	3	–
4	RA	500	0	35	0	30
5	RAEB	803	–	–	–	–
6	RAEB	47	0	1	0	0
7	RA	915	5	–	2	–
8	RA	167	4	–	1	–
9	RA	380	4	5	0	0
10	AMM	77	2	2	2	1
11	RAEB	33	7	4	4	5

CFU-E in semisolid culture increased in only two cases (Table 2). Responders were almost evenly distributed in each FAB group.

Although the lack of correlation between levels of Epo and hemoglobin concentrations might be due to the small number of cases in the study, from our data and the data of Jacobs et al. [12] we can draw the conclusion that Epo is sometimes inadequate to the degree of anemia in patients with MDS.

No obvious explanation can be found for this inadequacy, but several factors could account for it. One explanation finds support from two lines of evidence that bone marrow concentration and utilization of Epo are critical for its renal production. (a) Malignant erythroid cells are capable of Epo synthesis [15], thus an autocrine production by erythroblasts in MDS is conceivable. This, in turn, would limit the requirement for kidney-produced Epo. (b) Jacobs et al. [12] found that MDS patients with hypocellular bone marrow had the highest levels of the protein and speculated that Epo utilization by erythroid precursors is important in determining serum concentration. In our series the bone marrow erythroid component was rather hypercellular in all patients; therefore we cannot confirm this hypothesis.

A second possible explanation is as follows. As Erslev has pointed out in a recent review [16], the oxygen supply system is made up of several components which respond to hypoxic anemia together with the renal Epo feedback mechanism. Each component could be affected differently in a number of diseases, thus influencing Epo response.

Third, in vitro experiments suggest that cytokines such as interleukin-1 and tumor necrosis factor-A may inhibit Epo production [17].

In conclusion, Epo production may be lower than expected for several factors which can act either alone or, more likely, together. In addition, it is possible that, in the presence of adequate production of Epo for the degree of anemia, target cells are not stimulated because of a reduced number and/or function of receptors. While normal erythroid precursors have both high- and low-affinity receptors, in some clonal disorders only a single class of low-affinity receptors has been found [18].

References

1. Bennett JM, Catovsky D, Flandrin G, Galton DAG, Gralnick HR, Sultan C (1982) Proposal for the classification of the myelodysplastic syndromes. Br J Haematol 51:189–199
2. Chui DHK, Clarke BJ (1982) Abnormal erythroid progenitor cells in human preleukemia. Blood 60:362–367
3. Cazzola AM, Barosi G, Berzuini D, Dacco M, Orlandi D, Stefanelli M, Ascari B (1982) Quantitative evaluation of erythropoietic activity in dysmyelodyspoietic syndromes. Br J Haematol 50:55–62
4. Sherwood JB, Goldwasser E, Chilcote R, Carmichael LD, Nagel RL (1986) Sickle cell anemia patients have low erythropoietin levels for their degree of anemia. Blood 67:46–49

5. Hochberg MC, Arnold CM, Hogans BB, Spivak JL (1990) Serum immunoreactive erythropoietin in rheumatoid arthritis: impaired response to anemia. Arthritis Rheum 31:1318–1321
6. Baer AN, Dessypris EN, Goldwasser E, Krantz SB (1987) Blunted erythropoietin response to anaemia in rheumatoid arthritis. Br J Haematol 66:559–564
7. Miller CB, Jones RJ, Piantadosi S, Abeloff MD, Spivak JL (1990) Decreased erythropoietin response in patients with anemia of cancer. N Engl J Med 322:1689–1692
8. Ludwig H, Fritz E, Kotzmann H, Hocker A, Gisslinger H, Barnes U (1990) Erythropoietin treatment of anemia associated with multiple myeloma. N Engl J Med 322:1693–1699
9. Fischl M, Galpin JE, Levine JD, Groopman JE, Henry DH, Kennedy P, Miles S, Robins W, Starret B, Zalusky R, Abels RI, Tsai HC, Rudnick SA (1990) Recombinant human erythropoietin for patients with AIDS treated with Zidovudine. N Engl J Med 322:1488–1493
10. Vadhan-Raj S, Hittelman WN, Lepe-Zuniga JL, Gutterman JU, Broxmeyer HE (1990) Regulation of endogenous erythropoietin levels in anemia associated with myelodysplastic syndromes. N Engl J Med 75:1749–1750
11. Bessho M, Jinnai I, Matsuda A, Saito M, Kunitake H (1990) Improvement of anemia by recombinant erythropoietin in patients with myelodysplastic syndromes and aplastic anemia Int J Cell Cloning 8:445–458
12. Jacobs A, Janowska-Wieczorek A, Caro J, Bowen DT, Lewis T (1989) Circulating erythropoietin in patients with myelodysplastic syndromes. Br J Haematol 73:36–39
13. Lichtman MA, Brennan JK (1990) Preleukemia and oligoblastic leukemia (myelodysplastic disorders) In: Williams WJ, Beutler E, Erslev AJ, Lichtman MA (eds) Hematology. McGraw Hill, New York pp 175–187
14. Tichelli A, Gratwhol A, Driessen A, Mathys S, Pfefferkorn E, Regenass A, Schumacher P, Stebler C, Wernli M, Nissen C, Speck B (1990) Evaluation of the Sysmex R-1000, an automated reticulocyte analyzer. Am J Clin Pathol 93:70–78
15. Mitjavila MT, Dubart A, Casadevall N, Villeval JL, Breton-Gorius J (1988) Synthesis of erythropoietin by human erythroleukaemic cells. Blood 72 (Suppl):216a
16. Erslev AJ (1991) Erythropoietin. N Engl J Med 324:1339–1344
17. Jelkmann W, Wolff M, Fandrey J (1990) Modulation of the production of erythropoietin by cytokines: in vitro studies and their clinical implications. Contrib Nephrol 87:68–77
18. Means TR, Krantz SB, Sawyer ST, Gilbert HS (1989) Erythropoietin receptors in polycythemia vera. J Clin Invest 84:1340–1344

Antigenic Characteristics of Erythropoietin-Dependent and -Independent Erythroid Progenitors in Chronic Myeloproliferative Disorders Defined by Monoclonal Antibodies

T. Robak

Introduction

The most mature erythroid progenitor cells – colony forming units – erythroid (CFU-E) – give rise to the smallest colonies in the semisolid culture system of methylcellulose, plasma clots, or agar. More primitive cells – burst forming units–erythroid (BFU-E) – give rise to progressively larger colonies (called bursts) that hemoglobinize at correspondingly later times than CFU-E [6, 9]. The development of these colonies from normal bone marrow and peripheral blood requires the addition of erythropoietin (Epo) to the cultures. Erythroid colonies derived from bone marrow and peripheral blood cells of patients with polycythemia vera (PV) and some patients with idiopathic myelofibrosis (MF) or chronic granulocyte leukemia (CGL) can develop in semisolid culture medium also without the addition of Epo [1, 3, 8].

In this study we measured the reactivity of nine monoclonal antibodies (MAbs) directed against myeloid differentation antigens, with Epo-dependent and Epo-independent erythroid progenitor cells from PV, MF and CGL patients as well as from normal bone marrow by means of complement-mediated cytoxicity and colony assays in vitro.

Materials and Methods

Specimens of peripheral blood or bone marrow were collected from 12 patients with PV, 8 with MF, 18 with CGL in chronic phase and 16 normal subjects.

The separation of mononuclear cells (MNCs) from bone marrow and peripheral blood was performed by layering over lymphoprep (Nyegaard, Oslo) followed by centrifugation (400 *g*) for 25 min. Light-density MNCs were then harvested from the interface and washed twice with Hank's balanced salt solution. The cells were then resuspended in Iscove's modified Dulbecco minimum essential medium (IDMEM; Gibco Biocult Laboratories, UK).

We selected a panel of eight MAbs directed against different stages of differentiation of myeloid cells (Fig. 1; Table 1). We also used one MAb with HLA-DR specificity (McL243 Becton Dickinsonn, USA). The other MAbs

Pagel/Weiss/Jelkmann (Eds.)
Pathophysiology and Pharmacology
of Erythropoietin

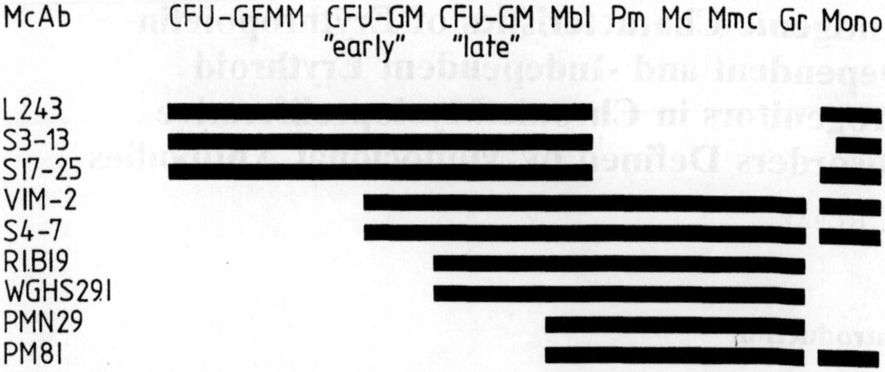

Fig. 1. Schematic presentation of the myeloid surface antigens recognized by the panel of MAbs. *Bars*, nonquantitative representation of the reactivity described for each antibody. *CFU-GEMM*, Colony forming unit–granulocyte, erythrocyte, megakaryocyte, macrophage; *CFU-GM*, colony forming unit–granulocyte-macrophage; *Mbl*, myeloblasts; *Pm*, promyelocytes; *Mc*, myelocytes; *Mmc*, metamyelocytes; *Gr*, granulocytes; *Mono*, monocytes. (For details see [7, 9])

were gifts and were kindly supplied as follows: MAbs S3-13, S17-25, S4-7, R.1B19, and WGHS.29.1 were given by Dr. G. Rovera (Wistar Institute, Philadelphia, USA), MAb VIM-2 by Dr. W. Knapp (University of Vienna, Vienna, Austria), and MAbs PMN29 and PM81 by Dr. E.D. Ball (Darmouth Medical College, Hannover, USA). All these MAbs were lytic in the presence of complement.

We used lyophylized baby rabbit complement (C'; Gibco Biocult Laboratories, USA) for cytotoxicity studies at a dilution of 1:4. MNCs (1×10^6) were suspended in 200 µl IDMEM, and 50 µl volume of MAB in optimal concentration was added to the suspension for 30 min. Subsequently, an equal volume of C' (250 µl) was added, and incubation was continued for

Table 1. Characteristics of MAbs used in this study

MAb	Immunogen	Isotype	Antigen	Dilution
L243	B-lymphoblastoid cell line	IgG 2b	HLA-DR	HS 1:40
S3-13	AML cells	IgM	p 29	AF 1:10
S17-25	AML cells	IgM	NK	AF 1:10
VIM-2	THP-1 cells	IgM	Glycolipids	1:100
S4-7	AML cells	IgM	gp 145-105	AF 1:10
R1.B19	AML cells	IgM	gp 145-105	AF 1:10
WGHS29.1	Gastric adeno-carcinoma cells		Lacto-*N*-fuco pentaose	AF 1:10
PMN29	Neutrophils	IgM	Glycolipids	HS.ND
PM81	HL 60 cells	IgM	NK	HS.ND

NK, Not known; HS, hybridoma supernatant; AF, ascitic fluid; NS, not diluted; CGL-BT, chronic granulocytic leukemia in blastic transformation; AML, acute myeloid leukemia. (For details see [7, 9].)

further 60 min. Control cultures with complement alone and MAbs alone were included in all experiments.

The assay for BFU-E and CFU-E was based on the method described by Iscove et al. [4] and modified by Goldman et al. [2]. We added 2 IU Epo (Terry Fox Laboratories, BC Cancer Research Center, Vancouver, Canada) to each plate on day 0 for BFU-E and CFU-E colony growth. In each experiment, similar cultures were established in plates with no Epo added. The reactivity of MABs with BFU-E and CFU-E were expressed as percentage inhibition of colony growth compared with control colony number after incubation with C' only.

Results and Discussion

We determined the antigenic phenotype of BFU-E and CFU-E from patients with PV, MF, and CGL by means of 9 MAbs directed against differentiation antigens of myeloid lineage cells. We considered both "endogenous" erythroid precursor cells forming colonies without the addition of Epo to the culture and cells forming erythroid colonies after the addition of Epo.

The growth of BFU-E in the presence of Epo was observed in all patients with PV, MF, and CGL and from normal bone marrow. The CFU-E colonies in the presence of Epo were obtained from peripheral blood in all patients with MF and CGL and from cultures of normal, PV, and CGL bone marrow. In all 12 patients with PV the growth of BFU-E and CFU-E colonies was observed without the presence of Epo. In contrast, "endogenous" BFU-E and CFU-E (without the addition of Epo) were found in only 4 out of 8 patients with MF and 4 out of 18 patients with CGL. No growth of "endogenous" BFU-E was observed in the cultures of normal bone marrow.

The reactivity of MAbs against Epo-dependent and Epo-independent BFU-E and CFU-E is presented in Tables 2 and 3. L243, S3-13, and S17-25 antibodies recognized antigens in over 50% of Epo-dependent as well as Epo-independent BFU-E. The percentage of Epo-dependent and Epo-independent BFU-E reacting with R1.B19, WGHS29.1, PMN29, and PM81 antibodies was small in the majority of patients with PV, MF, CGL and in normal individuals. Antigens recognized by these antibodies appear on more differentiated cells of the granulocyte-macrophage lineage [7, 9]. Both the results of our previous investigations and those of our present experiments reveal that committed precursor cells of erythroid lineage both in normal persons and in chronic myeloproliferative disorders do not have differentiation antigens appearing on promyelocytes and more mature myeloid cells. Our investigations and those of other authors indicate that the antigens present on committed cells of granulocyte-macrophage lineage (CFU-GM), and especially on pluripotent stem cells (CFU-GEMM) can also be present on more primitive erythroid progenitor cells (BFU-E). However, they are usually absent on CFU-E.

Table 2. Reactivity of MAbs against Epo-dependent (Epo+) and -independent (Epo–) BFU-E from patients with MF, PV, and CGL and from normals: percentage inhibition of colony growth compared with control colony numbers

MAb	MF		PV		CGL		Normals
	Epo+ (*n*=8)	Epo– (*n*=4)	Epo+ (*n*=12)	Epo– (*n*=12)	Epo+ (*n*=18)	Epo– (*n*=4)	Epo+(*n*=16)
L243	79.6±13.2	56.9±16.1	85.2±14.5	39.6± 4.8	97.5± 0.8	51.2± 6.9	98.2± 3.9
S3-13	51.0±10.1	56.2±12.8	70.0±16.3	73.4±10.2	70.± 9.9	62.9±10.2	92.6± 4.5
S17-25	60.3±12.8	78.3±11.6	59.7±18.6	68.8±12.3	62.2±11.8	78.4±12.6	95.7± 3.8
VIM-2	16.9± 3.1	68.3±13.7	12.4± 4.6	48.6± 4.1	24.5±11.2	69.7±18.2	75.8± 8.9
S4-7	12.9± 2.8	52.1±10.5	10.8± 3.9	28.2±10.3	33.4± 8.6	59.2±16.1	78.6±11.7
R1.B19	8.1± 1.8	10.8± 3.1	6.9± 2.8	8.4± 3.5	20.6± 4.7	11.4± 2.8	48.7± 6.2
WGHS29.1	3.8± 1.0	5.8± 3.0	3.8± 1.5	6.2± 1.9	23.2± 8.3	14.3± 7.3	32.4± 8.2
PMN29	2.1± 1.2	4.2± 1.8	1.0± 0.5	4.8± 2.4	1.9± 0.8	3.2± 1.2	3.8± 1.8
PM81	5.1± 2.1	1.2± 0.5	2.8± 1.0	1.6± 0.2	5.6± 3.8	2.8± 0.6	5.9± 2.3

100% represents total suppression of colony growth, and 0% represents the control value (inhibition with medium and C').

Table 3. Reactivity of MAbs against Epo-dependent (Epo+) and -independent (Epo–) CFU-E from patients with MF, PV, and CGL and from normals: percentage inhibition of colony growth compared with control colony numbers

MAb	MF		PV		CGL		Normals
	Epo+ (*n*=8)	Epo– (*n*=4)	Epo+ (*n*=12)	Epo– (*n*=12)	Epo+ (*n*=18)	Epo– (*n*=4)	Epo+(*n*=16)
L243	5.2± 1.0	26.9± 1.3	4.8± 2.0	19.6± 8.2	15.0± 3.5	10.4± 2.3	21.3± 3.6
S3-13	85.2±12.8	72.9±24.3	57.8±12.5	42.6± 7.8	70.2±10.2	45.4±12.6	72.8± 8.4
S17-25	72.6±14.3	72.9±24.3	52.8± 6.8	48.9±12.4	72.8±12.4	46.6±15.1	68.4±11.2
VIM-2	28.3± 8.5	36.8± 4.2	28.9± 9.1	32.4±12.6	48.4± 8.6	36.8±10.2	63.5±10.4
S4-7	18.5± 6.1	16.8±10.9	32.8± 8.2	32.3± 8.0	39.6± 8.2	20.8± 4.8	21.4± 8.2
R1.B19	4.2± 1.6	21.3±10.3	4.8± 1.6	3.8± 2.1	4.8± 1.2	3.6± 1.5	26.3± 4.1
WGHS29.1	5.9± 2.8	6.7± 4.0	5.9± 2.1	4.2± 1.9	6.8± 1.4	7.2± 2.3	25.3± 3.8
PMN29	5.2± 2.0	4.1± 1.9	5.0± 2.0	4.0± 1.6	4.2± 1.2	3.8± 1.5	5.4± 1.8
PM81	5.0± 2.1	1.7± 1.3	4.8± 1.6	1.9± 0.8	6.8± 1.9	7.2± 1.6	3.9± 1.3

Only a small percentage of CFU-E reacted with L243 antibody recognizing HLA-DR antigen. These results are in agreement with those of other authors who have shown that HLA-DR antigen is present in the majority of BFU-E and does not appear on CFU-E.

In our investigations we found no qualitative differences in the antigenic expression between Epo-dependent and Epo-independent BFU-E and CFU-E. However, the percentage of cells reacting with some antibodies showed significant differences. These results are in agreement with the observations of some authors who have shown that the increase of "endogenous" erythroid colonies is not completely independent of Epo, and that the cells forming these colonies are a subpopulation of precursor cells more sensitive to this hormone [6]. The hypresensitivity to Epo is not a constant feature of all "endogenous" erythroid precursor cells, but it decreases gradually, and their various fractions react to different though very small amounts of Epo that may be present in the serum added to the culture.

Our experiments have shown the heterogeneity of antigen phenotype of both Epo-dependent and Epo-independent BFU-E and CFU-E. In most cases the reactivity of these cells with MAbs was not complete. These results may be concluded with the different proliferative state of the investigated cells. It has been shown, for instance, that the expression of HLA-DR antigen is found in phase S of the cell life cycle [9]. It is possible that other antigens, depending on the division phase or the metabolic state of the hematopoietic cells, may also be found.

In the majority of PV, MF, and CGL patients a larger percentage of Epo-independent BFU-E reacting with S4-7 and VIM-2 antibodies was found ($p<0.01$). These antibodies recognize antigens appearing later in the process of differentiation of myeloid lineage cells than S17-25 and S3-13 antibodies. Their presence may point to a greater phenotype maturity of "endogenous" BFU-E than BFU-E colonies forming in the presence of Epo.

In our study we have not shown qualitative differences in the reactivity of MAbs with BFU-E and CFU-E in chronic myeloproliferative diseases or in normal persons (Tables 2, 3). We have found, however, quantitive differences in the percentage of reacting erythroid progenitor cells in these myeloproleferative syndromes and their normal counterparts. S3-13, VIM-2, S4-7, R1.B19, and WGHS.29.1 antibodies reacted with a larger percentage of BFU-E from normal bone marrow ($p<0.01$). These results may depend on the differences in the proliferative state of hemopoietic cells in normal persons and in patients with chronic myeloproliferative diseases.

References

1. Cashman J, Henkelman D, Humphries K, Eaves C, Eaves A (1983) Individual BFU-E in polycythemia vera produce both erythropoietin dependent and independent progeny. Blood 61:876–884

2. Goldmann JM, Hibbin J, Kearney L, Orchard K, Th'ng KH (1982) HLA-DR monoclonal antibodies inhibit the proliferation of normal and chronic granulocytic leukaemia myeloid progenitor cells. Br J Haematol 52:411–420
3. Greaves MF, Myers CD, Katz FE, Schneider C, Sutherland DR (1984) Cell surface structures involved in haemopoietic cell differentiation and proliferation. Br Med Bull 40:224–228
4. Iscove NN, Sieber F, Winterhalter KH (1974) Erythroid colony formation in cultures of mouse and human bone marrow: analysis of the requirements of erythropoietin by gel filtration and affinity chromatography on agarose-concanavalin A. J Cell Physiol 83:309–320
5. Lu L, Broxmeyer HE, Mayers PA, Malcolm AS, Thaler HT (1983) Association of cell cycle expression of Ia-like antigenic determinants on normal human multipotential (CFU-GEMM) and erythroid (BFU-E) progenitor cells with regulation in vitro by acid isoferritins. Blood 61:250–256
6. Reid CD (1987) The significance of endogenous erythroid colonies (EEC) in haematological disorders. Blood Rev 1:133–140
7. Robak T, Dowding C, Gareval G, Hibbin J, Th'ng KH, Goldman JM (1986) Antigenic determinants on myeloid leukaemia colony forming cells resemble those of normal progenitor cells and differ from those of circulating blast cells. Br J Haematol 64:133–148
8. Robak T, Hast R, Goldman JM (1986) Endogenous erythroid colony formation in myeloproliferative diseases does not depend on T-cells. Exp Hematol 14:197–201
9. Robak T (1987) Expression of myeloid differentiation antigens on normal and chronic granulocytic leukaemia erythroid progenitor cells identified by monoclonal antibodies. Arch Immunol Ther Exp (Warsz) 35:519–529

Chemotherapy and Erythropoietin Production

C. Pohl, A. Moter, D. Hasenclever, E.M. Woll, M. Löffler, and C.A. Baldamus

Erythropoietin (Epo) is a glycoprotein growth factor which controls erythropoiesis in mammals. It is well established that renal Epo production is stimulated solely by anemia and low arterial oxygen tension. While the Epo-deficient anemia of chronic renal insufficiency can now be treated effectively with recombinant human Epo [15], the usefulness of Epo therapy for hyporegenerative nonrenal anemia of malignant or chronic inflammatory disease is still under discussion [1, 5, 10–13, 18]. With regard to these conflicting data more information on Epo levels during chemotherapy is required. In the present study the correlation between the degree of chemotherapy-induced anemia and Epo levels was examined systematically in 24 patients with gastrointestinal neoplasia, high-grade non-Hodgkin's lymphoma or testicular cancer.

Materials and Methods

Patients were unselected and consecutive in presentation at the Medical Clinic I of the University Hospital of Cologne. Patients were observed over an average of three cycles of chemotherapy. In group 1 six patients with gastrointestinal carcinoma were followed over 24 cycles of the FAMtx or FEMtx regimen [8] consisting of 1500 mg/m^2 5-fluorouracil plus either 30 mg/m^2 doxorubicin (Adriamycin) or 60 mg/m^2 epirubicin, and 1500 mg/m^2 methotrexate (total intravenous dose administered over 8 days, every 4 weeks). In group 2, 12 patients with high-grade non-Hodgin's lymphoma were followed over 28 cycles of the MACOP-B regimen [9] consisting of 400 mg/m^2 methotrexate, 100 mg/m^2 doxorubicin (Adriamycin), 700 mg/m^2 cyclophosphamide, 2.8 mg/m^2 vincristine (Oncovin), 75 mg/day prednisone and 10 mg/m^2 bleomycin (total intravenous/oral dose administered over 28 days, every 4 weeks). In group 3, six patients with testicular cancer were followed over a total of 19 cycles of the PEB regimen [17] consisting of 100 mg/m^2 cisplatin, 500 mg/m^2 etoposide, and 60 mg/m^2 bleomycin (total intravenous dose administered over 8 days, every 4 weeks). Sera from 215 healthy individuals or patients with nonrenal anemia from acute blood loss served as controls.

Serum and urine specimens were collected during therapy at least every 3 days and after therapy in weekly intervals. Laboratory parameters in serum

Pagel/Weiss/Jelkmann (Eds.)
Pathophysiology and Pharmacology
of Erythropoietin

and/or urine including hemoglobin, hematocrit, white blood cell, reticulocyte, and platelet counts, electrolytes, transaminases, creatinine, urea, proteins, ß-microglobulin, iron, transferrin, and ferritin were determined by conventional methods at the Department of Clinical Chemistry of the University of Cologne. Samples for Epo determination were collected at the same occasion in the morning. Epo was measured in duplicate by an radioimmunoassay using radiolabeled recombinant human Epo (tissue culture grade, Amersham, UK) and polyclonal rabbit anti-Epo antibodies as described elsewhere [3]. The interassay variation was less than 10%.

Peak Epo values were compared to Epo levels prior to chemotherapy by the Wilcoxon rank test.

Results

In otherwise healthy individuals, i.e., with intact regulation of hematopoiesis, serum Epo concentration is strictly correlated with the degree of anemia at that time. Variations in Epo concentration thus can be interpreted only in conjunction with the respective degree of hemoglobin. Figure 1 presents the correlations between serum Epo concentration and hemoglobin values between 5 and 17 g/dl in 215 healthy individuals and in patients with acute blood loss from traumatic lesions or after surgery. An increase in Epo at hemoglobin values below 10 g/dl represents a normal feedback mechanism of an intact hematopoietic regulation. Since no evidence for impaired renal function, iron metabolism, or bone marrow function was found in these

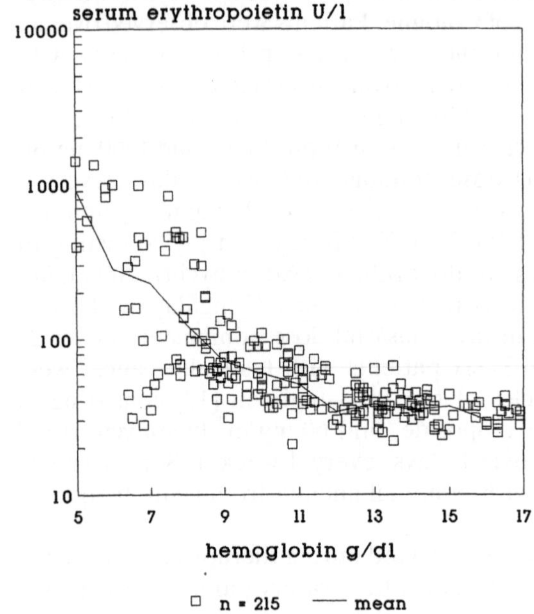

Fig. 1. Correlation of hemoglobin and erythropoietin in healthy adults and patients with acute blood loss

individuals, mean Epo concentrations within 3 standard deviations of the mean (Fig. 1) were considered as the hemoglobin-dependent normal reference values for our further studies.

Prior to chemotherapy all patients had appropriate Epo levels and normal renal and hepatic function. Epo responded appropriately to the degree of anemia induced by the FAMtx or FEMtx regimen in patients with gastrointestinal neoplasia (group 1), suggesting an intact feedback mechanism of Epo regulation. Similarly, patients treated with MACOP-B (group 2) developed a sometimes severe anemia and demonstrated correspondingly elevated peak Epo values (data not shown). Patients with testicular cancer, treated according to the PEB regimen (group 3), developed only a mild anemia (hemoglobin 10–12 g/dl). In contrast to the other groups, PEB-treated patients developed three- to four fold higher Epo levels during chemotherapy than predicted from the degree of anemia (68–354 U/l Epo, median 130; compared to 21–65 U/l Epo, median 38, in the control group at hemoglobin values of 10–12 g/dl). Peak Epo levels (Fig. 2) were measured 6–8 days after the start of the therapy protocol and always declined to normal or near normal concentrations prior to the next chemotherapy cycle.

The difference between Epo level before chemotherapy and the highest Epo level of the individual therapy cycle was statistically significant while hemoglobin remained unchanged. In patients 2 and 5 (Fig. 3) reactivity in terms of Epo increase was markedly different among individual patients. However, intraindividually, the Epo increase was seen repeatedly in each individual patient and therapy cycle. The effect of repetitive chemotherapy

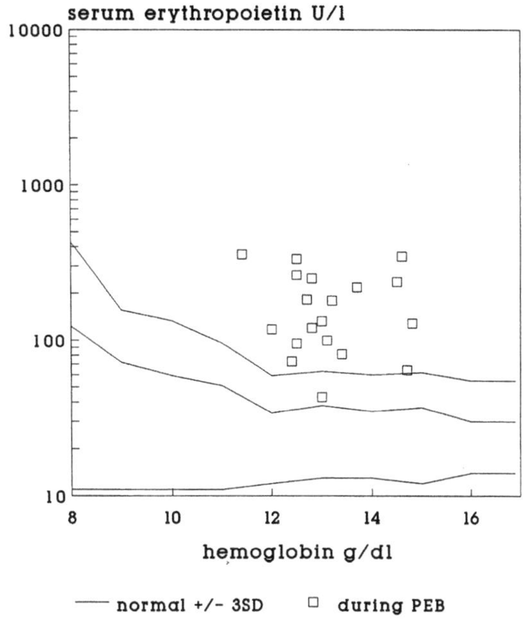

Fig. 2. Peak erythropoietin levels 6–8 days after start of PEB chemotherapy (six patients, 19 cycles)

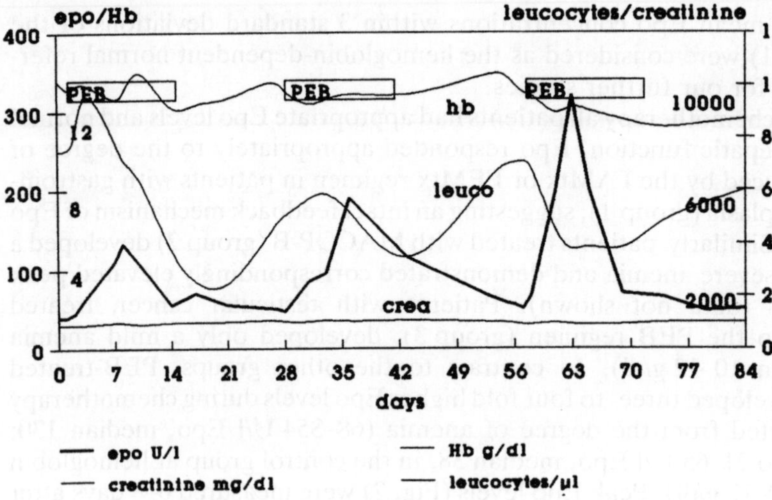

Pat. #2

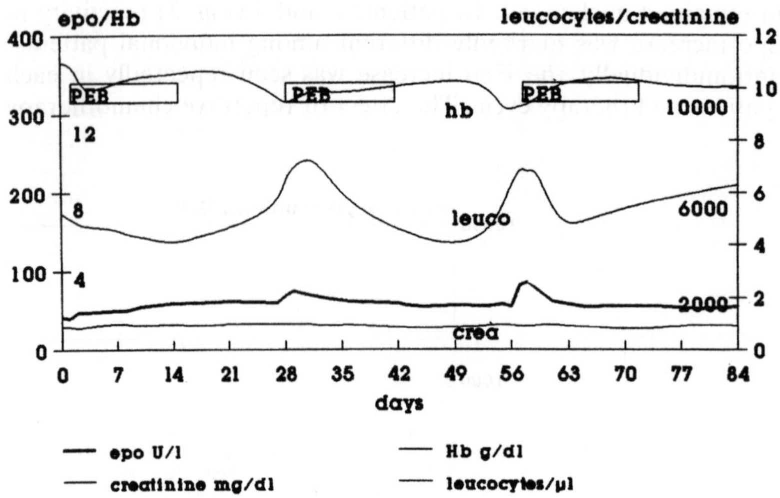

Pat. #5

Fig. 3. Influence of three cycles of PEB treatment in two patients on Epo, hemoglobin, creatinine, and leucocyte counts

appeared to be additive on the Epo increase; a distinctive increase in Epo concentrations was observed from cycle to cycle in all but one patient.

To explain this inappropriately high increase in Epo during PEB chemotherapy other laboratory parameters of renal of bone marrow function

Table 1. Patient groups

	Tumor	*Pts*	*Cycles*	*Length*
Group 1	gastrointestinal carcinoma	6	24	4 weeks
Group II	high grade NHL	12	28	4 weeks
Group III	testicular cancer	6	19	4 weeks
Control	healthy/acute blood loss	215		

were compared with Epo concentrations prior and during chemotherapy. No impairment of renal or bone marrow function was observed during or after PEB chemotherapy; especially renal function as measured by creatinine (Fig. 3), urea, and protein excretion was not affected. White blood cell counts (Fig. 3) and platelet counts declined during chemotherapy. This myelotoxic effect of chemotherapy became manifest only when peak Epo levels had already developed.

Discussion

A statistically significant increase in Epo before only a mild anemia developed was demonstrated after combination chemotherapy with PEB. No comparable effect was seen with the other chemotherapy regimens. While there was considerable interindividual variation, the overall pattern of PEB treatment showed a consistent Epo increase, with a maximum 3–4 times above normal 6–8 days after start of chemotherapy and a slow decrease over 2–3 weeks to normal values thereafter.

This observation is in accordance with recently published data of a similarly inappropriate Epo increase under the influence of a variety of cytostatic agents [1, 5, 13]. Also, parameters of the standard curve of normal Epo response to anemia are comparable to those published in the literature [7]. Contrasting data [6, 11, 16, 18] reported in regard to the effect of cisplatin and other chemotherapeutic agents may not be comparable to our data since a blunted Epo response was observed there at much lower hemoglobin levels, indicating a more cytotoxic and myelosuppressive therapy.

It is generally accepted that Epo production is stimulated by anemia and low arterial oxygen tension in the peritubular interstitium of the kidney. No other mechanism for the regulation of Epo production has been demonstrated so far [4]. As a consequence of this regulation mechanism anemia develops in Epo-deficient chronic renal disease, while hyporegenerative or hyperregenerative bone marrow disorders are associated with increased and decreased Epo secretion, respectively. In this study, however, Epo levels increased prior to a measurable suppression of hematopoiesis that would have triggered this increase. Possible explanations for this increase include

(a) chemotherapy-induced direct damage to the Epo-producing peritubular cells (e.g., cisplatin) in a manner which mimics hypoxia; (b) alteration of the pulmonary oxygenation of blood (e.g., bleomycin) or blood supply of the kidneys and thus inducing hypoxia in Epo-producing cells; (c) the selective disruption of the usual Epo degradation pathways and thus extending Epo half-life [5, 13]; (d) stimulation of extrarenal Epo production in liver or macrophages [2, 14]; and (e) decreased utilization of Epo by bone marrow aplasia [1]. Our data provide no support for any of these possibilities since no clinically overt renal toxicity or hepatic or pulmonary injury was observed, and bone marrow suppression was detected only after Epo levels had begun to increase. Although cisplatin, due to its well-known nephrotoxic effect [16], appears the most probable drug responsible for the observed Epo increase, our data do not permit attributing this effect to a single drug.

We conclude from our data that cytostatic treatment with cisplatin, bleomycin, and etoposide or a combination thereof induces an increased synthesis or decreased metabolism of Epo by an unknown mechanism that appears to be independent from hypoxia induced by anemia. A rationale for the administration of Epo in chemotherapy-induced anemia requires more research, including the systematic study of these and other cytostatic drugs in anemic and nonanemic states of various malignant diseases.

References

1. Birgegard G, Wide L, Simonsson B (1989) Marked erythropoietin increase before fall in Hb after treatment with cytostatic drugs suggest mechanism other than anemia for stimulation. Br J Haematol 72:462–466
2. Brown S, Caro J, Erslev AJ, Murray T (1980) Rise in erythropoietin and hematocrit associated with transient liver enzyme abnormalities in an anephric hemodialysis patient. Am J Med 68:280
3. Eckardt KU, Kurtz A, Hirth P, Scigalla P, Wieczorek L, Bauer C (1988) Evaluation of the stability of human erythropoietin in samples for radioimmunoassay. Klin Wochenschr 66:241–245
4. Goldberg MA, Gaut CC, Schapira L, Antin JH, Ransil BJ, Antman KH, Eder JP, McGarigle CJ, Bunn HF (1991) Erythropoietin gene regulation: from the laboratory to the bedside. In: Gurland HJ, Moran J, Samtleben W, Scigalla P, Wieczorek L (eds) Erythropoietin in renal and non renal anemias. Contrib Nephrol 88:35–45
5. Hellebostad M, Marstrander J, Slordahl SH, Cotes PM, Refsum HE (1990) Serum immunoreactive erythropoietin in children with acute leukemia at various stages of disease – and the effects of treatment. Eur J Haematol 44:159–164
6. Fisher JW, Roh BL (1964) Influence of alkylating agents on kidney erythropoietin production. Cancer Res 24:983–988
7. Jelkmann W, Wolff M (1991) Bestimmung der Erythropoietin-Aktivität im Serum. Dtsch Med Wochenschr 6:230–234
8. Klein, HO (1986) 5-Flourouracil (5-FU), Adriamycin (ADM) and methotrexate (MTX) – a combination protocol (FAMtx) for treatment of metastasized stomach cancer. Proc ASCO 84:325
9. Klimo P, Connors J (1985) MACOP-B chemotherapy for the treatment of diffuse large-cell lymphoma. Ann Intern Med 102:596–602

Ludwig H, Fritz E, Kotzmann H, Höcker P, Gisslinger H, Barnas U (1990) Erythropoietin treatment of anemia associated with multiple myeloma. N Engl J Med 322:1693–1699
Miller CB, Jones RJ, Piantadosi S, Abeloff MD, Spivak JL (1990) Decreased erythropoietin response in patients with the anemia of cancer. N Engl J Med 322:1689–1692
Oster W, Herrmann F, Cicco A, Gamm H, Zeile G, Brune T, Lindemann A, Schulz G, Mertelsman R (1990) Erythropoietin prevents chemotherapy-induced anemia. Blut 60:88
Piroso E, Erslev A, Caro J (1989) Inappropriate increase in erythropoietin titers during chemotherapy. Am J Hemtol 32:248–254
Rich I (1987) Erythropoietin production by macrophages: cellular response to physiological oxygen tension and detection of erythropoietin gene expression by in situ hybridization. In: Rich I (ed) Molecular and cellular aspects of erythropoietin and erythropoiesis. Springer, Berlin Heidelberg New York, pp 291–310 (NATO ASI series H, vol 8)
Scigalla P, Wieczorek L, Bicker U (1991) Treatment of renal anemia with recombinant human erythropoietin: European experience. In: MB Garnick (ed) Erythropoietin in clinical applications. Dekker, New York, pp 141–182
Smith DH, Guarneri CM, Whaling SM, Vokes EE (1988) Erythropoietin response in cancer patients receiving cisplatin. Proc Am Assoc Cancer Res 29:52
William S, Einhorn L (1988) Disseminated germ cell tumors: a comparison of cisplatin plus bleomycin plus either vinblastine (PVB) or VP16 (PEB). Proc ASCO 4:C–390
Wood P, Nygaand S, Hrushesky WJM (1988) Cisplatin-induced anemia is correctable with erythropoietin. Blood 72 (Suppl):52

Production of Erythropoietin During and After Thiamphenicol Treatment Occurs Independently of Peripheral Blood Cell Numbers*

W. Nijhof, H. Goris, and H. Hartsuiker

Introduction

Our understanding of the regulation of hematopoiesis has vastly increased during the past decade with the discovery of an extensive network of growth factors and interleukins that is able to stimulate or inhibit at different levels along the differentiation pathway from the pluripotent stem cells to the mature blood cell. Nevertheless questions concerning the way in which the production of these factors is regulated still remain open to debate. While no one questions that the level of the hormone erythropoietin (EPO) is dictated mainly by the oxygen supply to the tissues, early doubts have been raised whether this is the only regulating system.

Polycythemia, induced by non-oxygen-transporting cells (methemoglobinized), severely decreases Epo levels, as does transfusion with normal cells. This suggests that hematocrit as such and not the oxygen transport capacity under these conditions is a part of the regulation of EPO production (Kilbridge et al. 1969). The results of other experiments support this idea (Misago et al. 1986). On the other hand, Barcelo and Bozzini (1982) showed that EPO production induced under hypoxic conditions decreased without a change in hematocrit. Their experiments support the hypothesis that EPO production is inversely related to the erythroid activity of the marrow (Fried et al. 1971). Negative effects on EPO serum levels by hyperproliferative marrow were further shown after recovery from phenylhydrazine treatment (Cynshi et al. 1990), after treatment of pernicious anemia (De Klerk et al. 1981), and in sickle cell anemia (Sherwood et al. 1986).

Increased EPO production (apart from hypoxic conditions) could be linked to conditions with hypoproliferative marrow, such as aplastic anemia (De Klerk et al. 1981), leukemia (Jelkmann and Wiedemann 1990), thalassemia (McGonigle et al. 1985), and during early phases of cytostatic drug treatments (Birgegard et al. 1989; Piroso et al. 1989).

We investigated whether the antibiotic thiamphenicol (TAP) – a well-known bone marrow depressant eliminating the late erythroid progenitors, colony-forming units – erythroid (CFU-E) and all erythroblasts (Nijhof et al. 1982; Goris et al. 1989) – also produces an EPO response under conditions of

---

* Supported by the Deutsche Forschungsgemeinschaft grant Lo 342/1–2.

Pagel/Weiss/Jelkmann (Eds.)
Pathophysiology and Pharmacology
of Erythropoietin

near-normal blood parameters. In previous experiments with TAP and severely anemic mice, very high EPO levels were obtained, showing that its production is not affected by TAP (Nijhof et al. 1977). This study compared results of TAP treatment alone, one with a blood transfusion to compensate the hematocrit, and a mild bleeding.

Materials and Methods

Mice (C57bl) were treated with TAP (Inpharzam SA, Zambon Group, Cadempino, Switzerland) via a subcutaneously implanted dialysis bag, as described elsewhere (Nijhof et al. 1982). Mice were bled by puncturing the orbital plexus, under light ether anesthesia, with a hematocrit capillary. On day 0 0.3 ml blood was taken. Transfusion of 0.3 ml normal donor blood was carried out via the same route with a 1-ml syringe. For each experimental point two mice were used, and the blood and marrow specimens were mixed before the assays.

A bioassay with enriched murine splenic CFU-E was used. CFU-E from spleens of bled and TAP-pretreated mice were isolated by centrifugal elutriation as described elsewhere (Nijhof and Wierenga 1983). These partially enriched CFU-E (40%–50% pure) were cultured in the presence of 0–500 mU recombinant human EPO (Boehringer) or 0.1 ml test serum in a volume of 1 ml methylcellulose medium (Iscove and Sieber 1975). Benzidine-positive colonies were evaluated after 2 days of culture at 37°C. A

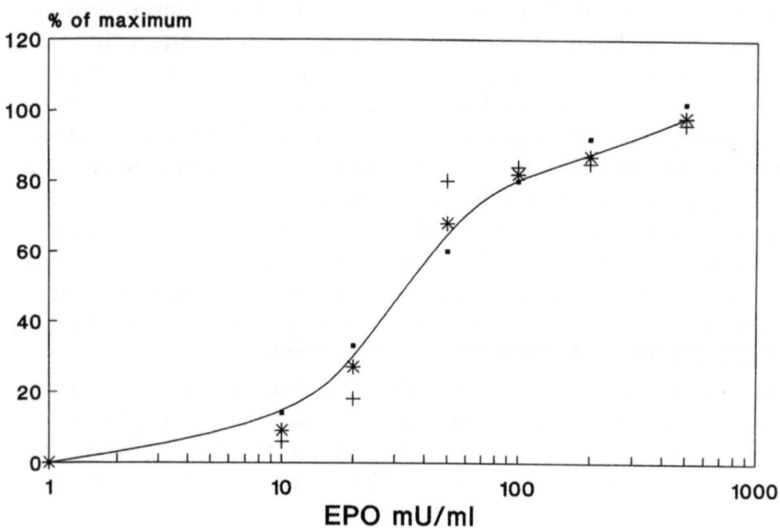

Fig. 1. Recombinant human EPO dose-response curve with splenic murine CFU-E. Three CFU-E preparations (■, *, +)

dose-response curve was constructed, and the serum Epo levels were calculated (Fig. 1).

The hematocrit and reticulocyte counts were carried out according to standard methods. Erythroid precursors were counted as a percentage of total cells on a May-Grünwald-Giemsa stained cytospin preparation of a bone marrow specimen from the femur contents.

Results

Effects of Thiamphenicol on Erythropoiesis. After a 4-day treatment of TAP, murine erythropoiesis was severely inhibited (Table 1). Whereas CFU-E, erythroid precursors, and reticulocyte numbers were practically absent from the marrow, this led to only a slight decrease in hematocrit after the 4-day period. Burst-forming units – erythroid (BFU-E) and CFU – granulocyte-macrophage (CFU-GM) numbers increased at the expense of the CFU – spleen (CFU-S). Thus the most prominent effect of TAP is on late erythroid progenitors and its descendants.

EPO Serum Levels During and After a 4-Day Treatment. Effects of TAP on erythropoiesis was measured as total morphologically recognizable erythroid precursor cells in the femur and the hematocrit and reticulocyte counts of the peripheral blood (Fig. 2). The hematocrit minimum was reached 2 days after removal of TAP at 37%, and the value increased slightly thereafter without reaching normal values within 6 days after the treatment. The erythroid precursor numbers were low after TAP (day 4) and peaked transiently on days 7–8. The peripheral reticulocyte numbers increased after day 7. The highest Epo levels (500 mU/ml) were attained 1 day before removal of TAP and gradually decreased to 20% of the maximum level 5 days after TAP removal. This decrease occurred while hematocrit still was below normal.

EPO Serum Levels During and After a 4-Day TAP Treatment and Blood Transfusions. To exclude the contribution of the small decrease in hematocrit as a possible cause for the rise in EPO we repeated the previous experiment simultaneously keeping the hematocrit close to its normal value

Table 1. Femoral hematopoietic cell numbers, reticulocytes, and hematocrit after 4 days of TAP treatment

	CFU-S	CFU-GM	BFU-E	CFU-E	E-prec	Ret (%)	Hct
Normal	6000	20 000	4500	50 000	6×10^6	3	44
TAP	3500	55 000	6000	0	$<0.3 \times 10^6$	0	40

E-prec, Erythroid precursors; Ret, reticulocytes; Hct, hematocrit.

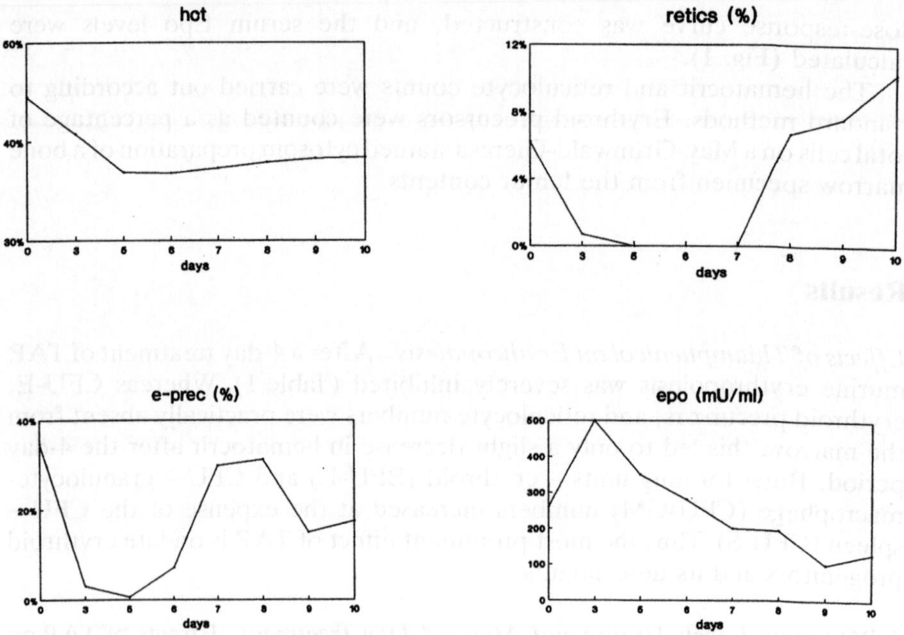

Fig. 2. Time courses of hematocrit (*Hct*), reticulocytes (*retics*), erythroid precursors (*e-prec*), and EPO serum level (*epo*) during and after 4 days of TAP treatment

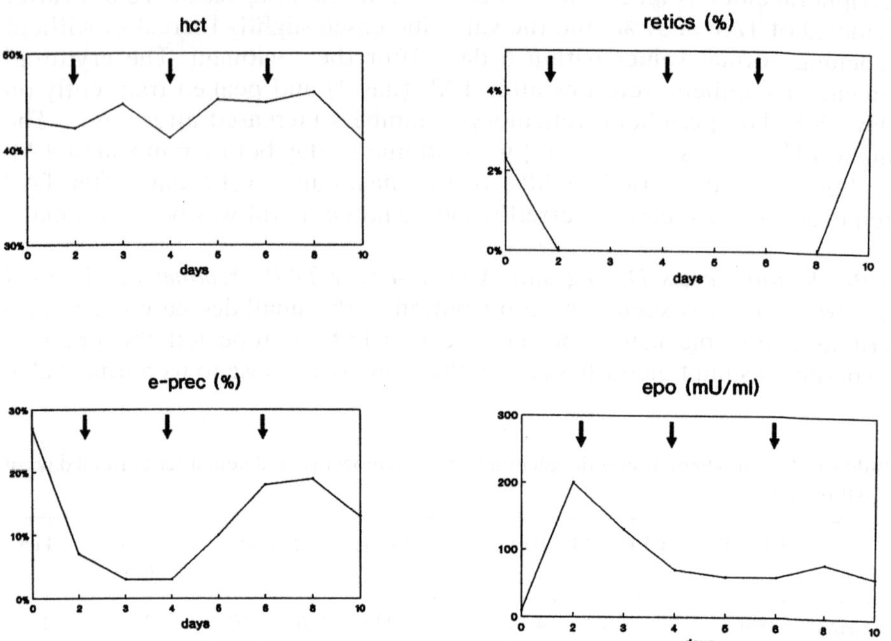

Fig. 3. Time courses of hematocrit (*Hct*), reticulocytes (*retics*), erythroid precursors (*e-prec*), and EPO serum level (*epo*) during and after TAP and three blood transfusions

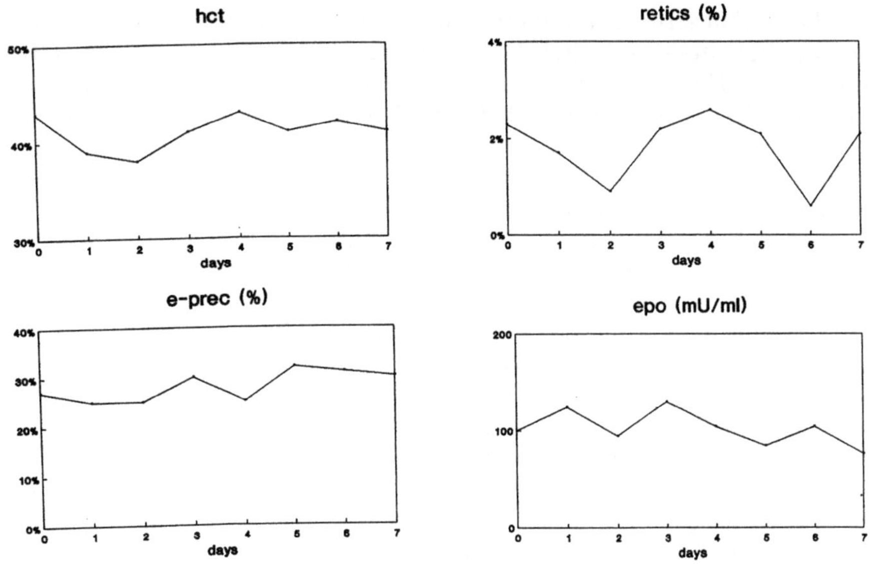

Fig. 4. Time courses of hematocrit (*Hct*), reticulocytes (*retics*), erythroid precursors (*e-prec*), and EPO serum level (*epo*) after a single bleeding

by three blood transfusions on the indicated days (Fig. 3). The peripheral reticulocytes disappeared after 4 days of TAP. Their numbers started to increase after day 8. Erythroid precursors were minimal for 2 days and recovered afterwards. The latter two parameters were similar as in the experiments without blood transfusion. The EPO time course was also similar, being high on the last day of the TAP treatment and declining afterwards to normal levels. The absolute peak value (200 mU/ml) was somewhat lower.

EPO Serum Levels After Bleeding. The previous experiments were performed with hypoproliferative bone marrow. We also investigated the effect of a small decrease in hematocrit to a value comparable to that reached after a TAP treatment by bleeding. The marrow activity was not exogenously impaired. The bleeding on day 0 led to a hematocrit of 37%. This decrease hardly influenced the reticulocyte count or the erythroid precursor compartment. The Epo level also did not change by this treatment and fluctuated around its baseline value.

Discussion

TAP, an antibiotic causing a reversible bone marrow depression, enabled us to study EPO production during a period in which erythroid precursors sharply declined, as well as during increasing numbers of erythroid cells. It was shown that serum levels increased during the TAP treatment regardless of whether hematocrit fell slightly, as was the case with a compensating blood transfusion. Therefore the induction of the EPO production could not be related to decreasing peripheral blood cell numbers. A simple bleeding to the same hematocrit value also had no effect on EPO levels. This phenomenon is comparable to results of Piroso et al. (1989) and Birgegard et al. (1989) from leukemic patients treated with chemotherapy, in whom EPO titers rose without a change in hemoglobin content. In these studies no bone marrow precursor numbers were presented. We take this increase to indicate the novo synthesis of EPO instead of a decreased consumption by the diminishing EPO-responsive cell numbers. The number of EPO receptors on these cells is rather small (<1000), and probably only a few must be triggered by EPO for a proper function (Sawyer 1990). Piroso et al. (1991) also rejected the EPO consumption theory; they found no differences in EPO half-lives under hyper- or hypoproliferative conditions.

After the highest Epo levels a continuous decrease to normal values was observed. This decrease started while hematocrit was still at the lowest level, as was the bone marrow precursor content. Assuming a half-life of about 3 h as Piroso et al. (1991) found in rats, this means that production of EPO continues in this period but at a lower rate. Several authors found low Epo levels in subjects with hyperproliferative marrow in comparison with hypoproliferative states (Fried et al. 1971: normal anemic mice versus WW_v mice; De Klerk et al. 1981: sickle cell anemia versus Fanconi's anemia; Jelkmann and Wiedemann 1990; Johannsen et al. 1989: ulcerative colitis versus leukemia).

Our experiments, however, show that this link is not absolute. The decrease in EPO started before a rise in the erythroid precursors could be ovserved. Nevertheless, such a hypothetical regulatory link could be advantageous to the system. The production of the hormone starts already before aplastic effects become visible in the peripheral blood. The mechanism of this type of regulation remains fully obscure. Despite the fact that the kidney is the main EPO plant, reacting to the oxygen supply by the blood, it remains possible that under our conditions extrarenal sites become activated. Rich et al. (1980) showed that stimulated macrophages of different origin, including marrow and spleen, produce EPO in vitro. Production in the latter two organs would be very interesting, because a much more efficient use of EPO could be achieved. This is precisely where it is needed. Macrophages have also the potential to produce negative regulators of erythropoiesis such as tumor necrosis factor-A and transforming growth factor-ß.

It is challenging to relate different populations of macrophages from different hematopoietic microenvironments to their ability to react differently to local stimuli, thereby producing negative or positive factors for erythropoiesis. Further experiments are necessary to give experimental support for these ideas.

References

Barcelo AC, Bozzini CE (1982) Erythropoietin formation during hypoxia in mice with impaired responsiveness to erythropoietin induced by irradiaction or 5-fluorouracil injection. Experientia 38:504–505

Birgegard G, Wide L, Simonsson B (1989) Marked erythropoietin increase before fall in Hb after treatment with cytostatic drugs suggests mechanism other than anemia for stimulation. Br J Haematol 72:462–466

Cynshi O, Shimonaka Y, Higuchi M, Imai N, Suzuki H, Togashi M, Okamoto MT, Hirashima K (1990) Effects of recombinant human erythropoietin on haemolytic anemia in mice. Br J Haematol 76:414–419

De Klerk G, Rosengarten PCJ, Vet RJWM, Goudsmit R (1981) Serum erythropoietin (ESF) titers in anemia. Blood 58:1164–1170

Fried W, Gregory SA, Knospe WH, Trobaugh Jr FE (1971) Regulation of plasma erythropoietin levels in mice with impaired responsiveness to erythropoietin. J Lab Clin Med 78:449–456

Goris H, Loeffler M, Bungart B, Schmitz S, Nijhof W (1989) Hemopoiesis during thiamphenicol treatment. I. Stimulation of stem cells during eradication of intermediate cell stages. Exp Hematol 17:957–961

Iscove NN, Sieber F (1975) Erythroid progenitors in mouse bone marrow detected by macroscopic colony formation in culture. Exp Hematol 3:32–43

Jelkmann W, Wiedemann G (1990) Serum erythropoietin level: relationship to blood hemoglobin concentration and erythrocytic activity of the bone marrow. Klin Wochenschr 68:403–407

Johannsen H, Jelkmann W, Wiedemann G, Otte M, Wagner T (1989) Erythropoietin/haemoglobin relationship in leukaemic and ulcerative colitis. Eur J Haematol 43:201–206

Kilbridge TM, Fried W, Heller P (1969) The mechanism by which plethora suppresses erythropoiesis. Blood 33:104–113

McGonigle RJS, Ohene-Frempong K, Lewy JE, Fisher JW (1985) Erythropoietin response to anaemia in children with sickle cell disease and Fanconi's hypoproliferative anaemia. Acta Haematol (Basel) 74:6–9

Misago M, Chiba S, Kikuchi M, Tsukada J, Suzuki H (1986) Effect of absolute and relative changes of hematocrit on erythropoiesis in mice. Int J Cell Cloning 4:320–330

Nijhof W, Wierenga PK, Kardaun S (1977) The effect of thiamphenicol on the production of immature red blood cells under anaemic conditions. Br J Haematol 36:29–40

Nijhof W, Wierenga PK, Goldwasser E (1982) The regeneration of stem cells after bone marrow depression induced by thiamphenicol. Exp Hematol 10:36–43

Nijhof W, Wierenga PK (1983) Isolation and characterization of the erythroid progenitor cell: CFU-E. J Cell Biol 96:386–392

Piroso E, Erslev AJ, Caro J (1989) Inappropriate increase in erythropoietin titers during chemotherapy. Am J Hematol 32:248–254

Piroso E, Erslev AJ, Flaharty KK, Caro J (1991) Erythropoietin life span in rats with hypoplastic and hyperplastic bone marrows. Am J Hematol 36:105–110

Rich IN, Heit W, Kubanek B (1980) An erythropoietic stimulating factor similar to erythropoietin released by macrophages after treatment with silica. Blut 40:297–303

Sawyer ST (1990) Erythropoietin: synthesis of the hormone and its interaction with receptors on erythroid progenitors. Clin Biotechnol 2:77–85

Sherwood JB, Goldwasser E, Chilcote R, Carmichael LD, Nagel RL (1986) Sickle cell anemia patients have low erythropoietin levels for their degree of anemia. Blood 67:46–49

VI. Application of Recombinant Erythropoietin in Non-Renal Anemia

Effects of Inflammatory Cytokines on the Production of Erythropoietin

W. Jelkmann, J. Fandrey, and H. Pagel

Introduction

Chronic infections, autoimmune diseases, and malignancies are frequently associated with a mild hypoplastic anemia. The causes of this so-called anemia of chronic disorders (ACD) include impaired ability of erythrocytic precursors to utilize iron for hemoglobin synthesis, inhibition of the proliferation of myeloid cells by inflammatory cytokines, hemolysis, and often bleeding. Furthermore, earlier bioassay studies showed that the plasma level of erythropoietin (Epo) is relatively low for the degree of the anemia in many patients suffering from chronic inflammation or malignancy [24, 33, 34]. Radioimmunological investigations have confirmed that there is a relative lack of Epo in rheumatoid arthritis [1, 12, 29], acquired immunodeficiency syndrome [28], and cancer [20].

Antigens, bacterial and viral pyrogens such as endotoxin, and complement components induce the secretion of immunomodulatory peptides from macrophages, T lymphocytes and other cells. In turn then, cytokines are important in the control of the acute-phase response and of the specific immune reactions. However, the effects of inflammatory cytokines are not restricted to the immune system. There is significant interaction with endocrine glands. Certain cytokines inhibit hormone secretion from anterior pituitary cells [26, 32].

The purpose of this work is to summarize evidence suggesting that inflammatory cytokines influence the production of Epo.

Effects of Cytokines on In Vitro Epo Production

Hepatoma Cultures. The only cell cultures that have clearly been proven to produce Epo in a regulated fashion are the two human hepatoma-derived lines HepG2 and Hep3B [10]. As in vivo, Epo synthesis in these cultures increases at low O_2 tension or upon the addition of cobalt ions. When confluent HepG2 cultures (about 0.5×10^6 cells/cm^2; medium RPMI 1640, 10% fetal bovine serum) are maintained in commun culture dishes made of polystyrene, their Epo production is considerably increased above the

Pagel/Weiss/Jelkmann (Eds.)
Pathophysiology and Pharmacology
of Erythropoietin

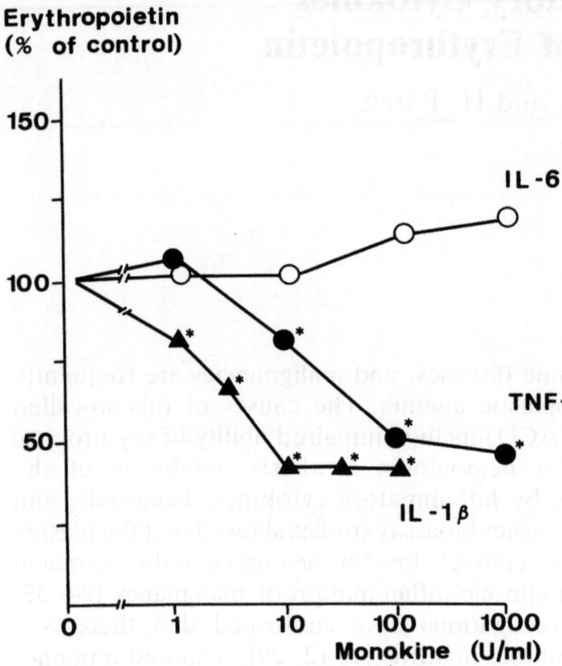

Fig. 1. Effects of the monokines interleukin-1β (*IL-1β*), tumor necrosis factor-α (*TNF-α*) and interleukin-6 (*IL-6*) on the 24-h rate of production of immunoreactive erythropoietin. Values are the mean of four subcultures. *, Significant difference compared to control cultures without added monokine. (Data from [8, 13])

constitutional level, because the cells are hypoxic due to the diffusion-limited supply with O_2 [9].

The addition of pure recombinant human cytokines to HepG2 cultures enables one to study the action of each of the single immunomodulatory peptides alone on hypoxia-induced production of Epo. Previously, experiments were carried out with interleukin (IL) 1α, 1ß, and 6, tumor-necrosis

Table 1. Synopsis of the effects of immunomodulatory peptides on erythropoietin production (Epo) in HepG2 cultures. (Data from [8, 13])

Peptide	Tested concentration	Effect on Epo
IL-1	1 – 100 U/ml	–
IL-6	1 –10000 U/ml	∅ (+)
TNF-α	1 – 1000 U/ml	–
TGF-β2	0.2– 100 ng/ml	∅
IFN-γ	5 – 5 000 U/ml	∅ (+)

–, Inhibition; ∅, no effect; +, stimulation.

factor-α (TNF-α), transforming-growth factor ß2 (TGF-ß2), and interferon γ (IFN-γ). Experimental details are described elsewhere [9, 13].

Figure 1 shows that the addition of IL-1ß and TNF-α resulted in a dose-dependent inhibition of the 24-h rate of production of immunoreactive Epo in HepG2 cultures. IL-1α also inhibited. None of the other cytokines tested lowered the production of Epo in hepatoma cells (Table 1). The action of IL-1 and TNF-α on Epo production was specific, since there was no overall decrease in protein synthesis, as assessed by determinations of the production of α-fetoprotein [13].

Isolated Perfused Rat Kidneys. Since IL-1ß was found to be the most potent cytokine inhibiting Epo synthesis in HepG2 cultures, its effects were also studied in isolated perfused rat kidneys. The perfusion system has been described in detail in [23]. In brief, rat kidneys were perfused at constant pressure (100 mmHg) and 37°C in a closed recirculation system for 3 h. The perfusion medium (0.2 l, Krebs-Henseleit solution supplemented with 60 g/l albumin and 5% human red cells) was dialyzed continuously against 5 l protein-free medium, which was equilibrated with 5% CO_2 and varied mixtures of O_2 and N_2.

Figure 2 shows that the rate of production of Epo increased from 0.1 to 0.9 U/g kidney in 3 h when the perfusion medium was equilibrated with 3% O_2 instead of 20% O_2. The addition of IL-1ß (10 U/ml) completely suppressed this in vitro renal production of Epo at lowered O_2 tension.

Clinical Relevance of Cytokine Inhibition of Epo Production

IL-1 and TNF-α are regarded as important mediators of the anemia of chronic disorders [15, 27, 30]. Treatment of mice with IL-1 [14] or endoxin [31] and treatment of humans [4], mice [15], and rats [30] with TNF-α resulted in suppressed erythropoiesis. With respect to the underlying mechanism, previous studies have focused on the myeloid action of monokines. It has been established that IL-1 and TNF-α inhibit the proliferation of erythrocytic progenitors.

The present in vitro findings indicate that pure recombinant human IL-1 and TNF-α may additionally suppress hepatic and renal synthesis of Epo. The role played by these monokines in the control of Epo production in vivo still remains to be elucidated. Little is known, for instance, about the interaction of the various cytokines, which can be synergistic or antagonistic. In addition, there is lack of knowledge with regard to the significance of the local versus the systemic actions of stimulated cytokine-synthetizing cells and their secretory products. Nevertheless, it may be asked whether there are disorders in which increased secretion of the monokines IL-1 and TNF-α could result in lowered Epo production. Such consideration is not only of interest with a

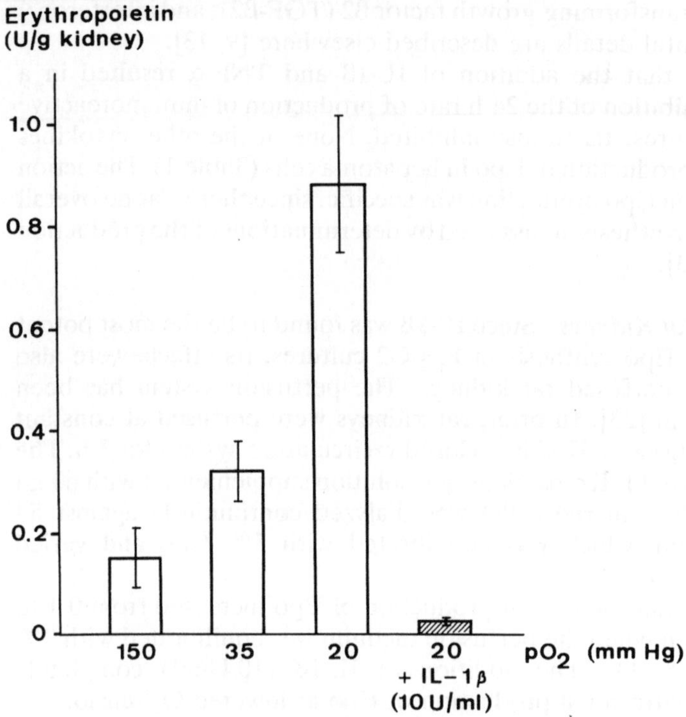

Fig. 2. Effect of interleukin-1ß (*IL-1β*); human, recombinant) on hypoxia-induced production of erythropoietin in isolated perfused rat kidneys (3-h perfusion period). Values are the mean ± SEM of five or six experiments

view to treatment of anemia with recombinant human Epo. It also raises questions as to the changes in Epo production following pharmacologic immunosuppression by glucocorticoids or antibodies to immunocompetent cells and to specific cytokines and their receptors.

IL-1 and TNF-α could contribute to the defective Epo production in renal diseases. Regarding the Epo deficiency in acute renal failure [17, 22] monocytic macrophages but also mesangial cells of the renal corpuscles are a potential source of monokines [18]. With respect to chronic renal failure, elevated blood levels of IL-1 and TNF-α were measured in hemodialysis patients [11]. Note that the ability to produce Epo is not necessarily abolished in chronic renal failure. Instead, the Epo-synthesizing cells can operate at a lower set point and be stimulated by severe hypoxic stress [6]. The association between monokines and Epo production is perhaps most obvious in renal transplanted patients. Plasma Epo levels fall [2, 25] while TNF-α levels rise [19] during episodes of acute renal allograft rejection.

The Epo response has been reported to be blunted in chronic inflammatory conditions (rheumatoid arthritis; [1, 12, 24, 29, 33]), infectious [5, 34] and malignant [20, 34] diseases. However, it should be noted that some investigators [3, 7, 21] have disputed the significance of Epo deficiency in the pathogenesis of ACD. Indeed, the slope of the log Epo/hemoglobin regression line is only slightly flatter in ACD than in other nonrenal anemias. The possibility that certain inflammatory cytokines inhibit Epo production in some cases of ACD has not been investigated as yet. Clearly, chronic infectious, inflammatory, and malignant diseases are associated with the expression of monokines and other immunomodulatory peptides.

Furthermore, the plasma levels of certain monokines increase following acute trauma and major surgery. Recent studies have shown that Epo production is suppressed in patients after surgical procedures [16]. The administration of recombinant Epo to surgical patients is expected to prove useful in decreasing the severity and duration of postoperative anemia and the need for red cell transfusions.

In conclusion, the role of inflammatory cytokines in the pathophysiology of the production of Epo is only incompletely understood. IL-1 and TNF-α, which suppress in vitro synthesis of Epo, are generated in several renal and nonrenal diseases that are characterized by an anomalously low plasma Epo level.

References

1. Baer AN, Dessypris EN, Goldwasser E, Krantz SB (1987) Blunted erythropoietin response to anaemia in rheumatoid arthritis. Br J Haematol 66:559–564
2. Besarab A, Caro J, Jarrell BE, Francos G, Erslev AJ (1987) Dynamics of erythropoiesis following renal transplantation. Kidney Int 32:526–536
3. Birgegård G, Hällgren R, Caro J (1987) Serum erythropoietin in rheumatoid arthritis and other inflammatory arthritides: relationship to anaemia and the effect of anti-inflammatory treatment. Br J Haematol 65:479–483
4. Blick M, Sherwin SA, Rosenblum M, Gutterman J (1987) Phase I study of recombinant tumor necrosis factor in cancer patients. Cancer Res 47:2986–2989
5. Camacho J, Arnalich F, Zamorano AF, Vázquez JJ (1991) Serum erythropoietin levels in the anaemia of chronic disorders. J Intern Med 229:49–54
6. Chandra M, Clemons GK, McVicar M (1988) Relation of serum erythropoietin levels to renal excretory function: evidence for lowered set point for erythropoietin production in chronic renal failure. J Pediatr 113:1015–1021
7. Erslev AJ, Wilson J, Caro J (1987) Erythropoietin titers in anemic, nonuremic patients. J Lab Clin Med 109:429–433
8. Fandrey J, Jelkmann WEB (1991) Interleukin-1 and tumor necrosis factor-α inhibit erythropoietin production in vitro. Ann N Acad Sci 628:250–255
9. Fandrey J, Seydel FP, Siegers CP, Jelkmann W (1990) Role of cytochrome P_{450} in the control of the production of erythropoietin. Life Sci 47:127–134
10. Goldberg MA, Glass GA, Cunningham JM, Bunn HF (1987) The regulated expression of erythropoietin by two human hepatoma cell lines. Proc Natl Acad Sci USA 84:7972–7976
11. Herbelin A, Nguyen AT, Zingraff J, Ureña P, Descamps-Latscha B (1990) Influence of uremia and hemodialysis on circulating interleukin-1 and tumor necrosis factor α. Kidney Int 37:116–125

12. Hochberg MC, Arnold CM, Hogans BB, Spivak JL (1988) Serum immunoreactive erythropoietin in rheumatoid arthritis: impaired response to anemia. Arthritis Rheum 31:1318–1321
13. Jelkmann W, Wolff M, Fandrey J (1990) Modulation of the production of erythropoietin by cytokines. In vitro studies and their clinical implications. Contrib Nephrol 87:68–77
14. Johnson CS, Keckler DJ, Topper MI, Braunschweiger PG, Furmanski P (1989) In vitro hematopoietic effects of recombinant interleukin-1α in mice: stimulation of granulocytic, monocytic, megakaryocytic, and early erythroid progenitors, suppression of late-stage erythropoiesis, and reversal of erythroid suppression with erythropoietin. Blood 73:678–683
15. Johnson RA, Waddelow TA, Caro J, Oliff A, Roodman GD (1989) Chronic exposure to tumor necrosis factor in vivo preferentially inhibits erythropoiesis in nude mice. Blood 74:130–138
16. Levine EA, Rosen AL, Sehgal LR, Gould SA, Egrie JC, Moss GS (1991) Erythropoietin deficiency after coronary artery bypass procedures. Ann Thorac Surg 51:764–766
17. Lipkin GW, Kendall RG, Russon LJ, Turney JH, Norfolk DR, Brownjohn AM (1990) Erythropoietin deficiency in acute renal failure. Nephrol Dial Transplant 5:920–922
18. Matsumoto K, Dowling J, Atkins RC (1988) Production of interleukin 1 in glomerular cell cultures from patients with rapidly progressive crescentic glomerulonephritis. Am J Nephrol 8:463–470
19. Maury CPJ, Teppo AM (1987) Raised serum levels of cachectin/tumor necrosis factor α in renal allograft rejection. J Exp Med 166:1132–1137
20. Miller CB, Jones RJ, Piantadosi S, Abeloff MD, Spivak JL (1990) Decreased erythropoietin response in patients with the anemia of cancer. N Engl J Med 322:1689–1692
21. Nielsen OJ, Andersen LS, Ludwigsen E, Bouchelouche P, Hansen TM, Birgens H, Hansen NE (1990) Anaemia of rheumatoid arthritis: serum erythropoietin concentrations and red cell distribution width in relation to iron status. Ann Rheum Arthr 49:349–353
22. Nielsen OJ, Thaysen JH (1990) Erythropoietin deficiency in acute tubular necrosis. J Intern Med 227:373–380
23. Pagel H, Jelkmann W, Weiss C (1991) Isolated serum-free perfused rat kidneys release immunoreactive erythropoietin in response to hypoxia. Endocrinology 128:2633–2638
24. Pavlovic-Kentera V, Ruvidic R, Milenkovic P, Marinkovic D (1979) Erythropoietin in patients with anaemia in rheumatoid arthritis. Scand J Haematol 23:141–145
25. Rejman ASM, Grimes AJ, Cotes PM, Mansell MA, Joekes AM (1985) Correction of anaemia following renal transplantation: serial changes in serum immunoreactive erythropoietin, absolute reticulocyte count and red-cell creatine levels. Br J Haematol 61:421–431
26. Schettini G, Landolfi E, Grimaldi M, Meucci O, Postiglione A, Florio T, Ventra C (1989) Interleukin 1 beta inhibition of TRH-stimulated prolactin secretion and phosphoinositides metabolism. Biochem Biophys Res Commun 165:496–505
27. Schooley JC, Kullgren B, Allison AC (1987) Inhibition by interleukin-1 of the action of erythropoietin on erythroid precursors and its possible role in the pathogenesis of hypoplastic anaemias. Br J Haematol 67:11–17
28. Spivak JL, Barnes DC, Fuchs E, Quinn TC (1977) Serum immunoreactive erythropoietin in HIV-infected patients. JAMA 261:3104–3107
29. Takashina N, Kondo H, Kashiwazaki S (1990) Suppressed serum erythropoietin response to anemia and the efficacy of recombinant erythropoietin in the anemia of rheumatoid arthritis. J Rheumatol 17:885–887
30. Tracey KJ, Wei H, Manogue KR, Fong Y, Hesse DG, Nguyen HT, Kuo GC, Beutler B, Cotran RS, Cerami A, Lowry SF (1988) Cachectin/tumor necrosis factor induces cachexia, anemia, and inflammation. J Exp Med 167:1211–1227

31. Udupa KB, Lipschitz DA (1982) Endotoxin-induced suppression of erythropoiesis: The role of erythropoietin and a heme synthesis stimulating factor. Blood 59:1267–1271
32. Walton PE, Cronin MJ (1990) Tumor necrosis factor-α and interferon-α reduce prolactin release in vitro. Am J Physiol 259:E672–E676
33. Ward HP, Gordon B, Pickett JC (1969) Serum levels of erythropoietin in rheumatoid arthritis. J Lab Clin Med 74:93–97
34. Ward HP, Kurnick JE, Pisarczyk MJ (1971) Serum level of erythropoietin in anemias associated with chronic infection, malignancy, and primary hematopoietic disease. J Clin Invest 50:332–335

The Use of Recombinant Human Erythropoietin in the Treatment of the Anemia of Chronic Disorders

J.P. Kaltwasser, F. Praeve, R. Wigand, and T. Brune

Introduction

Common infectious diseases as well as numerous noninfectious disorders such as rheumatoid arthritis, Hodgkin's disease, and metastatic carcinomas are associated with a similar type of anemia. As defined by Cartwright in his classic 1966 review [5]: "The anaemia is usually mild in the degree and not progressive in severity and is characterized by a low plasma iron, decreased total iron binding capacity, decreased saturation of transferrin with iron, decreased bone marrow sideroblasts, normal or increased reticuloendothelial iron, increased plasma copper and increased free erythrocyte protoporphyrin." Being roughly proportional to the severity of the underlying disease, hemoglobin (Hb) concentrations usually range from 7.0 to 11.0 g/dl [5, 14]. A variety of names have been proposed for this anemia, but since etiology and pathogenesis are still far from being clarified the most commonly used name is anemia of chronic disorders (ACD). ACD is thought to be a common feature in the broad spectrum of infections and inflammatory diseases [8]. There is in the literature, however, surprisingly little in the way of controlled data about its incidence. In an analysis conducted at the Rheumatology Unit of the University Hospital of Frankfurt, we calculated an incidence of 7.3% for ACD in a population of 218 patients with various rheumatic diseases, including rheumatoid arthritis (RA), systemic lupus erythematosus (SLE), and mixed connective tissue disease (MCT; Fig. 1).

Pathogenesis of ACD

The pathogenesis of ACD is still a matter of controversy. The erythropoietic activity of the bone marrow is hypoplastic in nature although the red blood cell life span is slightly to moderately shortened. Low serum levels of adequate iron stores indicate a profound disturbance in iron metabolism. A defective iron release from the macrophage-monocyte system and from hepatic cells (RES block) has been thought to be responsible for the changes in the iron transport and storage compartment [8, 13, 14]. Recent experimental and clinical observations indicate, however, that cytokines may play a

Pagel/Weiss/Jelkmann (Eds.)
Pathophysiology and Pharmacology
of Erythropoietin

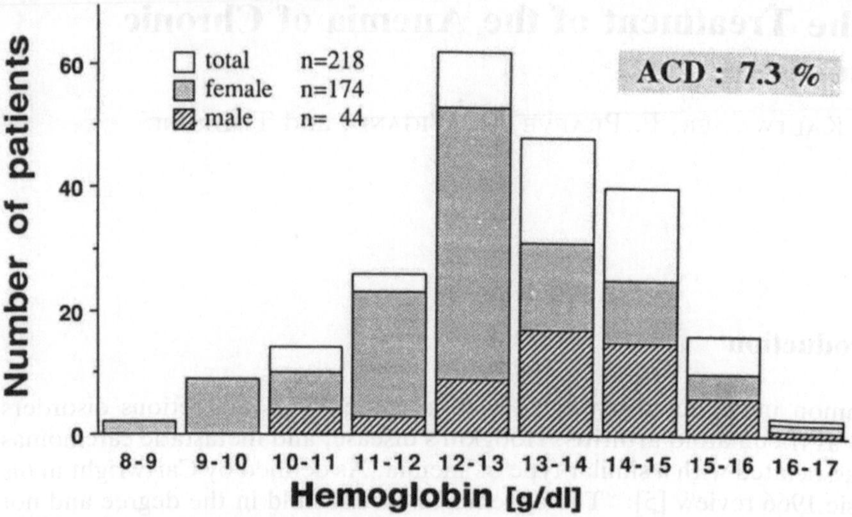

Fig. 1. Histogram of the Hb concentration in 218 patients with various chronic inflammatory diseases of the outpatient Rheumatology Unit of the University Hospital of Frankfurt. Excluding iron deficiency, B 12 and folate deficiencies and hemolysis, 7.3% remain as ACD, with hemoglobin less than 11 g/dl

key role in the pathogenesis of ACD. Among these intercellular mediators the monokine interleukin-1 (IL-1) is of particular interest. IL-1 induces fever, leukocytosis, acute-phase protein synthesis, and hypoferremia, which are all typical features of chronic infectious and inflammatory diseases [10, 16]. Recombinant IL-1ß also markedly suppresses colony formation of erythroid progenitor cells [15] and seems to be identical with the leukocyte endogenous mediator described by Lee [14] 8 years ago as probably the "final common pathway" that connects the various diseases associated with ACD.

Role of Erythropoietin in the Pathogenesis of ACD

Since normal bone marrow easily compensates for the shortened red blood cell life span and Hb reduction in ACD, the hypoplastic erythropoietic marrow response has also been assumed to be caused by a decreased release of, or a decreased reponse to erythropoietin (EPO) [8, 14, 20]. Studies of the release of EPO in patients with ACD have, however, produced conflicting results. Some have shown a blunted EPO response compared to the degree of anemia [2, 11, 19]; others have found appropriate EPO-concentrations [3, 7]. In animal experiments IL-1α and IL-1β have been shown to antagonize the capacity of EPO to stimulate the proliferation of mouse spleen and bone marrow erythroid precursor cells, and it has been suggested that the IL-1 mediated antagonism of the effects of EPO on erythroid precursor cells may be a factor in the pathogenesis of ACD [18].

We report here on a pilot study of the possible efficacy of recombinant human EPO (rHu-EPO) in the treatment of ACD due to various rheumatic diseases.

Patients and Methods

Eleven patients (ten women, one man) were included in the study. Eight patients had RA, two SLE, and one MCT (Table 1). Three of these, were lost to follow-up for various reasons, so that the following analysis includes the remaining eight patients. Patients with iron, folate of rB_{12} deficiency as well as those with hypertension, acute infectious diseases, malignant tumors, pregnancy, transfusion of packed red cells in the last 3 months or age below 18 years were excluded. All RA and SLE patients fulfilled the criteria of the American Rheumatism Association [1]. Each patient underwent hematologic evaluation, including blood cell counts, reticulocytes, blood sedimentation rate, C-reactive protein, serum iron, total iron-binding capacity, serum ferritin and plasma EPO according to the treatment schedule (Fig. 2). Measurements were carried out by standard methods.

All patients were outpatients of the Rheumatology Unit of the University Hospital of Frankfurt. Informed consent was obtained in the study protocol which was approved by the local committee on medical ethics. The administration of EPO was preceded by a 2-week placebo period (Fig. 2).

Table 1. Pretreatment laboratory data on patients included in the study

Patient	Age (years)	Sex	Disease	Hb (g/dl)	MCH (pg)	Retic (%)	S-iron (μg/dl)	% Satur. of TIBC	Ferritin (μg/l)	S-EPO (mU/ml)
V.I.	43	F	SLE	10.7	26.0	2.8	40	12	63	42
S.D.	59	F	RA	9.4	23.4	0.9	80	23	27	22
E.D.	66	F	RA	8.6	25.1	1.8	57	18	65	93
H.G.	45	F	MCT	10.4	31.0	0.9	95	32	198	114
K.G.	74	F	RA	8.7	27.2	1.4	44	15	557	52
M.G.	71	F	RA	9.8	34.0	0.3	70	26	595	48
A.G.[a)]	19	F	SLE	10.1	34.1	1.1	110	37	293	49
H.H.[a)]	55	F	RA	9.7	24.2	0.8	23	9	47	26
R.L.[a)]	79	F	RA	8.1	33.3	0.9	54	24	502	165
C.R.	54	F	RA	10.3	24.6	0.6	13	6	93	48
U.F.	27	M	RA	11.3	22.1	1.7	39	13	105	20
Mean	48			9.7	27.7	1.2	57	20	229	62
±SD	21			1.1	4.5	0.7	30	11	222	45

Hb, Hemoglobin; MCH: mean corpuscular hemoglobin, Retic, reticulocytes; S-iron, serum iron; TIBC, total iron binding capacity; SLE, systemic lupus erythrematosus; RA, rheumatoid arthritis; MCT, mixed connective tissue disease.

[a]Drop-outs during the study (A.G, due to hemolysis; H.H. and R.L., due to transfusion requirement).

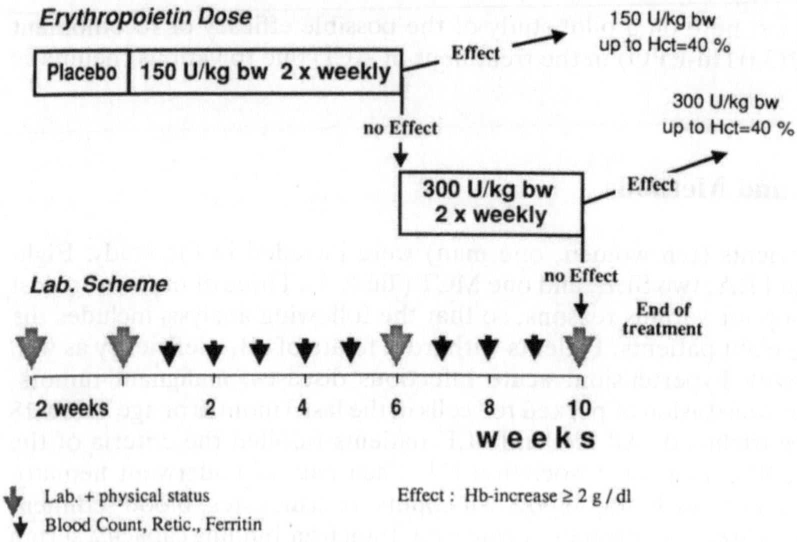

Fig. 2. Schedule of the EPO administration. *Arrows*, laboratory and physical status, including blood counts

EPO was administered twice a week subcutaneously in a dose of 150 U/kg body weight. Hb, serum ferritin, and reticulocyte counts were determined at weekly intervals. After 6 weeks of EPO treatment the net increase in Hb was calculated. If the increase was less than 2.0 g/dl, the EPO dose was doubled to 300 U/kg for a maximum of an additional 4 weeks. Treatment was stopped after 10 weeks if the increase in Hb was less than 2,0 d/dl, or when hemocrit reached 40%. The treatment goal was to achieve an Hb increase of at least 2.0 d/dl.

We used 62.5 μg rHu EPO (kindly provided by Behringwerke, Marburg, FRG) dissolved in 1 ml phosphate buffer, equivalent to 5000 IU rHu-EPO. The phosphate buffer used for the dissolution of rHu-EPO was also used as placebo during the first 2 weeks of the study.

Results

All patients except one (N.R.) responded by an increase in Hb concentration and reticulocyte count and decrease in serum ferritin concentration. The mean Hb increase was 2.1 g/dl (Fig. 3) and was achieved within 6 weeks using the starting dose of 300 U/kg per week of rHu-EPO. One female patient (K.G.), who was repeatedly transfused by packed red cells because of the severity of her anemia, became transfusion independent and could be brought to a necessary joint replacement surgery. Serum ferritin decreased during rHu-EPO treatment from 156 ± 57 μg/l to 120 ± 60 μg/l (Fig. 4).

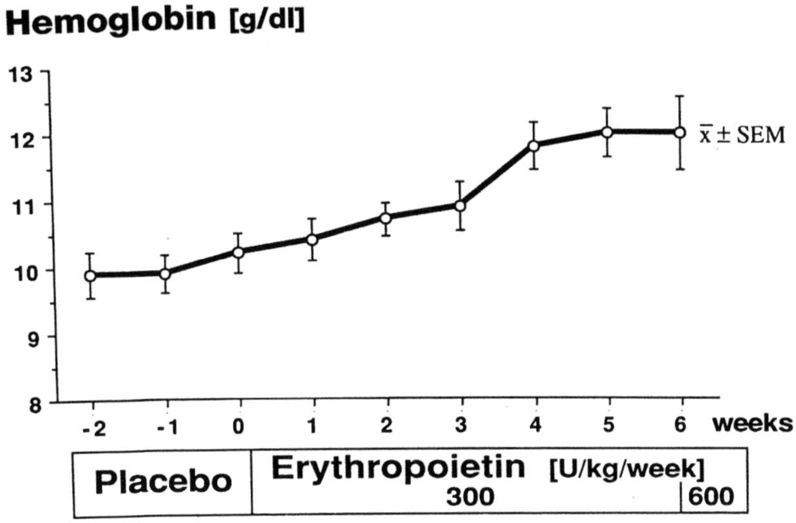

Fig. 3. Course of Hb concentration in eight patients with ACD during placebo and EPO administration

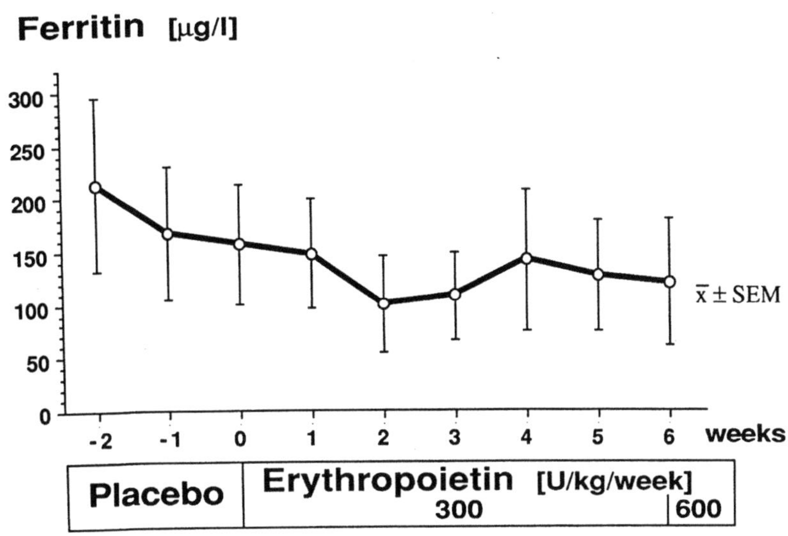

Fig. 4. Course of serum ferritin concentration in eight patients with ACD during placebo and EPO administration

There was, however already a decrease in ferritin concentration during the preceding 2 weeks of placebo. Reticulocytes increased from 1.7% ± 0.6% to 2.3% ± 0.8% after 5 weeks of EPO treatment. The rHu-EPO injections were well tolerated. There were no increases in blood pressure or frequency of

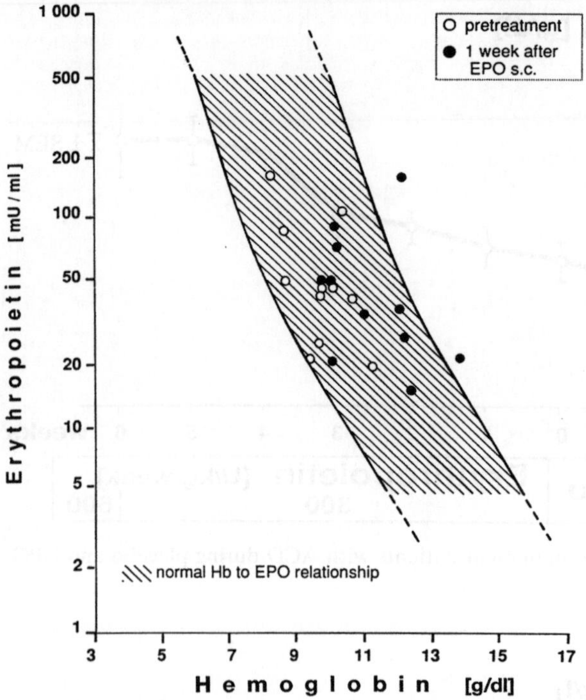

Fig. 5. Relationship between serum erythropoietin and hemoglobin concentration in the 11 patients originally included in the study prior to rHu-EPO treatment and 1 week after first subcutaneous EPO administration

thromboses. Serum EPO concentrations prior to rHu-EPO treatment and 1 week after the start of treatment (Fig. 5) were – except in two – all within the normal Hb-to-EPO relationship, as given by Erslev et al. [9].

Discussion

EPO deficiency is usually not regarded as the primary cause of ACD. Accordingly, Cotes et al. [6], Erslev et al. [9] and Birgegard et al. [3] have reported that serum EPO levels in patients with RA are essentially the same as those in patients with anemias of other etiologies. Other investigators have, however, described a rather blunted response to anemia, such as in RA compared to other forms of anemia not related to chronic inflammatory disorders [2, 11]. More recently, Takashina et al. [19] have shown that serum EPO levels and Hb concentration in anemic patients with RA and concomitant iron deficiency were significantly negatively correlated whereas this was not the case in those with RA anemia without iron deficiency (serum ferritin of 20 μg/l or more). Noé et al. (this volume) using a sensitive enzyme-linked immunosorbent assay for EPO also demonstrate a blunted EPO response to anemia of patients with RA as compared to anemias of other etiologies. Inadequate production or accelerated catabolism of EPO may therefore be a factor in the pathogenesis of ACD. There is also evidence that treatment with rHu-EPO may improve erythropoietic response to anemia in patients with ACD, as shown in two cases by Means et al. [17] in four by Takashina et al. [19], in 5/7 by Birgegard et al. [4], and in 7/8 patients of the study presented here. As has been shown by Birgegard et al. [4], the duration of Hb response is limited after cessation of treatment and may be influenced by the course of the inflammatory activity in ACD. In the few cases evaluated so far it is not possible to determine whether the pretreatment EPO levels are of importance for the response to treatment.

In our study, it was not investigated whether the improvement in Hb concentration was associated with an improvement in quality of life. Five of our seven patients, however, have stated that they felt better with higher Hb levels. No meaningful changes were seen in scores for activities of daily living and pain in the two cases treated by Means et al. [17]. On the other hand, Takashina et al. [19] described a slight reduction in the number of painful joints after EPO therapy associated with a reduction in erythrocyte sedimentation rate, which may, however, also be caused by the improvement in the anemia itself. Of particular interest in this context is a recent observation by Kimball and Kerman [12] that EPO may also modulate the immune response by decreasing IL-1 production.

In conclusion, we have found in accordance with others, that ACD responds to subcutaneous rHu-EPO in a considerable proportion of patients. Whether restoration of a normal Hb concentration by rHu-EPO in ACD also increases quality of life requires further evaluation. With regard to a reasonable price of rHu-EPO it is important to differentiate between an expensive cosmetic procedure and a true benefit for the quality of life of the patient. Controlled clinical trials are therefore necessary to determine the therapeutic value of rHu-EPO in ACD.

References

1. Arnett FC, Edworthy SM, Bloch DA et al. (1988) The American Rheumatism Association 1987. Revised criteria for the classification of rheumatoid arthritis. Arthritis 31:315–324
2. Baer AN, Dessypris N, Goldwasser E, Krantz SB (1987) Blunted erythropoietin response to anemia in rheumatoid arthritis. Br J Haematol 66:559–564
3. Birgegard G, Hällgren R, Caro J (1987) Serum erythropoietin in rheumatoid arthritis and other inflammatory arthritides: relationship to anemia and the effect of anti-inflammatory treatment. Br J Haematol 65:479–483
4. Birgegard G, Gudbjörnsson B, Hällgren R, Wide L (1991) Anemia of chronic inflammatory arthritides: treatment with recombinant human erythropoietin. In: Gurland HJ, Movan J. Samtleben W, Scigalla P, Wieczovek L (eds) Erythropoietin in renal and non-renal anemias. Contrib Nephrol 88:295–303
5. Cartwright GR (1966) The anemia of chronic disorders. Semin Hematol 3:351–375
6. Cotes, PM, Brozovic B, Mansell M, Samson DM (1980) Radioimmunoassay of erythropoietin in human serum. Exp Hematol 8:292

7. Erslev AJ, Caro J, Miller O, Silver R (1980) Plasma erythropoietin in health and disease. Ann Clin Lab Sci 10:250–259
8. Erslev AJ (1983) Anemia of chronic disorders. In: Williams WJ, Beutler E. Erslev AJ, Lichtman MA (eds) Hematology 3rd edn. McGraw-Hill, New York, pp 522–528
9. Erslev AJ, Wilson J, Caro J (1987) Erythropoietin titers in anemic, non-uremic patients. J Lab Clin Med 109:429–433
10. Gordeuk VR, Prithiraj P, Dolinart T, Brittenham GM (1988) Interleukin-1 administration in mice produces hypoferremia dispite neutropenia. J Clin Invest. 82:1934–1938
11. Hochberg MC, Arnold CM, Hogans BB, Spivak JL (1988) Serum immunoreactive erythropoietin in rheumatoid arthritis: impaired response to anemia. Arthritis 31:1318–1321
12. Kimball PM, Kerman RH (1991) Erythropoietin: a potential immunomodulator? Transplant Proc 23/1:336
13. Konijn AM, Hershko C (1977) Ferritin synthesis in inflammation. I. Mechanism of impaired iron release. Br J Haematol 37:7–16
14. Lee GR (1983) The anemia of chronic disease. Semin Hematol 20:61–80
15. Maury CPJ, Andersson LC, Teppo AM, Partanen S, en Juvon E (1988) Mechanism of anemia in rheumatoid arthritis: demonstration of raised interleukin-1-beta concentrations in anemic patients and of interleukin-1 mediated supression of normal erythropoiesis and proliferation of human erythroleucaemia (HEL) cells in vitro. Ann Rheum Dis 47:972–978
16. Maury CPJ (1989) Anemia in rheumatoid arthritis: role of cytokines. Scand J Rheumatol 18:3–5
17. Means RT, Olsen NJ, Krantz SB, Dessypris EN, Graber SE, Stone WJ, O'Neil VL, Pincus T (1989) Treatment of the anemia of rheumatoid arthritis with recombinant human erythropoietin: clinical and in vitro studies. Arthritis Rheum 32:638–642
18. Schooley JC, Kullgren B, Allison AC (1987) Inhibition by interleukin-1 of the action of erythropoietin on erythroid precursors and its possible role in the pathogenesis of hypoplastic anemias. Br J Haematol 67:11–17
19. Takashina, N., Kondo H, Kashiwazki S (1990) Suppressed serum erythropoietin response to anemia and the efficacy of recombinant erythropoietin in the anemia of rheumatoid arthritis. Rheumatol 17:885–887
20. Wallner, SF (1983) The anemia of chronic disorders: clinical and pathological features. In: Dunn, CDR (ed) Current concepts in erythropoiesis. Wiley, Chichester, pp 209–231

Recombinant Human Erythropoietin in the Treatment of Anaemia in Rheumatoid Arthritis

T. Pettersson, K. Rosenlöf, A. Mickos, C. Friman, and F. Fyhrquist

Introduction

Rheumatoid arthritis (RA) is often accompanied by anaemia, the severity of which correlates with the activity of the disease. This particular kind of anaemia, anaemia of chronic disease (ACD) is characterised by decreased serum iron and decreased transferrin saturation despite the presence of normal or increased bone marrow iron stores (Cartwright and Lee 1971). The reticulocyte count is low for the degree of anaemia.

The pathogenesis of ACD remains incompletely understood (Baer et al. 1990; Vreugdenhil and Swaak 1990). A slight reduction in erythrocyte life span, the alteration in iron metabolism and a reduced bone marrow response to anaemia have been thought to contribute to the development of ACD (Cartwright and Lee 1971). A blunted erythropoietin production in response to anaemia has been suggested by some authors (Baer et al. 1987), while others have found increased concentrations of erythropoietin in patients with ACD (Birgegård et al. 1987; Erslev et al. 1987). Another possibility is that inflammation induces a decreased erythroblast sensitivity to erythropoietin. There is evidence that mediators of inflammation, such as interleukin-1 and tumour necrosis factor, have a suppressant effect on erythropoiesis in vitro (Schooley et al. 1987; Maury et al. 1988).

In patients with RA, correction of anaemia is achieved by successful treatment with disease-modifying drugs. Corticosteroids are also known to induce an increase in haemoglobin values in RA patients. To evaluate the usefulness of erythropoietin in the treatment of anaemia in RA we started a 24-week open clinical study in which recombinant human erythropoietin (rHuEPO) was given to 12 RA patients. A few earlier reports have shown that rHuEPO can be successfully used in the treatment of anaemia in patients with RA (Means et al. 1989; Pincus et al. 1990; Tauchi et al. 1990; Takashina et al. 1990).

Patients and Methods

The study population comprised 11 women and one man with active, erosive RA, diagnosed according to the 1987 American Rheumatism Association

Pagel/Weiss/Jelkmann (Eds.)
Pathophysiology and Pharmacology
of Erythropoietin

revised criteria for RA (Arnett et al. 1988). The mean age of the patients was 53 years (range 33–72) and the mean duration of RA 15 years (range 5–28). Ten patients had seropositive disease. Two had secondary amyloidosis, and one of them had mild renal insufficiency. All 12 patients received nonsteroidal anti-inflammatory drugs, ten low-dose prednisone or equivalent, three intramuscular gold, two penicillamine, three sulphasalazine, one hydroxychloroquine and three methotrexate. Treatment with disease-modifying antirheumatic drugs had been unchanged for at least 12 weeks prior to the study.

All patients had stable initial haemoglobin values of 110 g/l or less. At baseline all patients underwent thorough clinical examination including evaluation of joint disease activity by a joint index scoring system (Thompson et al. 1987), by a visual analog pain scale score and by questions regarding duration of morning stiffness. The following laboratory investigations were performed at the beginning of the study: routine haematology and serum chemistry, reticulocyte count, serum iron, transferrin and ferritin concentrations, serum vitamin B_{12} and erythrocyte folate concentrations, serum haptoglobin concentration, a direct Coombs' test and thyroid function tests. Exclusion criteria included anaemia due to haemorrhage or haemolysis, vitamin B_{12} or folic acid deficiency, uncontrolled hypertension, clinically relevant impairment of renal or liver function and high-dose systemic corticosteroid therapy (prednisone 10 mg daily at maximum).

rHuEPO (Eprex 10 000 U/ml, Cilag, Sweden) was given as a subcutaneous injection twice weekly. The starting dose was 150 U/kg body weight. The dose was adjusted higher or lower to maintain a target haemoglobin value of at least 20 g/l higher than the initial haemoglobin level. Dose adjustments were not made at intervals less than 4 weeks. If the haemoglobin value exceeded 140 g/l the rHuEPO injections were discontinued for at least 4 weeks.

Monitoring of haematological indices and C-reactive protein (CRP) was conducted weekly, and serum iron, transferrin and ferritin concentrations were measured every second week. Blood pressure was measured before every rHuEPO injection. All patients except one received oral iron preparations throughout the study.

Results

At the beginning of the study the patients' haemoglobin values were 90–110 g/l (mean 102 g/l), erythrocyte sedimentation rates 13–72 mm/h (mean 46 mm/h), CRP values 7–64 mg/l (mean 33 mg/l), ferritin values 9–348 μg/l (mean 63 μg/l) and erythropoietin levels 5–32 pmol/l (mean 13 pmol/l; reference range 8–43 pmol/l).

At present eight patients have completed the study, and the remaining four have been treated for at least 12 weeks. Eight of the 12 patients have shown a response defined as an increase in haemoglobin level of at least 20 g/l

(Table 1). The doses of rHuEPO required to maintain the target haemoglobin level ranged from 50 to 200 U/kg body weight, two times weekly. As a group, the nonresponders initially had a higher mean joint index, longer mean duration of morning stiffness and higher mean erythrocyte sedimentation rate and mean CRP value than the responders (Table 2). There was no difference in the maximal absolute reticulocyte counts between the nonresponders (mean $0.17 \times 10^{12}/l$, range 0.14–0.23) and the responders (mean $0.17 \times 10^{12}/l$, range 0.11–0.21). No significant changes in leukocyte or platelet counts were seen during the study. Serum ferritin concentrations showed a decrease during the study despite oral iron supplementation. Bone marrow aspiration was performed in five patients before and after the study. Initially, normal or reduced bone marrow iron stores were present in all five

Table 1. rHuEPO treatment in 12 patients with rheumatoid arthritis and anaemia

Patient	Initial Hb (g/l)	Initial MCV (fl)	Weeks to ΔHb=20 g/l	Highest Hb (g/l)	Highest EPO dose (U/kg)	Maintenance EPO dose (U/kg)
Responders						
1	97	90	3	137	150	125
2	104	89	8	135	150	100
3	110	76	9	140	150	50
4	99	77	5	141	150	50
5	90	82	7	136	150	150
6	100	92	18	124	200	200
7	106	99	13	128	200	150
8	106	85	17	128	200	150
Nonresponders						
1	108	81		116	200	
2	100	79		108	200	
3	105	83		110	300	
4	93	83		107	300	

Hb, Haemoglobin concentration; MCV, mean corpuscular volume.

Table 2. Baseline clinical data on 12 patients with anaemia and rheumatoid arthritis treated with rHuEPO

	Responders (n=8) Mean	Range	Nonresponders (n=4) Mean	Range	p
Joint index (range 0–534)	55	0–152	142	74–190	0.02
Morning stiffness (h)	0.6	0– 2	4	1– 7	0.004
Visual analog pain scale score (range 0–100 mm)	39	10– 71	36	25– 50	NS
Erythrocyte sedimentation rate (mm/h)	43	13– 72	53	40– 66	NS
C-reactive protein (mg/l)	29	7– 64	44	32– 52	NS

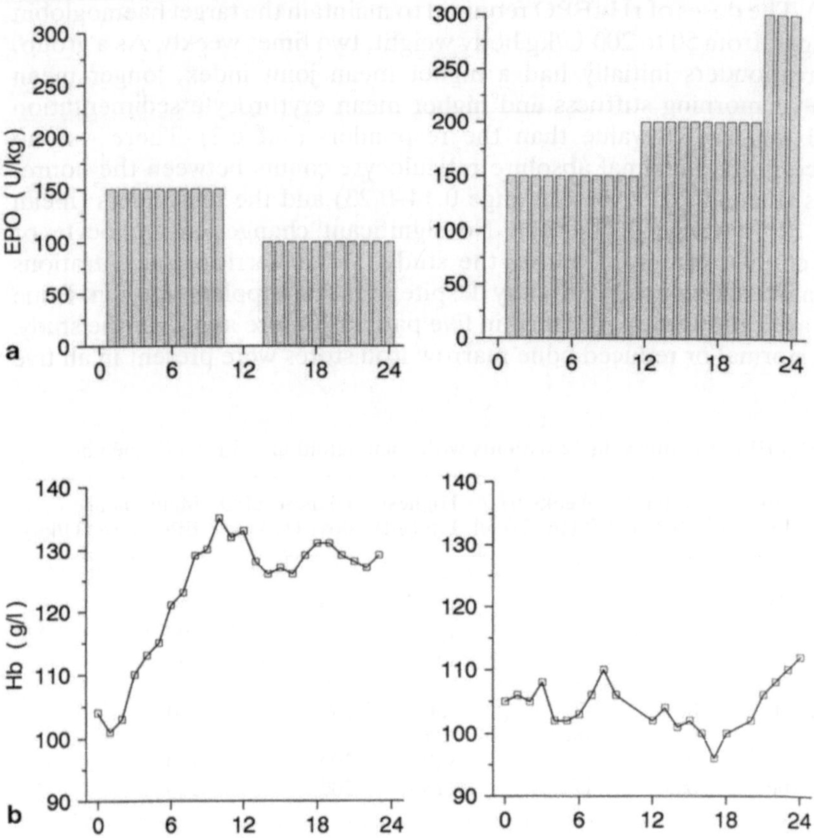

Fig. 1a–d. Comparison of rHuEPO treatment in one responder (*left*) and one nonresponder (*right*). **a** rHuEPO dosage. **b** Haemoglobin concentration.

patients, but after the study iron was absent from the bone marrow aspirates of four of these patients.

Figure 1 shows the rHuEPO doses used, haemoglobin concentrations, CRP values, and ferritin concentrations in one responder and one nonresponder.

No side effects of rHuEPO injections have been noted. There has been no increase in blood pressure. No significant change in the patients' joint status has been noted during the study.

Discussion

Our results agree with those obtained by previous investigators (Means et al. 1989; Pincus et al. 1990; Tauchi et al. 1990; Takashina et al. 1990) that anaemia in RA can respond to rHuEPO. We observed a correction of the

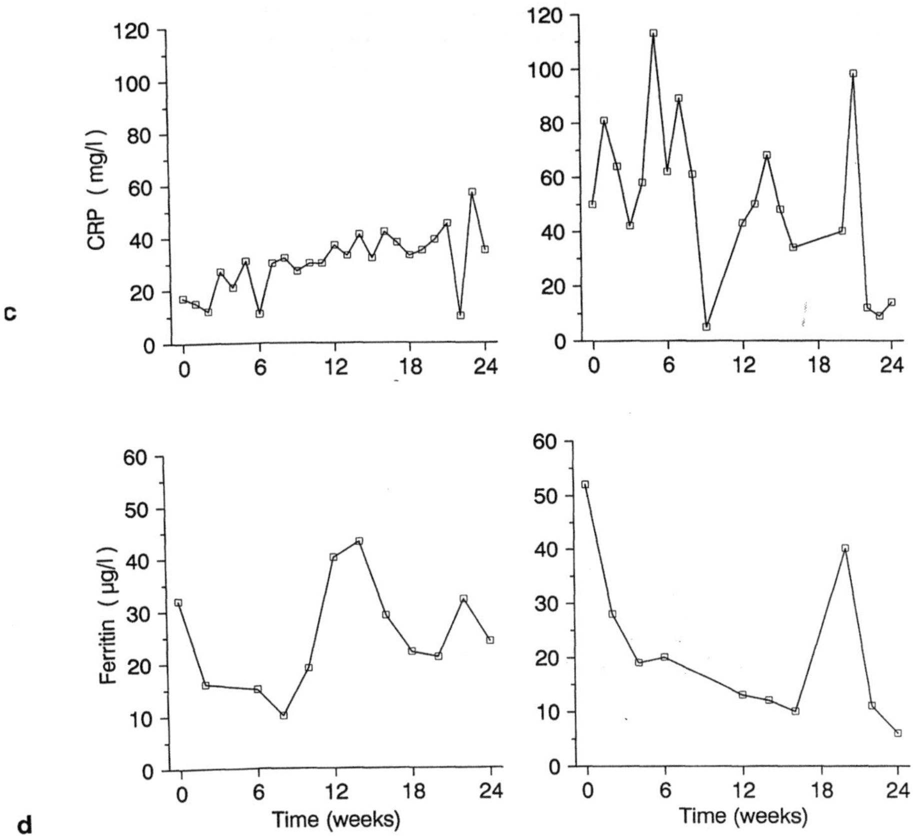

Fig. 1. c CRP concentration. **d** Ferritin concentration

haemoglobin value in 8 out of 12 patients with RA. There was a great variation in response, both with respect to the time required to achieve the target haemoglobin and to the dose of rHuEPO needed to maintain the increase in haemoglobin level. The most rapid response was seen in the patient with RA and secondary amyloidosis who also had mild renal insufficiency. Three patients did not respond until after 13–18 weeks of treatment, when the rHuEPO dose had been increased to 200 U/kg twice weekly. Administration of rHuEPO was well tolerated; no side effects were noted, including no increase in blood pressure.

We have been concerned with the problem of concomitant iron deficiency in some of our patients. Four of them, three responders and one nonresponder initially had low serum ferritin values suggesting iron deficiency. To prevent the iron deficiency that follows stimulation of erythropoiesis the patients were given oral iron supplementation throughout the study. The responses cannot, however, be attributed to the administration of iron alone. Most patients had received oral iron preparations for several periods prior to administration of rHuEPO and no consistent response had occurred. We have also noted that the haemoglobin levels of the responders tended to decrease after discontinuation of rHuEPO despite continued iron supplementation, a finding which indicates that rHuEPO was of major importance for the correction of anaemia.

We observed no significant change in the patients' rheumatological status during the treatment (eight patients hitherto evaluated), but almost all patients, including two nonresponders, reported an enhanced level of energy during rHuEPO treatment.

According to our study the degree of inflammatory activity of RA seems to influence the responsiveness to rHuEPO. As a group the nonresponders had more active disease than the responders, as judged by articular indices, duration of morning stiffness, erythrocyte sedimentation rate and CRP values. From a pathogenetic point of view our results support the hypothesis that decreased erythroblast sensitivity to erythropoietin is a major factor in the development of anaemia in inflammatory diseases such as RA. If a deficiency in erythropoietin production had been the cause of anaemia a more uniform response among our patients would have been expected. From a clinical point of view, administration of rHuEPO may be useful for certain patients with RA, for example when patients are being prepared for elective surgery. However, patients with severe inflammation seem to show resistance to rHuEPO and may require higher doses than those used in this study.

References

Arnett FC, Edworthy SM, Bloch DA et al. (1988) The American Rheumatism Association 1987 revised criteria for the classification of rheumatoid arthritis. Arthritis Rheum 31:315–324

Baer AN, Dessypris EN, Goldwasser E, Krantz SB (1987) Blunted erythropoietin response to anemia in rheumatoid arthritis. Br J Haematol 66:559–564

Baer AN, Dessypris EN, Krantz SB (1990) The pathogenesis of anemia in rheumatoid arthritis: a clinical and laboratory analysis. Semin Arthritis Rheum 19:209–223

Birgegård G, Hällgren R, Caro J (1987) Serum erythropoietin in rheumatoid arthritis and other inflammatory arthritides. Relationship to anemia and the effect of anti-inflammatory treatment. Br J Haematol 65:479–483

Cartwright GE, Lee GR (1971) The anemia of chronic disorders. Br J Haematol 21:147–152

Erslev AJ, Wilson J, Caro J (1987) Erythropoietin titers in anemic nonuremic patients. J Lab Clin Med 109:429–433

Maury CPJ, Andersson LC, Teppo A.M et al. (1988) Mechanism of anemia in rheumatoid arthritis: demonstration of raised interleukin-1 concentrations in anemic patients and of interleukin-1 mediated suppression of normal erythropoiesis and proliferation of human erythroleukaemia (HEL) cells in vitro. Ann Rheum Dis 47:972–978

Means RT, Olsen NJ, Krantz SB, Dessypris EN, Graber SE, Stone WJ, O'Neil VJ, Pincus T (1989) Treatment of the anemia of rheumatoid arthritis with recombinant human erythropoietin: clinical and in vitro studies. Arthritis Rheum 32:638–642

Pincus T, Olsen NJ, Russell IJ, Wolfe F, Harris ER, Schnitzer TJ, Boccagno JA, Krantz SB (1990) Multicenter study of recombinant human erythropoietin in correction of anemia in rheumatoid arthritis. Am J Med 89:161–168

Schooley JC, Kullgren B, Allison AC (1987) Inhibition by interleukin-1 on the action of erythropoietin on erythroid precursors and its possible role in the pathogenesis of hypoplastic anemias. Br J Haematol 67:11–17

Takashina N, Kondo H, Kashiwazaki S (1990) Suppressed serum erythropoietin response to anemia and the efficacy of recombinant erythropoietin in the anemia of rheumatoid arthritis. J Rheumatol 17:885–887

Tauchi T, Ohyashiki JH, Fujeda H, Lin K-Y, Ohyashiki K, Toyama K (1990) Correction of anemia in rheumatoid arthritis by recombinant human erythropoietin. Br J Haematol 29:235–236

Thompson PW, Silman AJ, Kirwan JR, Currey HLF (1987) Articular indices of joint inflammation in rheumatoid arthritis. Correlation with the acute-phase response. Arthritis Rheum 30:618–623

Vreugdenhil G, Swaak AJG (1990) Anemia in rheumatoid arthritis: pathogenesis, diagnosis and treatment. Rheumatol Int 9:243–257

Erythropoietin in Neoplastic Diseases: Preliminary Results

M. Pecherstorfer, S. Janisch, and I. Zimmer-Roth

Introduction

Anemia is found in about 60% of patients with disseminated neoplastic disease (Ultmann and Golomb 1983). The definition of chronic anemia of cancer as a specific subset of anemia is based on the exclusion of iron depletion, vitamin deficiency, hemolysis, and renal failure as underlying causes (Cartwright 1971) and on the evidence of low or normal serum iron and transferrin levels combined with normal or increased ferritin concentration. So far there is no agreement on whether anemia due to myelophthisis and cytotoxic treatment accompanied by impairment of iron metabolism should be considered as chronic tumor associated.

Several groups have reported on successful treatment of chronic anemia in low-grade non-Hodgkin's lymphoma and multiple myeloma with recombinant human erythropoietin (Ludwig et al. 1990; Oster et al. 1990b; Pecherstorfer et al. 1990). However, the number of patients included in these studies was small, and the treatment of anemia in solid tumors has been described only in case reports (Oster et al. 1990a). Therefore an Austrian multicenter trial was initiated by H. Ludwig to evaluate the impact of recombinant human erythropoietin in chronic anemia associated with hematoligic as well as solid neoplasmas regardless of myelophthisis and cytotoxic treatment.

Patients and Methods

Since February 1991 ten patients have been enrolled in the trial (Table 1). The median age was 60.5 years (range 48–82). Inclusion criteria were histologically confirmed malignancy and chronic tumor-associated anemia. At entry into the study none of the patients showed clinical evidence of infection. Chronic anemia of cancer was defined by hemoglobin levels under 11 g/dl and the exclusion of other causes of anemia such as blood loss (e.g., stool positive for occult blood), iron deficiency (serum ferritin concentration < 30 ng/ml, sideropenia in bone marrow biopsy), vitamin B_{12} and folinic acid deficiency, hemolysis (absent haptoglobin), chronic inflammation, and renal failure (serum creatinine concentration > 2.5 mg/dl). Cyclic aggressive chemotherapy was not a reason for exclusion from the study. Patients with cerebral

Pagel/Weiss/Jelkmann (Eds.)
Pathophysiology and Pharmacology
of Erythropoietin

Table 1. Patient characteristics

Patient	Sex	Age	Diagnosis	Treatment[a]	Biopsy[b]
S.I.	F	61	Breast cancer	Chemotherapy	+
W.S.	F	68	Lung cancer	Chemotherapy	–
K.M.	F	65	Breast cancer	Chemotherapy	ND
P.E.[c]	F	79	Chronic lymphatic leukemia	No treatment	+
S.A.[c]	F	51	Breast cancer	Chemotherapy	+
S.M.[c]	F	48	Breast cancer	Chemotherapy	–
W.M.	F	62	Multiple myeloma	Chemotherapy	+
S.H.[c]	F	82	Multiple myeloma	Chemotherapy	+
M.L.	F	51	Breast cancer	Chemotherapy	+
G.M.	F	52	Hodgkin's disease	No treatment	–

[a] In the 1 month prior to or during erythropoietin trial.
[b] Results of bone marrow biopsy concerning evidence of tumor cells.
[c] Patient responding to erythropoietin treatment.

tumor involvement, history of seizures, and uncontrolled hypertension (diastolic blood pressure > 100 mmHg) were excluded.

Recombinant human erythropoietin (specific activity 110.048 U/mg hormone) was provided by Cilag (Vienna). For the first 6 weeks erythropoietin was administered three times weekly subcutaneously in a dose of 150 U/kg body weight. If the hemoglobin level of the 6th week was less than 1 g/dl above the baseline value, the administration interval of erythropoietin was maintained and the dose was increased to 300 U/kg; if the hemoglobin concentration was 1 g/dl or more above the initial level, the dose of 150 U/kg was maintained for another 6 weeks. Patients who required blood transfusions during the 12 weeks of erythropoietin treatment remained in the trial, depending on the report of late responses to erythropoietin in anemic patients with multiple myeloma (Ludwig et al. 1990).

The protocol was accepted by the local ethics committees. Informed consent was obtained from each patient.

A physical examination of the patient (including control of blood pressure) and complete blood counts including differential count of leukocytes and determination of reticulocytes were performed weekly. Prior to the initiation of erythropoietin treatment and after 4, 8, and 12 weeks we measured the concentrations of blood sugar, creatinine, blood urea nitrogen, sodium, potassium, chloride, calcium, iron, transferrin, ferritin, bilirubin and the serum activities of alkaline phosphate, glutamic-oxaloacetic and glutamic-pyruvic transaminase. At the time of baseline examination and after 12 weeks bone marrow biopsies were obtained from the posterior iliac crest. Prussian blue stained preparations were examined to determine iron stores. A response was defined as an increase of at least 2 g/dl in the initial hemoglobin concentration.

Preliminary Results

At present ten patients have been treated for more than 6 weeks with recombinant human erythropoietin and are therefore evaluable for this interval (Tables 1, 2). In four women (two with breast cancer, one with multiple myeloma, and one with chronic lymphatic leukemia) an increase in the hemoglobin concentration of more than 2 g/dl compared to the baseline examination was seen (mean initial hemoglobin concentration 9.2±0.5 g/dl, mean hemoglobin concentration after 6 weeks 11.7±1.87 g/dl; Fig. 1a). In two other women (multiple myeloma, small cell lung cancer) the increase in hemoglobin was less than 2 g/dl. Both cases are regarded as nonresponders for this preliminary evaluation although treatment with 150 U/kg erythropoietin was continued. The patient suffering from lung cancer fulfilled response criteria after 8 weeks of erythropoietin therapy. In 3 patients with breast cancer and one with Hodgkin's disease no increase in hemoglobin was found after 6 weeks. Therefore the dose of erythropoietin was increased to 300 mg/kg. The mean hemoglobin levels of nonresponding patients were 9.16±1.1 g/dl at the baseline examination and 8.93±1.66 g/dl after 6 weeks (Fig. 1b).

Table 2. Initial erythrocyte and reticulocyte counts, hemoglobin (Hb) concentration, and serum levels of transferrin and ferritin

Patient	Erythrocytes ($\times 10^6/\mu l$)	Reticulocytes (%)	Hb (g/dl)	Transferrin (mg/dl)	Ferritin (ng/ml)
S.I.	2.80	4.1	7.7	243	425
W.S.	2.93	12.0	10.4	226	208
K.M.	3.40	11.6	9.4	297	203
P.E.[a]	3.59	6.5	9.9	72	274
S.A.[a]	3.18	5.1	9.2	167	251
S.M.[a]	2.73	0.7	7.0	123	769
W.M.	3.34	1.7	10.3	225	225
S.H.[a]	2.51	1.5	8.7	188	122
M.L.	4.02	3.2	9.7	150	984
G.M.	2.81	2.6	7.9	156	839

[a] Patients responding to erythropoietin treatment.

Table 3. Median change in serum iron, transferrin, and ferritin in responders and nonresponders

	Serum iron (μg/dl)		Transferrin (mg/dl)		Ferritin (ng/ml)	
	Week 0	Week 6	Week 0	Week 6	Week 0	Week 6
Responders	40.0	61.2	148.5	228.5	313.5	199.5
Nonresponders	84.5	75.0	225.5	254.0	316.5	273.0

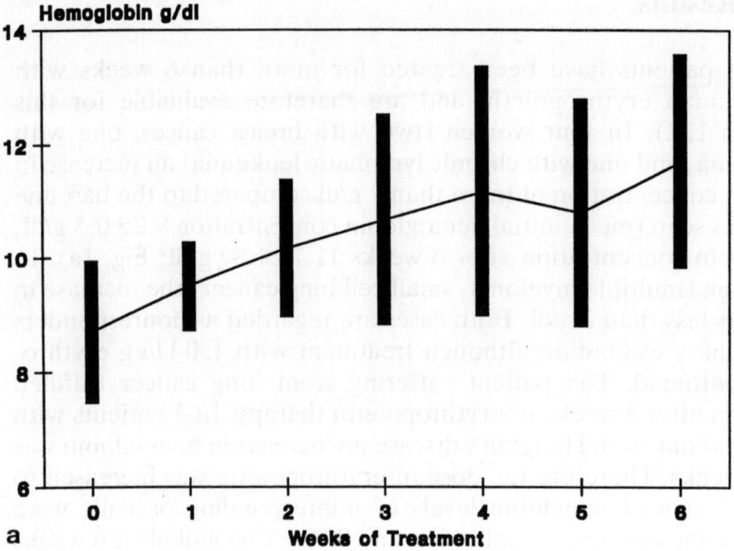

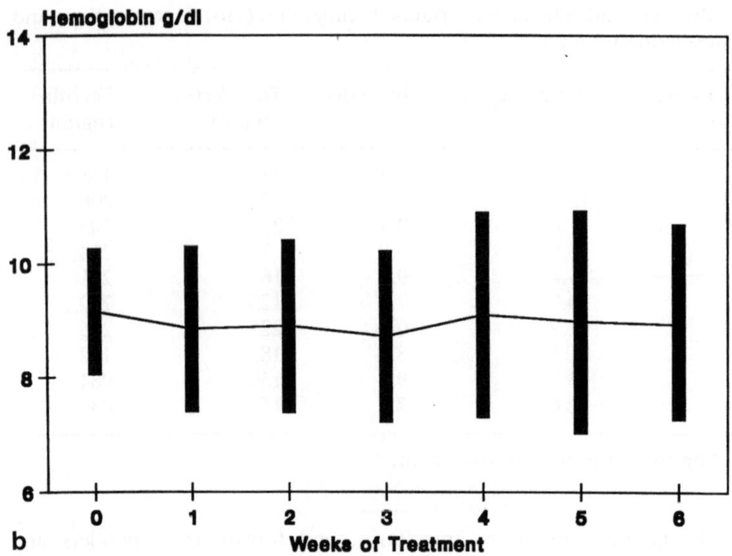

Fig. 1a, b. Change in hemoglobin concentrations (mean ± SD) during 6 weeks of erythropoietin treatment. **a** Responders (n=4). **b** Nonresponders (n=6)

Examinations of iron metabolism revealed in both responders and nonresponders a decrease in ferritin and an increase in transferrin whereas serum iron rose slightly in responders and fell in nonresponders (Table 3). Due to the small number of patients no statistical significance was shown.

Discussion

The pathogenesis of chronic tumor associated anemia is still controversial: In the 1960s and 1970s studies concentrated on impaired iron flow from reticuloendothelian cells to the plasma (Haurani et al. 1963) and on the uptake of iron by bone marrow cells (Douglas and Adamson 1975). More recently, suppression of erythropoiesis by cytokines (interferons, tumor necrosis factor; Johnson et al. 1989) and impaired erythropoietin secretion in chronic anemia of cancer were shown (Dainiak et al. 1983). At the same time, Dainiak found that erythroid progenitors (burst- and colony-forming units – erythroid) derived from anemic cancer patients proliferated normally in vitro in response to erythropoietin whether or not tumor cells were present in bone marrow. In 1990 Miller et al. demonstrated that for any given degree of anemia the serum concentration of immunoreactive erythropoietin was lower in 81 patients with solid tumors compared to patients with iron deficiency or hemolytic anemia. The erythropoietin response to anemia was further blunted in cytostatic-treated persons. This finding confirmed earlier experiments of Fisher and Roh (1964), who perfused dog kidneys with alkylating agents and observed a negative influence on erythropoietin production.

The clinical importance of erythropoietin treatment of cancer-associated anemia is derived from the persistent threat of transmission of infectious diseases and the inability to restore a normal hemoglobin concentration by blood transfusions. Moreover, immunologic abnormalities in patients receiving multiple transfusions have been reported (Gascon et al. 1984).

The initial analysis of patients participating in our multicenter trial suggests that erythropoietin is favorable for anemic patients with solid and hematologic neoplasmas. In the four patients responding to the erythropoietin dosage of 150 U/kg the decrease in ferritin levels and the increase in transferrin and iron indicated a mobilizing effect of erythropoietin on stored iron. In one patient the ferritin concentration reached the lower limit (< 30 ng/ml), and therefore oral iron substitution was started. The fall in ferritin among nonresponders was due to two patients who showed an increase in the hemoglobin concentration of less than 2 g/dl after 6 weeks.

The number of cases discussed here is small. So far the results indicate a trend which seems to persist in eight further patients who have been enrolled but are not evaluable at the present time.

References

Cartwright GE (1971) The anemia of chronic disorders. Br J Haematol 21:147–152

Dainiak N, Kulkarni V, Howard D, Kalmanti M, Dewey MC, Hoffmann R (1983) Mechanisms of abnormal erythropoiesis in malignancy. Cancer 51:1101–1106

Douglas SW, Adamson JW (1975) The anemia of chronic disorders: studies of marrow regulation and iron metabolism. Blood 45:55–65

Fisher JW, Roh BL (1964) Influence of alkylating agents on kidney erythropoietin production. Cancer 24:983–988

Gascon P. Zoumbos NC, Young NS (1984) Immunologic abnormalities in patients receiving multiple blood transfusions. Ann Intern Med 100:173–177

Haurani FI, Young K, Tocantins LM (1963) Reutilization of iron complicating malignant neoplasmas. Blood 22:73–81

Johnson RA, Waddelow TA, Caro J, Oliff A, Roodman GD (1989) Chronic exposure to tumor necrosis factor in vivo preferentially inhibits erythropoiesis in nude mice. Blood 74:130–138

Ludwig H, Fritz E, Kotzmann H, Höcker P, Gisslinger H, Barnas U (1990) Erythropoietin treatment of anemia associated with multiple myeloma. N Eng J Med 322:1693–1699

Miller CB, Jones RJ, Piantadosi S, Abeloff MD, Spival JL (1990) Decreased erythropoietin response in patients with the anemia of cancer. N Eng J Med 322:1689–1692

Oster W, Herrmann F, Cicco A, Gamm H, Zeile G, Brune T, Lindemann A, Schulz G, Mertelsmann R (1990a) Erythropoietin prevents chemotherapy-induced anemia: case report. Blut 60:88–92

Oster W. Herrmann F, Gamm H, Zeile G, Lindemann A, Müller G, Brune T, Kraemer HP, Mertelsmann R (1990b) Erythropoietin for the treatment of anemia of malignancy associated with neoplastic bone marrow infiltration. J Clin Oncol 8:956–962

Pecherstorfer M, Schmoranzer F, Vesely M, Fortelny A, Baumgartner G (1990) Erythropoietin in multiple myeloma. Dtsch Med Wochenschr 115:1096–1099

Ultmann JE, Golomb HM (1983) Principles of neoplasia: approach to diagnosis and management. In: Petersdorf RG, Adams RD, Braunwald E, Isselbacher KJ, Martin JB, Wison JD (ed) Harrison's principles of internal medicine, 10th edn. McGraw Hill, New York, p 784

Erythropoietin Effects on Acute Anemia Induced by Antineoplastic Therapy in the Case of a Uremic Patient with Lymphoma

A. Di Benedetto, F. Nuzzi, and P. Sorice

Introduction

Anemia occurs with high incidence in patients affected by chronic renal failure (CRF) [1]. Many factors have been described as being responsible for the anemic status in these patients [2–5]. Erythropoietin deficiency seems to be the main cause of anemia in CRF [6]. Blood transfusions are frequently needed to counteract this complication, with the risk of hepatitis, AIDS, iron overload, and sensitization of hystocompatibility antigens, with reduction in the success probability rates for renal transplantation [7]. Nowadays, the possibility of having sufficient amounts of human erythropoietin, produced by DNA-recombinant methods (rHuEPO) [8, 9] has made the therapeutic use of this compound available for the treatment of anemia in CRF patients [7, 10–14].

We report here on the effects of rHuEPO on chemotherapy-induced acute anemia in a CRF patient affected by non-Hodgkin's lymphoma.

Case Report

A 35-year-old woman suffering from CRF who had undergone hemodialysis for 48 months presented with an asymmetric inguinal-crural lymph node swelling 2 cm in size and aching on palpation. Histopathological examination of the material obtained from removed lymph nodes led to the diagnosis of a centrocytic and centroblastic non-Hodgkins's lymphoma (Fig. 1). Thoracic and abdominal computed tomography carried out to determine the lymphoma stage showed lymph node swellings in the anterior-superior region of mediastinum and in the pancreas head. Thus, the patient was classified as being in the third stage in terms of the Ann Arbor classification [15].

It was decided to administer five chemotherapy cycles, at 21-day intervals, using the CEOP protocol. We modified drug doses according to the dialytic clearances of the drugs involved, so that each cycle consisted of the sequential administration of: 20 mg vincristine intravenously, 30 mg/m^2 epirubicin intravenously, 300 mg/m^2 cyclophophamide intravenously, and 60 mg/m^2 prednisone orally, for 5 days. Prednisone was reduced to 40 mg/m^2 on the 6th day, 20 mg/m^2 on the 7th, and 10 mg/m^2 on the 8th. The presence of CRF anemia together with the well-known marrow-inhibitory effect induced by

Pagel/Weiss/Jelkmann (Eds.)
Pathophysiology and Pharmacology
of Erythropoietin

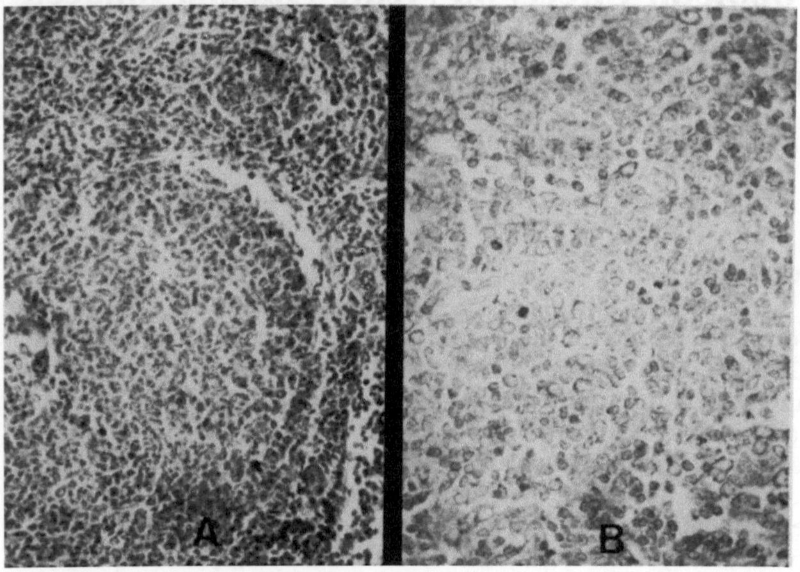

Fig. 1. *Left*, lymphonodular section: nodular aspects (EE, × 375); *right*, lymphonodular section: neoplastic population of centrofollicolar type (Giemsa-Lennert, × 600)

these drugs, suggested a hematological consultation. The specialists consulted recommended blood transfusion (300 ml) for each chemotherapy cycle, in relation to the variations in the blood count. Since following the first blood transfusion there were no effects on the blood count or the blood indices, we decided – upon obtaining the patient's informed consent – to use rHuEPO at the initial dose of 75 U/kg at the start of the second chemotherapy cycle. This a dose derived from balancing the need to counteract the chemotherapy-induced marrow inhibition and the patient's concomitant hypertensive state, which was not always pharmacologically well controlled.

Erythropoietin (Eprex, Cilag Biotech, Italy) was administered intravenously three times a week, at the end of hemodialytic treatment. The patient also received trivalent iron therapy at a dose of 62.5 mg pa dialysis, her serum iron level being about 40 μg/100 ml and ferritin level about 279 μg/l. During the study period the patient's vital signs (oral temperature, blood pressure, heart and pulmonary rates) were well checked by recording each sign before and 60 min after rHuEPO administration.

Blood count, performed immediately before and 14 days after each chemotherapy cycle, was recorded by Coulter counter, while reticulocytes were counted by the usual standard methods. At the same time, serum iron and ferritin levels were also determined using routine methods.

Since after the second and third chemotherapy cycles, no significant variations in blood pressure had occurred, the rHuEPO dose was increased to 100 U/kg to obtain further improvement in blood count and blood indices.

Results

Table 1 shows that the beginning of rHuEPO therapy, started at the second chemotherapy cycle, not only prevented a reduction in the blood count and blood indices induced by the marrow-inhibitory effect of chemotherapy but also led to a progressive increase in hemoglobin concentration and in hematocrit. The increase in rHuEPO dose to 100 U/kg, administered from the fourth chemotherapy cycle, induced a further improvement in blood parameters, counteracting the marrow-inhibitory effect of chemotherapy without any need for blood transfusions (Fig. 2).

Erythropoietin administration did not bring about blood pressure increase during the follow-up period. The increase from 75 to 100 U/kg was made without any significant problems; moreover, the patient reported feeling well and having a more effective strain endurance.

Thoracic and abdominal computed tomography was performed as control at the end of the chemotherapy protocol, and these no longer showed the pancreatic and mediastinic lymphonodular swellings that had previously been observed.

Table 1. Variations in the analyzed parameters over the follow-up period

Parameters	1st cycle[a]	2nd cycle[b]	3rd cycle[b]	4th cycle[c]	5th cycle[c]
Before chemotherapy					
Red blood cells ($\times 10^6$)	2.18	2.10	2.46	2.51	2.65
Hematocrit (%)	21.4	20.3	24.4	24.7	26.8
Hemoglobin (g/100 ml)	6.6	6.0	7.1	8.0	9.5
Mean corpuscular volume (μ^3)	96	90	99	100	101
Mean corpuscular hemoglobin (pg)	30	30	30.5	32.3	32.1
Iron serum (μg/100 ml)	38	40	90	140	148
Ferritin (μg/l)	279	279	42	24	24
14 Days after chemotherapy					
Red blood cells ($\times 10^6$)	2.09	2.41	2.48	2.56	3.18
Hematocrit (%)	19.4	23.3	24.6	26.2	32.4
Hemoglobin (g/100 ml)	6.0	6.8	7.8	9.1	11.0
Mean corpuscular volume (μ^3)	93	97	99	102	102
Mean corpuscular hemoglobin (pg)	29.7	31.3	32.2	32.4	32.5
Iron serum (μg/100 ml)	40	46	90	140	146
Ferritin (μg/l)	280	276	40	26	22

[a] + blood transfusion (300 ml), [b] + 75 U rHuEPO/kg pa dialysis
[c] + 100 U rHuEPO/kg pa dialysis

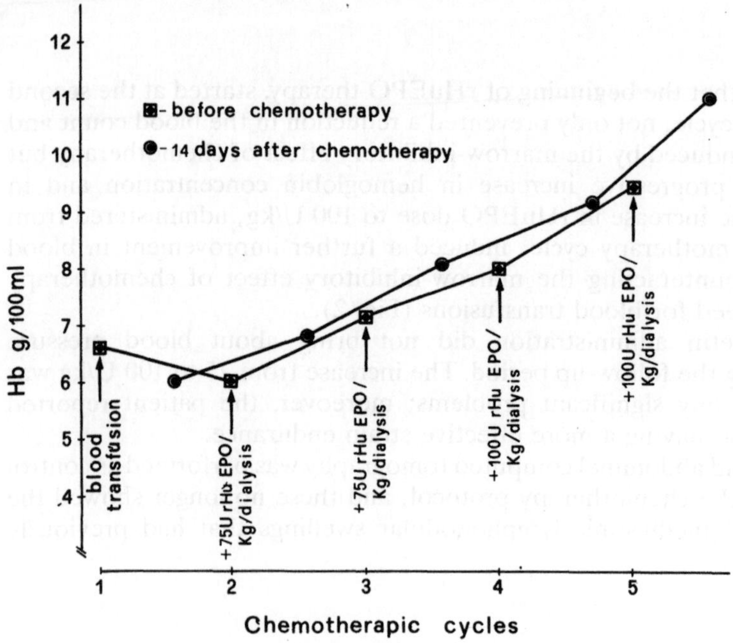

Fig. 2. Variations in hemoglobin concentration before and 14 days after chemotherapy, in relation to blood transfusion and rHuEPO administration

Discussion

Our results demonstrate that rHuEPO is able to improve CRF-induced anemia and to counteract effectively marrow-inhibitory effects of chemotherapy. Blood transfusion, performed at the first chemotherapy cycle, was not able to prevent the blood count from dropping secondary to marrow inhibition. rHuEPO administration at the start of the second chemotherapy cycle led to a continuous increase in blood indices before and after cytostatic therapy. In this way, it was possible to complete the CEOP protocol in a patient already suffering from CRF-induced anemia and to obtain a full regression of the mediastinic and pancreatic matastases. An increase in the rHuEPO dose administered at the start of the fourth cycle proved without side effects, even on arterial blood pressure, while we observed a further improvement in the anemic status.

The occurence of non-Hodgkins's lymphoma in a patient suffering from CRF thus led us to extend the field of application for a drug currently used only to correct the anemic status in uremic patients undergoing hemodialysis [9–14] or in the predialytic phase [16–20]. Our findings suggest the possibility of employing rHuEPO also in oncology, where it is important to counteract the marrow-inhibitory effect of chemotherapy. Further studies are needed to establish the rHuEPO doses in relation to cytostatic protocols, patient's age, and concomitant diseases, etc.

References

1. Eschbach JW (1983) Hematologic problems of dialysis patients. In: Drukker W, Parson FM, Maher JF (eds) Replacement of renal function by dialysis. Nijhoff, Boston, pp 630–645
2. Eschbach JW, Adamson JW (1985) Anemia of end-stage renal disease (ESRD). Kidney Int 28:1
3. Adamson JW, Eschbach JW, Finch CA (1968) The kidney and erythropoiesis. Am J Med 44:725
4. Jacobson LO, Goldwasser E, Fried W, Plzak L (1957) Role of the kidney in erythropoiesis. Nature 179:633
5. Caro J, Brown S, Miller O, Murray T, Erslev AJ (1979) Erythropoietin levels in uremic nephric and anephric patients. J Lab Clin Med 93:449
6. Erslev AJ (1969) Erythropoietin function of the kidney. In: Wesson LG (ed) Physiology of the human kidney. Grune and Stratton, New York, pp 521–534
7. Winearls CG, Pippard MJ, Downing MR et al. (1986) Effect of human erythropoietin derived from recombinant DNA on the anaemia of patients maintained by chronic haemodialysis. Lancet ii:1175
8. Jacobs K, Shoemaker C, Roudersdorf R et al. (1985) Isolation and characterization of genomic and cDNA clones of human erythropoietin. Nature 313:806
9. Lin FK, Suggs S, Lin CH et al. (1985) Cloning and expression of the human erythropoietin gene. Proc Natl Acad Sci USA 82:7580
10. Eschbach JW, Egrie JC, Downing MR et al. (1987) Correction of the anaemia of end stage renal disease with recombinant human erythropoietin. N Engl J Med 316:73
11. Eschbach JW, Adamson JW (1988) Recombinant human erythropoietin: implications for nephrology. Am J Kidney Dis 3:203
12. Casati S, Passerini P, Campise MR, Graziani G, Cesana B, Perisic M, Ponticelli C (1988) Benefici e rischi del trattamento protratto con eritropoietina umana ricombinante in pazienti emodializzati. Br Med J (Ital Edn) 8:104
13. Schaefer RM, Hörl WH, Massry SG (1989) Treatment of renal anaemia with recombinant human erythropoietin. Am J Nephrol 9:353
14. Ponticelli C, Casati S (1989) Correction of anaemia with recombinant human erythropoietin. Nephron 52:201
15. Hermanek P, Sobin LH (eds) (1987) TNM classification of malignant tumors. Springer, Berlin Heidelberg New York
16. Stone WJ, Graber SE, Krantz SB et al. (1988) Treatment of the anaemia of predialysis patients with recombinant human erythropoietin: a randomized, placebo-controlled trial. Am J Med Sci 296:171
17. Koene RAP (1988) Recombinant human erythropoietin in predialysis patients: results from Belgium and the Netherlands. Dialysis and Transplantation 17:639
18. Lim VS, De Gowin RL, Zavala D et al. (1989) Recombinant human erythropoietin treatment in pre-dialysis patients. Ann Intern Med 110:108
19. Eschbach JW, Kelly MR, Haley NR et al. (1989) Treatment of the anaemia of progressive renal failure with recombinant human erythropoietin. N Eng J Med 321:158
20. Kleinman KS, Schweitzer SU, Perdue ST (1989) The use of recombinant human erythropoietin in the correction of anaemia in predialysis patients and its effects on renal function: a double-blind, placebo-controlled trial. Am J Kidney Dis 6:486

The Therapeutic Effect of Recombinant Human Cytokines (GM-CSF, Interleukin-3, Erythropoietin) in Patients with Myelodysplastic Syndromes

A. Ganser, M. Eder, G. Geissler, U. Hess, O. G. Ottmann, G. Seipelt, H. Vogt, and D. Hoelzer

Introduction

Myelodysplastic syndromes (MDS) are characterized by progressive refractory cytopenias with cellular dysfunction and defective myeloid maturation involving at least two and generally three hematopoietic cell lineages. According to the definition of the French-American-British Group, patients with less than 5% blast cells in the bone marrow are classified as refractory anemia (RA) or RA with ring sideroblasts, those with 5%–20% blast cells in the marrow as RA with excess of blasts (RAEB), and those with 20%–30% bone marrow blasts as RAEB in transformation (RAEB-T) [1]. Patients with chronic myelomonocytic leukemia (CMML) have a significant proportion of monocytes in the circulation (>1000/μl), and up to 20% blast cells in the marrow. Although depending on the initial blast cell load 10%–40% of these patients will proceed to acute myeloid leukemia (AML), infections due to neutropenia and thrombocytopenic bleeding are the most frequent causes of death [2].

Uncoupling between proliferative and differentiative programs in hematopoietic stem cells of patients with MDS has been proposed as the basic lesion within their clonal hematopoietic cells. Analyses of biological parameters, for example, the in vitro clonogenic growth of the progenitor cells, cytogenetics, and restriction fragment length polymorphism of X-linked genes, have indicated the presence of an abnormal clone of hematopoietic cells. In vitro culture studies have demonstrated subnormal clonal growth in vitro in response to granulocyte (G-CSF) and granulocyte-macrophage colony-stimulating factors (GM-CSF), and defective cellular maturation of myeloid and erythroid precursors in most patients with MDS. These parameters deteriorate further with progression toward AML. The defective proliferation of hematopoietic precursors in the marrow of these patients has been suggested to be due to either decreased responsiveness to and/or decreased production of hematopoietic growth factors.

Treatment of anemia and severe thrombocytopenia still depends mainly on red blood cell and platelet transfusions eventually resulting in iron overload or refractoriness to donor platelets. Treatment with cytotoxic or differentiation-inducing agents has been only partially effective, and only younger patients might benefit from allogeneic bone marrow transplantation. With the recent availability of recombinant human hematopoietic CSFs and

Pagel/Weiss/Jelkmann (Eds.)
Pathophysiology and Pharmacology
of Erythropoietin

erythropoietin, stimulation of hematopoiesis in vivo has evolved as a new, potentially efficacious treatment modality. In theory, the hematopoietic growth factors could exert their action at various levels: (a) they could stimulate the proliferation and maturation of residual normal hematopoietic progenitor cells; (b) they could induce proliferation and maturation of the myelodysplastic cell clone leading to clonal extinction; or (c) they could sensitize the malignant clone for the cytostatic action of chemotherapeutic agents, such as cytosine arabinoside.

GM-CSF in MDS

Several clinical trials with recombinant human GM-CSF treatment alone in patients with MDS have been published on a total of 213 patients [3–13] (Table 1). Of these patients, who were treated with different schedules, GM-CSF doses, and routes of administration, 77% (83% after exclusion of

Table 1. Effect of GM-CSF in patients with MDS

	Schedule				Increase in			
Study	Dosage μg/m²	Route	Duration (days)	*n*	N	P	R	B
Vadhan-Raj et al. 1988 [3]	60–500 (24 h)	IV	14	8	8	3	7	1
Antin et al. 1988 [4]	15–240 (1–12 h)	IV	14	7	6	0	7	0
Ganser et al. 1989 [5]	15–150 (8 h)	IV	7–14	11	9	1	2	5
Herrmann et al. 1989 [6]	5–750 (0.5 h)	IV	5	4	4	0	0	4
Hoelzer et al. 1990	250	SC	3× 14	9	7	2	2	0
Estey et al. 1989 [9]	120	IV	21–84	22	22	2	0	4
Thompson et al. 1989 [7]	0.3 – 10 μg/kg	SC	10	16	12	2	3	2
Schuster et al. 1990	3 μg/kg	SC	>180	26	26	0	0	3
Willemze et al. 1990 [11]	108 (2x/day)	SC	56	33	22			1
	216(2×/day)		56	35	22			4
Dunbar et al. 1990 [12]	500	SC	42	13	13	0	11	0
Estey et al. 1991 [13]	5– 10	SC	14–18	29	14	0	0	3
				213	(77%)	(5%)	(15%)	(13%)

N, Neutrophils; P, platelets; R, reticulocytes; B, blasts

those receiving low-dose GM-CSF [13]) showed a dose-dependent increase in neutrophil counts. An increase in reticulocyte counts was observed in 15% of the patients. Platelets increased above baseline values in 5%, but decreases of platelet counts were also reported. In 14 patients (13%) a transient increase in the marrow and/or peripheral blood blast cells was noted in some of the patients, particularly those with over 15% bone marrow blasts progressing to acute leukemia [5].

In the only randomized trial testing GM-CSF versus supportive care, patients with RA, RAEB, or CMML received either 3 μg/m² GM-CSF subcutaneously per day or were observed [10]. With more than 25 patients in each arm and treatment periods in excess of 6 months, all patients receiving GM-CSF had a sustained increase in neutrophils coupled with a decrease in infection rate. No effect was seen on the platelet or reticulocyte counts. Progression to acute leukemia was comparable in the two arms (about 10% each). These results indicate that GM-CSF has a beneficial effect, with a lowering of the infection rate, while the previously observed progressions to AML appear to have been due mainly to the natural course of the disease.

Interleukin-3 in MDS

As part of phase I/II trials, 19 patients with MDS were treated with recombinant human interleukin-3 (IL-3) [12, 14]. Daily subcutaneous bolus injections of 125–500 μg/m² induced an increase in bone marrow cellularity and in granulocytic precursor cells, resulting in a rise in circulating white blood cell counts in the majority of the patients (Table 2). Most patients showed a rise in neutrophils, eosinophils, basophils, monocytes and lymphocytes. After discontinuation of IL-3 treatment white blood cell counts gradually declined to baseline values.

While the leukocyte response to IL-3 was comparable in MDS patients and in patients with normal hematopoiesis [15], the increase in platelet and

Table 2. Effect of interleukin-3 in patients with MDS

Study	Schedule			n	Increase in			
	Dose μg/m²	Route	Days		N	P	R	B
Ganser et al. 1989 [5]	250–500	SC	15	9	7	5	2	1
Dunbar et al. 1990 [12]	125	SC	28	10	6	4	5	0
				19	(68%)	(47%)	(37%)	(5%)

N, Neutrophils; P, platelets; R, reticulocytes; B, blasts

reticulocyte counts induced by IL-3 was only moderate, although occurring in about one-half and one-third of patients, respectively (Table 2). Nonetheless, several transfusion-dependent patients had a clinical benefit from the modest rise in platelet counts, not requiring further platelet transfusions for a period of 4 months. With regard to its action on thrombopoiesis in MDS patients, IL-3 appears to be at least as effective as GM-CSF or G-CSF.

In one patient with secondary RA the proliferative effect of IL-3 resulted in an increase of blast cells in the bone marrow and peripheral blood. In the remaining patients, there were no increased blast numbers in the bone marrow or in the peripheral blood, indicating that the malignant cell populations remained under differentive control. Dysplastic features in the bone marrow persisted under IL-3 treatment [16].

Erythropoietin in MDS

The rationale for treating MDS patients who have an adequate increase in plasma erythropoietin level (Fig. 1) [17] with recombinant erythropoietin lies in the possibility of overcoming the defective proliferation and maturation of the erythroid precursors by pharmacological doses, thereby increasing the hematocrit or reducing the transfusion requirements. At our own institution, 12 patients with MDS (ten RA, two RAEB) were treated with recombinant

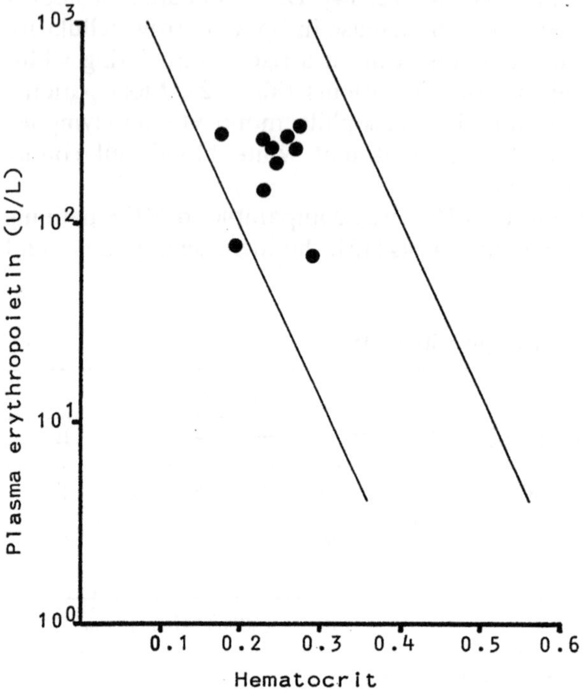

Fig. 1. Plasma erythropoietin and hematocrit in ten patients with myelodysplastic syndromes. *Lines*, 95% confidence limits for anemic persons with normal renal function

man erythropoietin (patients 1–5: 450 U/kg twice weekly as intravenous lus injection; patients 6–11: 10 000 U five times per week as subcutaneous lus injection; Table 3). One patient with cytogenetically confirmed RA (a q-abnormality) had a marked response in reticulocyte counts, with a rise in moglobin from 9.1 g/dl to above 12 g/dl when treated intravenously with 0 U/kg twice weekly for 12 weeks. After a treatment-free interval, during ich the hemoglobin levels dropped to 7.9 g/dl, erythropoietin was again ministered subcutaneously at a dosage of 10 000 U five times per week. is led to a rise in the hemoglobin level above 14 g/dl, while serum-ferritin vels decreased from 547 to 154 μg/l. The serum level of erythropoietin was U/ml, the lowest of all patients studied. Although a study by Bowen et al. 7] showed no correlation between erythropoietin concentration and moglobin level in MDS patients, indicating that not only a decreased sponsiveness to erythropoietin plays a role in MDS, a low serum ythropoietin level could be of predictive value for the response to ythropoietin therapy [18].

Combining our own and published data on a total of 64 MDS patients, travenous or subcutaneous bolus injections of comparable amounts of ythropoietin led to stimulation of erythropoiesis with an increase in ticulocyte counts in 23% of patients, an increase in hemoglobin levels in %, and a reduction in red blood cell transfusion requirements in 22%, ile the platelet transfusion requirements were reduced in two patients 8–24] (Table 4). Since the studies were small and produced diverse response

ble 3. Patient characteristics

ımber of patients	12
en/women	7/5
ge (years)	
Median	59
Range	30–79
agnosis	
RA	9
RA with excess blasts	2
Secondary RA	1
emoglobin (g/dl)	
Median	7.4
Range	5.7–9.1
rum erythropoietin (mU/ml)	71–336
ansfusion dependency	11/12

Table 4. Effect of erythropoietin in treatment in patients with MDS

Study	Schedule Dosage (U/kg)	Route	Duration (weeks)	*n*	Increase in Hb	Ret
von Kamp et al. 1990 [20]	50–250 3 times/week	SC	15	10	0	1
Besa et al 1990 [21]	150 3 times/week or 150 daily[a]	SC	12	6	5	5
Stebler et al, 1990 [22]	50–500 3 times/week	IV/SC	56	2	1	1
Bowen et al. [17]	60– 90 6 times/week	SC	16	10	2	1
Hirashima et al. 1991 [23]	50–200 3 times/week	IV	4–54	8	1	1
Laporte et al. 1991 [24]	40–300 3 times/week	SC	12	6	5	
Oster et al. 1990 [19]	450 2 times/week or 10 000[b] 5 times/week	IV SC	12	10	0	4
Present study	450 2×/week or 10 000[b] 5×/week	IV SC	12	12	1	2
				64	(23%)	(23%)

[a] Combined with 13-*cis*-retinoic acid and tocopherol. [b] total dose
Hb, Hemoglobin; Ret, reticulocytes.

rates, a larger controlled trial seems warranted to evaluate the value of erythropoietin therapy in MDS.

Conclusion

Further attempts are required with either biological or pharmacological agents to suppress the abnormal clone to the extent that it will allow normal hematopoiesis to expand. Aggressive chemotherapy has generally been considered ineffective and contraindicated in elderly MDS patients because of the assumption that the bone marrow of these patients is incapable of recovering normal function, even if the dysplastic clone could be eradicated [25]. However, complete remissions have been reported with the use of aggressive chemotherapy in about 55% of patients [26, 14]. As clinical studies in patients with RAEB and RAEB-T have shown, GM-CSF is capable of recruiting leukemic blast cells into proliferation in vivo. Possibly a combination of hematopoietic growth factors such as G-CSF, GM-CSF or IL-3 with a more aggressive cytotoxic drug regimen could lead to more promising results in patients with RAEB/RAEB-T, and this needs to be further evaluated in these subgroups. In this setting the hematopoietic growth factors could not only be used to recruit leukemic cells but also to shorten the period of cytopenia after aggressive chemotherapy, which is characterized by severe and even fatal infections in more than one-third of patients.

The treatment of RA patients with GM-CSF showed that the neutrophil count can be increased in all patients without stimulation of blast cells. GM-CSF, but also G-CSF, given as long-term maintenance therapy in MDS patients with severe neutropenia could therefore lower the frequency of infectious episodes in these patients.

During IL-3 treatment the platelets increased in a number of patients whereas the observed rise in reticulocytes did not result in an increase in hemoglobin levels. Long-term treatment with IL-3 might prove to be more effective in elevating the platelet and hemoglobin levels. Since treatment with erythropoietin alone is also insufficient in the majority of patients with MDS, a combination of erythropoietin with IL-3 might have superior stimulatory effects on erythro- and megakaryopoiesis. The long-term effects of G-CSF, GM-CSF, IL-3, and erythropoietin, given either alone or in combination, on the peripheral blood count, frequency of infectious episodes, transfusion requirements, and natural course of this disease should therefore be evaluated in phase III studies.

References

1. Bennett JM, Catovsky D, Daniel MT, Flandrin G, Galton DAG, Granlnick HR, Sultan C (1982) Proposals for the classification of the myelodysplastic syndromes. Br J Haematol 51:189
2. Mufti GJ, Galton DAG (1986) Myelodysplastic syndromes: natural history and features of prognostic significance. Clin Haematol 15:953
3. Vadhan-Raj S, Kellagher MJ, Keating M et al. (1988) Phase I study of recombinant human granulocyte-macrophage colony stimulating factor in patients with myelodysplastic syndrome. N Engl J Med 317:1545–1552
4. Antin JH, Smith BR, Holmes W et al. (1988) Phase I/II study of recombinant human granulocyte-macrophage colony stimulating factor in aplastic anemia and myelodysplastic syndrome. Blood 72:705–713
5. Ganser A, Völkers B, Greher J et al. (1989) Recombinant human granulocyte-macrophage colony-stimulating factor in patients with myelodysplastic syndromes – a phase I/II trial. Blood 73:31–37
6. Herrmann F, Lindemann A, Klein H et al. (1989) Effect of recombinant human granulocyte-macrophage colony-stimulating factor in patients with myelodysplastic syndrome with excess blasts. Leukemia 3:335–338
7. Thompson JA, Douglas JL, Kidd P et al. (1989) Subcutaneous granulocyte-macrophage colony-stimulating factor in patients with myelodysplastic syndrome: toxicity, pharmacokinetics, and hematological effects. J Clin Oncol 7:629–637
8. Hoelzer D, Ganser A, Ottmann OG et al. (1990) Effect of treatment with rhGM-CSF and low-dose cytosine arabinoside on leukemic blast cells in patients with myelodysplastic syndromes. Haematol Blood Transfus 33:763–769
9. Estey E, Kurzrock R, Talpaz M, Beran M, Kantarjian H, Keating M, McCredie K, Freireich E, Deisseroth A, Gutterman JU (1989) Therapy of myelodysplastic syndromes (MDS) with GM-CSF. Proc Am Soc Clin Oncol 8 (Suppl):200
10. Schuster MW, Thompson JA, Larson R, Allen SL, O'Laughlin R, Israel R, Fefer A (1990) Randomized trial of subcutaneous granulocyte-macrophage colony-stimulating

factor (GM-CSF) versus observation in patients (pts) with myelodysplastic syndrome (MDS). J Cancer Res Clin Oncol 116 (Suppl):1079

11. Willemze R, Visani G, de Witte T, Zwierzina H, Lebar B, Jacobs A, Gerhartz H, Haak HL, Stern AC, Suciu S, Solbu G, Zittoun R (1990) A randomized phase I/II study with recombinant human GM-CSF in patients (pts) with myelodysplastic syndromes at a relatively low risk of developing acute leukemia. Blood 76 (Suppl 1):337a
12. Dunbar CE, Smith D, Kimball J, Garrison L, Nienhuis AW, Young NS (1990) Sequential treatment with recombinant human growth factors to compare activity of GM-CSF and IL-3 in the treatment of primary myelodysplasia. Blood 76 (Suppl 1):141a
13. Estey EH, Kurzrock R, Talpaz M, McCredie KB, O'Brien S, Kantarjian M, Keating MJ, Deisseroth AB, Gutterman JU (1991) Effects of low doses of recombinant human granulocyte-macrophage colony-stimulating factor (GM-CSF) in patients with myelodysplastic syndromes. Br J Haematol 77:291–295
14. Aul C, Schneider W (1989) The role of low-dose cytosine arabinoside and aggresive chemotherapy in advanced myelodysplastic syndromes. Cancer 64:1812–1818
15. Ganser A, Lindemann A, Seipelt G et al. (1990) The effect of recombinant human interleukin-3 in patients with normal hemopoiesis and in patients with secondary bone marrow failure. Blood 76:666–676
16. Ganser A, Seipelt G, Lindemann A et al. (1990) Effects of recombinant human interleukin-3 in patients with myelodysplastic syndromes. Blood 76:455–462
17. Bowen DT, Jacobs A, Cotes MP, Lewis TC (1990) Serum erythropoietin and erythropoieses in patients with myelodysplastic syndromes. Eur J Haematol 44:30–32
18. Bowen D, Culligan D, Jacobs A (1991) The treatment of anemia in the myelodysplastic syndromes with recombinant human erythropoietin. Br J Haematol 77:419–423
19. Oster W, Krumwieh D, Brune T, Kraemer HP, Herrmann F, Mertelsmann R (1990) Evaluation of erythropoietin (EPO) in the treatment of patients with myelodysplastic syndromes (MDS) – preclinical data supporting a rationale for sequential therapy with granulocyte-macrophage colonoy-stimulating factor (GM-CSF) and EPO. J Cancer Clin Oncol 116 (Suppl):280
20. Van Kamp H, Prinsze-Postema TC, Kluin PM, den Ottolander GJ, Willemze R, Fibbe WE (1990) Effect of subcutaneously administered human recombinant erythropoietin on erythropoiesis in patients with myelodysplasia. Blood 76 (Suppl 1):170a
21. Besa EC, Nathan F, Sylvester L, Catalano P (1990) Erythroid response of severely anemic or transfusion-dependent patients with myelodysplastic syndrome to recombinant human erythropoietin (Epo) does not cerrelate with baseline serum Epo levels. Blood 76 (Suppl 1):133a
22. Stebler C, Tichelli A, Dazzi H, Gratwohl A, Nissen C, Speck B (1990) High-dose recombinant human erythropoietin for treatment of anemia in myelodysplastic syndromes and paroxysmal nocturnal hemoglobinuria: a pilot study. Exp Hematol 18:1204–1208
23. Hirashima K, Bessho M, Jinnai I (1991) Improvement in anemia by recombinant human erythropoietin in patients with myelodysplastic syndrome and aplastic anemia. Contrib Nephrol 88:254–265
24 Laporte JPH, Isnard F, Fenaux P, Woler M, Najman A (1991) Recombinant human erythropoietin at high dose is effective for the treatment of the anemia of myelodysplastic syndromes: Contrib Nephrol 88:271–272
25. Armitage JO, Dick FR, Needlemann SW et al. (1981) Effect of chemotherapy for the dysmyelopoietic syndrome. Cancer Treat Rep 65:601–605
26. Tricot G, Boogaerts MA (1986) The role of aggressive chemotherapy in the treatment of the myelodysplastic syndromes. Br J Haematol 63:477-483

The Effect of Erythropoietin in Chemotherapy-Induced Anaemia in Patients with Osteosarcoma and Ewing's Sarcoma

C. WURNIG, R. WINDHAGER, T. ZETTL, and F. STOCKENHUBER

Introduction

Severe anaemia develops in patients with osteosarcoma and Ewing's sarcoma who are treated with chemotherapy before and after operation. In most cases patients require multiple blood transfusions. The aetiology of anaemia in these tumour patients may be multifactorial. It may be caused in part by the cytotoxic effect of chemotherapeutic agents on the bone marrow. Inappropriately low erythropoietin levels, reduced responsiveness to erythropoietin, elevated levels of tumour necrosis factor leading to a decrease in erythropoietin responsiveness may be further potential causes [1–5].

No matter what are the underlying causes, anaemia is a major problem in the handling of tumour patients. Severe anaemia can cause a delay or even elimination of chemotherapy. The risk of blood transfusions has become well documented in patients with malignant disease [6–8]. Besides adverse effects such as haemolytic reactions and transmission of infectious diseases, more recently immunosuppression following blood transfusions have been described. A decrease in the natural killer cell function [9, 10], increase in the suppressor cell function [11], and decreased helper/suppressor cell ratio [10] following blood transfusions might be crucial problems in tumour patients leading perhaps to enhanced tumour progression. In spite of aggressive chemotherapy the reduction or even avoidance of blood transfusions in these patients should be a goal for the future. To evaluate whether recombinant human erythropoietin is of benefit in ameliorating tumour anaemia we studied patients with osteosarcoma or Ewing's sarcoma receiving aggressive chemotherapy who were treated additionally with high doses of recombinant human erythropoietin.

Patients and Methods

Seven patients with osteosarcoma or Ewing's sarcoma (four women and three men; median age 38 years, range 17–64) enrolled in this study. Patients with osteosarcoma (n=6) were treated according to the COSS 91 protocol: 12 g/m^2 methotrexate, 120 mg/m^2 cisplatin (Platinol), 30 mg/m^2 doxorubicin (Adriamycin) and 3 g/m^2 ifosfamide (Holoxan); the one patient with Ewing's

Pagel/Weiss/Jelkmann (Eds.)
Pathophysiology and Pharmacology
of Erythropoietin

sarcoma was treated according to the CESS 90 protocol: 30 mg/m^2 doxorubicin, 1.5 mg/m^2 vincristine (Oncovin), 3 mg/m^2 ifosfamide, and 0.5 mg/m^2 actinomycin D (Cosmegen-Lyovac). The period of chemotherapy was 27 weeks. Patients received three cycles of chemotherapy prior to tumourectomy and four cycles after. Tumourectomy was performed in week 10 after starting chemotherapy. No patient had hypertension, apoplectic insult, seizures, thrombophilia, evidence of blood loss, iron deficiency, or vitamin B_{12} or folic acid deficiency.

Recombinant human erythropoietin (Boehringer Mannheim, Vienna, Austria) was administered intravenously twice per week during the period of chemotherapy. The dose was 600 U/kg body weight. A routine evaluation was performed before each injection of erythropoietin during the first 2 weeks of treatment, every week until the end of chemotherapeutic treatment. Each evaluation comprised a physical examination, a complete blood count including leukocyte differential, red cell, reticulocyte, and platelet counts, measurement of haemoglobin and haematocrit, determination of serum ferritin and transferrin concentrations, and a blood chemistry profile including serum iron, electrolytes, and indexes of renal and liver function. If haemoglobin dropped to less than 9 g/dl, blood units were given. Serum erythropoietin concentration was measured by a specific radioimmunoassay [12]. Informed consent was obtained from all patients who participated in this trial.

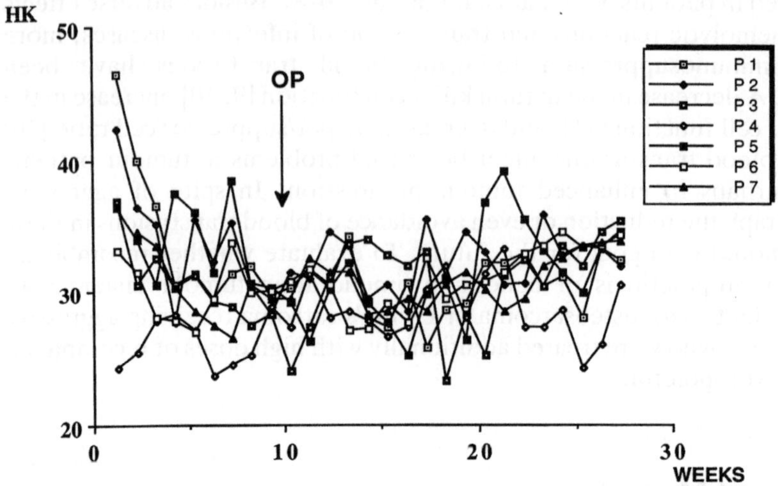

Fig. 1. Kinetics of haematocrit (*HK*) before and during erythropoietin therapy in seven patients (*P1–7*) with osteosarcoma or Ewing's sarcoma

Results

Pretreatment levels of haematocrit ranged from 25% to 46% (median 35%); the kinetics of haematocrit during the 27 weeks of erythropoietin therapy are shown in Fig. 1. A marked decrease in haematocrit levels appeared in five of seven patients in the first 3 weeks of chemotherapy. After this initial phase haematocrit values seemed to stabilize or to show a slight increase at the end of the treatment period. In all but one patients blood units had to be administered. The mean amount of blood units in these patients was 4.5. Compared to a matched control group this was a significant reduction from 7.5 to 4.5. Pretreatment serum levels of erythropoietin did not exceed 120 U/l in any of the patients. Serum erythropoietin concentrations after intravenous administration increased to 3000 U/l and showed a kinetic profile with a monoexponential decay and a half-life of 6.5 h. As ferrokinetics revealed initially rather low serum iron and ferritin concentrations, patients were treated orally with iron. With the substitution of iron no significant further decrease in serum iron or ferritin was observed (data not shown).

Discussion

The result of this pilot study shows that anaemia in this group of patients can be ameliorated and the demand of blood units reduced, but in the majority of patients not avoided. In this kind of tumour and chemotherapy-induced anaemia erythropoietin therapy under the described regimen does not seem to overcome the myelosuppressive effect completely.

No side effects occurred during our trial. In particular, none of our patients had hypertension, seizures or thrombohaemorrhagic complications. These symptoms have been described during erythropoietin treatment in patients with anaemia due to chronic renal failure [13]. This study of therapy with recombinant human erythropoietin in patients with osteosarcoma and Ewing's sarcoma-associated anaemia showed excellent tolerance of the treatment regimen. Patients treated with recombinant human erythropoietin also experienced subjective improvement in their quality of life.

This study cannot yet be considered representative of the effectness of recombinant human erythropoietin but does suggest that erythropoietin could be of some value in the treatment of tumour-associated anaemia in Ewing's sarcoma, and osteosarcoma. A double-blind, placebo-controlled trial is in progress to confirm these preliminary data.

References

1. Miller CB, Jones RJ, Piantadosi S, Abeloff MD, Spivak JL (1988) Decreased erythropoietin (EPO) response associated with the anemia of malignancy. Proc ASCO 8:182
2. Johnson CS, Cook CA, Furmanski P (1989) Reversal of tumor necrosis factor alpha (TNF-alpha) induced suppression of erythropoiesis in vivo by treatment with erythropoietin (EPO). Proc Am Assoc Cancer Res 30:409
3. Rothmann SA, Paul P, Weick JK, McIntyre WR, Fantelli F (1985) Effect of *cis*-diamminedichloroplatinum on erythropoietin production and hematopoietic progenitor cells. Int J Cell Cloning 3:415–423
4. Cox R, Musial T, Gyde OHB (1986) Reduced erythropoietin levels as a cause of anaemia in patients with lung cancer. Eur J Cancer Clin Oncol 22:511–514
5. Wood P, Hrushesky WJM (1984) Cisplatin-induced anemia: an erythropoietin deficiency syndrome? Proc ASCO 3:22
6. Blumberg N, Heal JM (1989) Transfusion and host defenses against cancer recurrence and infection. Transfusion 29:236–245
7. Rosenberg SA, Seipp CA, White DE, Wesley R (1985) Perioperative blood transfusions are associated with increased risks of recurrence and decreased survival in patients with high-grade soft-tissue sarcomas of the extremities. J Clin Oncol 3:698–709
8. Johnson JT, Taylor FH, Thearle PB (1987) Blood transfusion and outcome in stage III head and neck carcinoma. Arch Otolaryngol Head Neck Surg 113:307–310
9. Gascon P, Zoumbos NC, Young N (1984) Immunologic abnormalities in patients receiving multiple blood transfusions. Ann Intern Med 100:173–177
10. Kaplan J, Sarnaik S, Gitlin J, Rusher J (1981) Diminished helper/suppressor lymphocyte ratios and natural killer activity in recipients of repeated blood transfusions. Blood 64:308–310
11. Smith MD, Williams JD, Coles GA, Salaman JR (1981) The effect of blood transfusion on T-suppressor cells in renal dialysis patients. Transplant Proc 13:181–183
12. Eckardt KM, Kurtz A, Hirth P, Scigalla P, Wieczorek L, Bauer C (1988) Evaluation of the stability of human erythropoietin in samples for radioimmunoassay. Klin Wochenschr 66:241–245
13. Eschbach JW (1989) The anemia of chronic renal failure: pathophysiology and the effects of recombinant erythropoietin. Kidney Int 35:134–48

Erythropoietin in Bone Marrow Transplantation: Pathophysiology and Clinical Use

A. M. Vannucchi, A. Grossi, A. Bosi, D. Rafanelli, S. Guidi, R. Saccardi, G. Longo, A. R. Miele, and P. Rossi-Ferrini

Introduction

Both autologous (ABMT) and allogeneic (BMT) bone marrow transplantations are increasingly employed for the treatment of hematological and nonhematological malignancies. Following autologous rescue in ABMT, or graft infusion in BMT, a varying length of time is required for the infused progenitors to proliferate and mature until adequate levels of peripheral blood cells are restored (Atkinson 1990). Recent studies have established the usefulness of myeloid hematopoietic growth factors (namely, granulocyte and granulocyte-macrophage colony-stimulating factors) in accelerating myeloid cell recovery and reducing the incidence of infectious episodes following ABMTor BMT (Aurer et al. 1990). On the other hand, surprisingly little attention has been devoted up to now to the possible use of factors affecting erythroid reconstitution, although erythropoietin (Epo) was the first hematopoietic growth factor to be produced by genetic engineering (Jacobs et al. 1985) and used in the clinical setting (Eschbach et al. 1987). While red blood cell (RBC) unit transfusion is an effective and relatively inexpensive way to treat anemia, one must also consider the risks, both immediate and delayed, that are associated with multiple RBC transfusions in these patients. Of particular concern is the likelihood of viral infections in the immunocompromised host; among these, cytomegalovirus infection in patients treated with bone marrow transplantation may range from subclinical infections to fatal multiorgan diseases and may contribute to a delayed hematopoietic engraftment (Levin 1990).

We were therefore interested in evaluating the pathophysiology of Epo regulation in bone marrow transplant patients (Bosi et al. 1991a, b), and in determining whether the delivery of recombinant Epo (rhEpo) is effective in accelerating the erythroid reconstitution after graft infusion (Bosi et al. 1991c; Vannucchi et al. 1992).

Regulation of Epo Production in Bone Marrow Transplantation

We evaluated serum Epo (sEpo) levels in serial serum samples, obtained from the day before the institution of conditioning regimen up to at least day

Pagel/Weiss/Jelkmann (Eds.)
Pathophysiology and Pharmacology
of Erythropoietin

30, in 14 ABMT and 10 BMT patients. sEpo levels were determined with a commercially available radioimmunoassay kit (Epo-Trac, Instar, Minnesota, USA).

In all the 14 ABMT patients, a characteristic behavior of sEpo levels during the ABMT procedure was observed. sEpo levels increased markedly soon after the institution of conditioning regimen, reaching a peak on day 0 in 9/14 patients and still significantly higher than controls on day 45. During all the study period, sEpo and hemoglobin (Hb) levels were significantly correlated, and the parameters of the regression line were superimposable upon those of the control group (30 healthy subjects and 15 iron-deficiency anemic patients; Table 1).

In the BMT patients, on the other hand, sEpo and Hb levels were significantly correlated before the institution of conditioning regimen but not after graft infusion (Table 1). Moreover, the regression curve relating sEpo and Hb was remarkably different from that of controls (Table 1). In 60/67 serum samples available following graft infusion Epo levels fell below the 95% confidence limit of the controls' curve and were considered to be inappropriately low.

From these data it appears that while sEpo levels in ABMT are adequate for the severity of anemia, an inappropriate response occurs in BMT patients, who show lower than expected sEpo levels for any given degree of anemia in comparison with a control population of normal and iron-deficiency anemic subjects. Comparable results have been also reported by other researchers (Abedi et al. 1990; Ireland et al. 1990; Shapira et al. 1990; Beguin et al. 1991).

The mechanism(s) for the different behavior of Epo production in BMT versus ABMT patients is presently unknown. However, we recently investigated the role of the immunosuppressant agent cyclosporine A (CsA), which is used for the prevention of graft-versus-host disease in BMT and is nephrotoxic at doses very close to the therapeutic ones. Using an animal model, our findings suggest that CsA may be a (co)factor in the impaired Epo production in BMT. In fact, in mice treated with CsA in therapeutic dosages and later made anemic with phenylhydrazine, we observed sEpo levels significantly lower than those measured in control mice with comparable degrees of anemia; in addition, the parameters of the regression line correlating sEpo and hematocrit values in mice receiving CsA were significantly different from those of controls (Vannucchi et al. 1991). In

Table 1. Correlation values between Ln sEpo and hemoglobin levels

	r	p	Slope	Intercept
Controls	−0.867	<0.01	−0.400	8.50
ABMT	−0.819	<0.01	−0.382	8.35
BMT	−0.310	NS	−0.114	4.68

agreement with this, Abedi et al. (1990) observed significantly lower sEpo levels in patients treated with CsA plus methothrexate for prophylaxis against graft-versus-host disease than in recipients of a T-cell depleted graft, who did not receive an immunosuppressive agent. Further work is certainly needed to clarify the pathogenesis of the reduced Epo production in BMT.

Effects of rhEpo Administration on Erythroid Reconstitution After BMT

Because of the demonstrated defective Epo regulation in BMT, we started a pilot study of 8 BMT patients who received rhEpo and 13 comparable BMT patients who did not. The rhEpo (Eprex; Cilag, Milan, Italy) was administered intravenously at a dose of 50 U/kg three times a day from day 1 to day 30. rhEpo significantly reduced both the time to erythroid engraftment (defined as the time to reach an unsupported hematocrit value ≥35%) and the numer of RBC units before becoming transfusion independent following BMT (Table 2). Moreover, the number of reticulocytes measured on day 21 (Sysmex R-1000 automated reticulocyte flow cytometry analyzer; TOA Medical Electr., Hamburg, FR) was significantly greater in the rhEpo group than in controls (Table 2), as was the number of the most immature reticulocytes (cells with high fluorescence ratio due to the high RNA content). Bone marrow aspirates, obtained as early as day 15, revealed a conspicuous number of erythroid cells at various stages of maturation, which were virtually absent in the controls. Finally, the number of both burst- and colony-forming units – erythroid in the bone marrow was several times as high in the rhEpo group as in the controls (Vannucchi et al., unpublished data). Therefore, we suggest a role for rhEpo in accelerating erythroid reconstitution after BMT and in reducing RBC transfusion requirements. A double-blind, placebo-controlled trial is needed to confirm these preliminary, promising results.

Table 2. Effects of rhEpo in BMT

	rhEpo	Controls	p
RBC units	4.0± 2.0	11.2± 3.0	<0.01
Days until Htc ≥35%	58.0±23.0	123.0±117.0	<0.01
Reticulocytes on day 21 ($\times 10^9$/l)	109.1±63.6	27.5± 19.5	<0.01
HFR reticulocytes on day 21 ($\times 10^9$/l)	11.0± 5.8	1.8± 1.7	<0.001

Htc, Hematocrit; HFR, high fluorescence ratio.

References

Abedi MR, Backman L, Borstrom L, Lindback B, Ringden O (1990) Markedly increased serum erythropoietin levels following conditioning for allogeneic bone marrow transplantation. Bone Marrow Transplant 6:121–126

Atkinson K (1990) Reconstruction of the haemopoietic and immune systems after marrow transplantation. Bone Marrow Transplant 5:209–226

Aurer I, Ribas A, Gale RP (1990) What is the role of recombinant colony stimulating factors in bone marrow transplantation? Bone Marrow Transplant 6:79–87

Beguin Y, Clemons GK, Oris R, Fillet G (1991) Circulating erythropoietin levels after bone marrow transplantation: inappropriate response to anemia in allogeneic transplants. Blood 77:868–873

Bosi A, Vannucchi AM, Grossi A, Guidi S, Vannucchi L, Saccardi R, Bernabei PA, Longo G, Rafanelli D, Rossi Ferrini P (1991a) Serum erythropoietin levels in patients undergoing autologous bone marrow transplantation. Bone Marrow Transplant 7:421–425

Bosi A, Vannucchi AM, Grossi A, Guidi S, Saccardi R, Rafanelli D, Longo G, Rossi Ferrini P (1991b) Inadequate erythropoietin production in allogeneic bone marrow transplant patients. Haematologica 76:280–284

Bosi A, Vannucchi AM, Grossi A, Guidi S, Saccardi R, Rafanelli D, Rossi Ferrini P (1991c) Serum erythropoietin levels in allogeneic, non T-depleted bone marrow transplantation, and the effects of rhEpo administration. Bone Marrow Transplant 7 (Suppl 2):87

Eschbach JW, Egrie JW, Downing MR et al. (1987) Correction of the anemia of end-stage renal disease with recombinant human erythropoietin: results of the phase I and II clinical trial. N Engl J Med 316:73–78

Ireland RM, Atkinson K, Concannon A, Dodds A, Downs K, Biggs JC (1990) Serum erythropoietin changes in autologous and allogeneic bone marrow transplant patients. Br J Haematol 76:128–134

Jacobs K, Shoemaker C, Rudersdorf R et al. (1985) Isolation and characterization of genomic and cDNA clones of human erythropoietin. Nature 313:806–810

Levin MJ (1990) Current approaches to the prevention and treatment of cytomegalovirus disease after bone marrow transplantation: an overview. Semin Hematol 27:1–4

Schapira L, Antin JH, Ransil BJ, Antman KH, Eder JP, McGarigle CJ, Goldberg MA (1990) Serum erythropoietin levels in patients receiving intensive chemotherapy and radiotherapy. Blood 76:2354–2359

Vannucchi AM, Grossi A, Bosi A, Rafanelli D, Guidi S, Saccardi R, Alterini R, Rossi Ferrini P (1991) Impaired erythropoietin production in mice treated with Cyclosporine A. Blood 78:1615–1618

Vannucchi AM, Bosi A, Grossi A, Guidi S, Saccardi R, Lombardini L, Rossi Ferrini P (1992) Stimulation of erythroid engraftment by recombinant human erythropoietin in ABO-compatible, HLA-identical, allogeneic bone marrow transplant patients. Leukemia, In press

Experimental and Clinical Results of Perioperative Treatment with Recombinant Human Erythropoietin*

O.R.C. Busch, M.A.W. Hoynck van Papendrecht, R.L. Marquet, and J. Jeekel

Introduction

The side effects of allogeneic blood transfusions are nowadays well known [1]. Especially the risk of transmitting viral infections is a reason to seek methods to reduce the need for allogeneic blood transfusions. Many retrospective studies suggest that in patients undergoing resection for malignancy there is a further adverse effect of blood transfusions: perioperative transfusions may stimulate tumor growth and therefore worsen the prognosis of these patients [2, 3]. Additionally, the rate of postoperative infectious complications may be increased if blood transfusions are given [4]. All these effects can be avoided if the patient's own blood is used because this is the safest blood to be transfused. Autologous transfusions can be achieved in several ways: (a) by predeposit autologous blood donations, (b) by preoperative intentional hemodilution, and (c) by reinfusion of blood saved during the operation. In cancer surgery these methods may have their limitations due to logistical problems and anemia. In addition, intraoperative cell saving is not allowed in cancer surgery because of the possibility of introducing metastases.

Recombinant human erythropoietin (rHuEPO) has been shown to be a powerful stimulator of erythropoiesis in patients with chronic anemia due to renal failure [5]. To investigate whether rHuEPO can play a role in preventing blood transfusions in the postoperative period, we first performed dose-effect studies in rats. These studies revealed that there is a maximal effective dose, which produces the best effect when rHuEPO is administered daily. Importantly, far more red cells could be harvested due to rHuEPO treatment [6]. Others have confirmed these data in animal [7] and human studies [8]. An alternative way to reduce blood transfusion in the postoperative period may be the administration of rHuEPO in the perioperative period. Animal studies have indicated that rHuEPO is effective in reducing postoperative anemia in an extreme blood loss model [9, 10]. Since it has been reported that patients may become unresponsive to rHuEPO due to operation or infection [11], the clinical relevance of this extreme blood loss model in which no real surgical stress was imposed is questionable. In our rat experiments the animals did

* This research was funded by Cilag, Brussels, Belgium

Pagel/Weiss/Jelkmann (Eds.)
Pathophysiology and Pharmacology
of Erythropoietin

undergo surgery. An ileum resection was performed during which 20% of their blood volume was taken. When we investigated the efficacy of perioperative rHuEPO treatment in these rats, it was found that rHuEPO is effective only if given preoperatively [12].

Considering these experimental conclusions we conducted an open randomized clinical trial with the following aims: (a) to determine the effectiveness of rHuEPO on erythropoiesis if administered in the perioperative period, (b) to assess the safety of rHuEPO in elective cancer surgery, and (c) to determine whether rHuEPO can diminish or prevent postoperative anemia, thus reducing the need for postoperative blood transfusions.

Materials and Methods

Patients with a gastrointestinal malignancy who were scheduled for surgical resection, with an expected blood loss more than 500 ml, were asked to participate in this study. Patients were eligible if they were, except for their malignancy, in good general condition. Their hemoglobin (Hb) level was over 7.0 mmol/l, and they showed no evidence of folate, vitamin B_{12}, or iron deficiency. Patients were excluded if they had: (a) a history of any primary hematologic disease, (b) a history of seizures, (c) uncontrolled hypertension (i.e., diastolic blood pressure $>$ 110 mmHg), (d) cerebral metastases, or (e) any acute illness within 7 days of study entry. Patients were also excluded if they had received any therapy which could influence bone marrow activity (e.g., corticosteroid or other immunosuppressive therapy, chemo- or radiation therapy). Women of childbearing age were eligible only if they used adequate contraceptives and had a negative pregnancy test immediately prior to study entry.

Four of the original 30 patients who entered the study were excluded from analysis. Patients were excluded if blood loss was less than 500 ml, or if the treatment protocol was violated. We excluded these patients, because our main goals were the effectiveness and safety of rHuEPO administration. Two control patients were not resected and did not lose blood because at operation they appeared to have liver metastases. Two patients treated with rHuEPO were excluded because they were not treated completely according to protocol due to a postponement of the operation. Two patients were found to have a locally nonresectable tumor; because their blood loss exceeded 500 ml, they remained in the study. Characteristics of the 26 patients who were analyzed are given in Table 1. The two groups were similar with regard to age, sex, baseline hematologic values, tumor site, and blood loss.

After informed consent patients were randomized either to treatment with rHuEPO (Cilag, Brussels, Belgium) or to no treatment. No placebo was given. For 9 days treated patients received daily 200 U rHuEPO/kg body weight intravenously. They were treated from 5 days before operation (day –5) until 3 days (day 3) after operation. After each injection, body temperature, respiration rate, pulse rate, and blood pressure were measured.

Table 1. Patient characteristics

	EPO	Control
Total	13	13
Men	11	11
Women	2	2
Mean age (years)	60.9	62.4
Baseline values		
Hb (mmol/l)	9.0±0.2	8.9±0.2
Hct (vol%)	42.1±0.8	42.4±1.1
Tumor site		
Esophagus	9	8
Pancreas	4	4
Rectosigmoid	0	1
Blood loss (ml)		
Mean	1700	2000
Range	550–3800	1000–6500

Both groups were given daily 200 mg elemental iron orally. During operation blood loss was carefully measured, and blood transfusion was allowed only if blood loss exceeded 1000 ml, or the hemoglobin (Hb) level was under 6.0 mmol/l. The hematologic values – including Hb, hematocrit (Hct), erythrocytes, reticulocytes, leukocytes, and thrombocytes – were measured daily. Serum chemistry – including serum iron, transferrin, and ferritin levels – was assessed on days –5, –1, 1, and 5.

Non-parametric tests were used. Increases and decreases within the group were tested with the Wilcoxon matched-pairs test. In comparisons between the two groups the Mann-Withney *U* test was used.

Results

No adverse effects were seen in the 15 treated patients; no changes in blood pressure were noted.

Hct in the treated group was significantly elevated on day –1 in comparison with baseline values (p=0.0033). Hct did not differ significantly between the two groups at any time. The same was seen with regard to Hb and erythrocytes count. Hct levels reached their minimum 3 days after surgery (EPO 31.6%±1.2%, control 31.6%±1.4%). From this day until 5 days later it increased significantly ($p<0.05$) in both groups (EPO 34.9%±1.0%, control 34.5%±1.2%). The reticulocytes counts are given in Fig. 1. Significantly higher numbers of reticulocytes were found in the treated group on days –2, 1, 2, 3, and 4 compared to the controls ($p<0.02$). rHuEPO treatment resulted in an increased number of reticulocytes after only 3 days of treatment

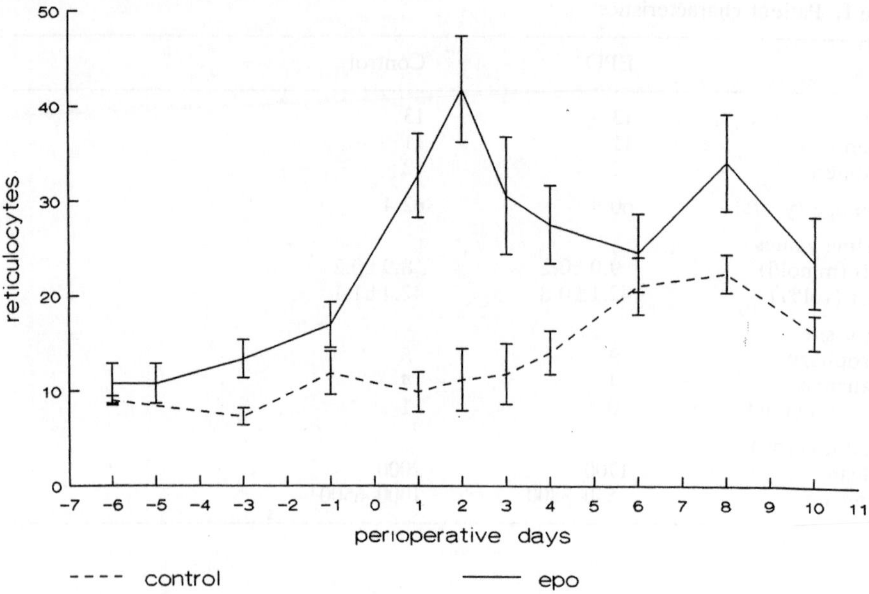

Fig. 1. Reticulocyte levels (mean±SEM) in the perioperative period of cancer patients treated with rHuEPO and control patients. Day 0 is the day of surgery. Reticulocytes are expressed as number of cells counted per 1000 erythrocytes

(p=0.01). The highest value was reached on day 2, after which it dropped significantly (p=0.046) on day 3 and remained at this level until day 10. Maximum values in the control group were reached on day 8; 2 days later it was again significantly lower (p=0.012). The maximum value among treated patients was significantly higher than the maximum value in the untreated group.

Figure 2 shows the need for blood transfusions in both groups. Excluding the two patients who needed over 3 units, a mean of 1.3±0.3 units of blood were transfused per control patient. Treated patients received a mean of 0.9±0.3 blood transfusions.

Discussion

The recent success of rHuEPO in abolishing the anemia of patients with different kinds of chronic diseases [13] has raised the question of whether perioperative rHuEPO treatment can reduce the need for postoperative blood transfusions. The rationale for perioperative treatment with rHuEPO is that the physiologic period between onset of anemia and red cell production is shifted forward, but also that higher erythropoietin levels may be reached postoperatively. The normal total marrow transit time of erythroid cells in humans is 7 days; however, at high erythropoietin concentrations, the

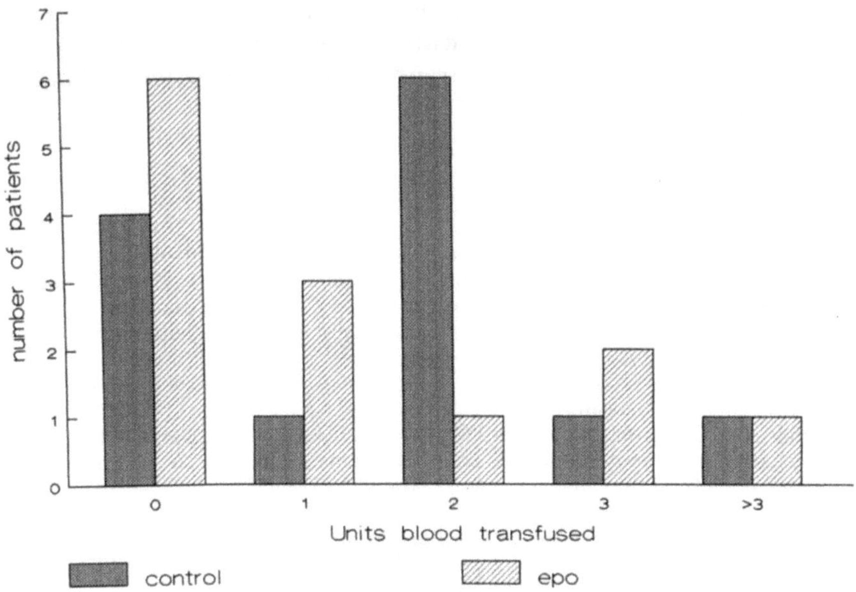

Fig. 2. Number of patients transfused with 0, 1, 2, 3, or > 3 units of blood after treatment with rHuEPO or no treatment (control)

reticulocyte pool is shifted to the peripheral blood within 1 day, and newly formed red cells appear in the circulation within 5 days. Therefore in the current trial rHuEPO therapy was started 5 days before surgery and was continued up to day 3 after. Results from our animal studies had indicated that daily administration is highly effective. We therefore treated the patients daily with 200 U rHuEPO/kg. Since at that time there was little clinical experience with subcutaneous administration, rHuEPO was administered intravenously. No adverse effects were seen by this short course of treatment. The only reason for excluding treated patients was violation of the treatment protocol due to events unrelated to the rHuEPO treatment.

With our dosing schedule, the reticulocytes were significantly elevated after only 2 treatment days, and the highest values were reached in the first 2 postoperative days. The time before reticulocyte were significantly elevated was 5–6 days in control patients; the maximum value was reached 2 days later and was significantly lower compared to the maximum value in treated patients. Hct was at no time significantly different between the two groups. However, fewer blood transfusions were given in the treated group. Because this study is not placebo controlled it is difficult to speculate on the significance of this issue.

In conclusion, perioperative treatment with a daily dose of 200 U rHuEPO/kg was given safely and was successful in stimulating erythropoiesis in elective cancer surgery. Although our results suggest that rHuEPO, given perioperatively, can reduce postoperative blood transfusions, placebo-controlled double-blind studies are needed to confirm this. By combining preoperative rHuEPO therapy with preoperative blood transfusion or hemodilution, allogeneic blood transfusions may become redundant in elective cancer surgery.

References

1. Walker RH (1987) Transfusion risks. Am J Clin Pathol 88:374–378
2. Marquet RL, de Bruin RWF, Dallinga RJ, Kumar S, Jeekel J (1986) Modulation of tumor growth by allogeneic blood transfusion. J Cancer Res Clin Oncol 111:50–53
3. Blumberg N, Heal JM (1989) Transfusion and host defences against cancer recurrence and infection. Transfusion 29:236–245
4. Tartter PI (1988) Blood transfusion and infectious complications following colorectal cancer surgery. Br J Surg 75:789–792
5. Winearls CG, Oliver DO, Pippard MJ, Reid C, Downing MR, Cotes PM (1986) Effect of human erythropoietin derived from recombinant DNA on the anemia of patients maintained by chronic haemodialysis. Lancet 2:1175–1178
6. Hoynck van Papendrecht MAW, Jeekel J, Marquet RL. Dosage of recombinant human erythropoietin in autologous blood donation programs; a study in rats (submitted)
7. Levine EA, Rosen AL, Gould SA et al. (1988) Recombinant human erythropoietin and autologous blood donation. Surgery 104:365–369
8. Goodnough LT, Rudnick S, Price TH et al. (1989) Increased preoperative collection of autologous blood with recombinant human erythropoietin therapy. N Engl J Med 321:1163–1168
9. Levine EA, Gould SA, Rosen AL et al. (1989) Perioperative recombinant human erythropoietin. Surgery 106:423–428
10. Krieter H, Brückner UB, Krumwieh D, Seiler FR, Messmer K (1990) Erythropoietin accelerates the recovery from extreme hemodilution: a randomized, placebo-controlled study in dogs. Eur Surg Res 22:121–127
11. Muirhead N, Hodsman AB (1990) Occult infection and resistance of anaemia to rHuEPO therapy in renal failure. Nephrol Dial Transplant 5:232–234
12. Hoynck van Papendrecht MAW, Jeekel J, Busch ORC, Marquet RL. The efficacy of recombinant erythropoietin to obviate the need for blood transfusions in surgery (in preparation)
13. Adamson JW, Eschbach JW (1990) The use of recombinant human erythropoietin (rHuEPO) in humans. Cancer Surv 9:157–167

Subject Index

MIX
Papier aus verantwortungsvollen Quellen
Paper from responsible sources
FSC® C105338

If you have any concerns about our products,
you can contact us on
ProductSafety@springernature.com

In case Publisher is established outside the EU,
the EU authorized representative is:
Springer Nature Customer Service Center GmbH
Europaplatz 3, 69115 Heidelberg, Germany

Printed by Libri Plureos GmbH
in Hamburg, Germany